Scientific Data Analysis with R

In an era marked by exponential growth in data generation and an unprecedented convergence of technology and healthcare, the intersection of biostatistics and data science has become a pivotal domain. This book is the ideal companion in navigating the convergence of statistical methodologies and data science techniques with diverse applications implemented in the open-source environment of R. It is designed to be a comprehensive guide, marrying the principles of biostatistics with the practical implementation of statistics and data science in R, thereby empowering learners, researchers, and practitioners with the tools necessary to extract meaningful knowledge from biological, health, and medical datasets.

This book is intended for students, researchers, and professionals eager to harness the combined power of biostatistics, data science, and the R programming language while gathering vital statistical knowledge needed for cutting-edge scientists in all fields. It is useful for those seeking to understand the basics of data science and statistical analysis, or looking to enhance their skills in handling any simple or complex data including biological, health, medical, and industry data.

Key Features:

- Presents contemporary concepts of data science and biostatistics with real-life data analysis examples.
- Promotes the evolution of fundamental and advanced methods applying to real-life problem-solving cases.
- Explores computational statistical data science techniques from initial conception to recent developments of biostatistics.
- Provides all R codes and real-world datasets to practice and competently apply into reader's own domains.
- Written in an exclusive state-of-the-art deductive approach **without any theoretical hitches** to support all contemporary readers.

Scientific Data Analysis with R

Biostatistical Applications

Azizur Rahman
Faruq Abdulla
Md. Moyazzem Hossain

CRC Press
Taylor & Francis Group
Boca Raton London New York

CRC Press is an imprint of the
Taylor & Francis Group, an **informa** business

A CHAPMAN & HALL BOOK

Designed cover image: © Shutterstock Stock Illustration ID 1925880932, Illustration Contributor vectorfusionart

First edition published 2025
by CRC Press
2385 NW Executive Center Drive, Suite 320, Boca Raton FL 33431

and by CRC Press
4 Park Square, Milton Park, Abingdon, Oxon, OX14 4RN

CRC Press is an imprint of Taylor & Francis Group, LLC

© 2025 Azizur Rahman, Faruq Abdulla and Md. Moyazzem Hossain

ISBN: 978-1-032-54692-6 (hbk)
ISBN: 978-1-032-54693-3 (pbk)
ISBN: 978-1-003-42618-9 (ebk)

DOI: 10.1201/9781003426189

Typeset in Palatino
by SPi Technologies India Pvt Ltd (Straive)

Access the Instructor and Student Resources: https://www.routledge.com/Scientific-Data-Analysis-with-R-Biostatistical-Applications/Rahman-Abdulla-Hossain/p/book/9781032546926

Contents

Preface

In an era marked by exponential growth in data generation and an unprecedented convergence of technology and healthcare, the intersection of biostatistics and data science has become a pivotal domain. This book is your companion in navigating the convergence of statistical methodologies and data science techniques with diverse applications implemented in the open-source environment of R. This book is designed to be a comprehensive guide, marrying the principles of biostatistics with the practical implementation of statistics and data science in R, thereby empowering researchers, and practitioners with the tools necessary to extract meaningful knowledge from biological, health, and medical datasets.

Biostatistics has long been an essential tool in deciphering biological phenomena, aiding researchers in drawing meaningful inferences from data. However, the rapid advent of data science has transformed the landscape, introducing novel methodologies and computational techniques that empower scientists to extract valuable insights from vast and diverse datasets. Moreover, this book bridges the traditional foundations of biostatistics with the cutting-edge methodologies of data science, providing readers with a holistic perspective on the application of statistical principles with data science notions in the context of biological and medical sciences research. Each chapter is meticulously crafted to guide readers through fundamental concepts, methodologies, and real-world applications. The key features of this book include exploring the fundamental principles of biostatistics, incorporating an introduction to data science, measures of locations and variability, summarizations and advanced level presentations of contemporary and complex data, probability theory, statistical estimation, inferential techniques, hypothesis testing, measures of statistical associations, regression modeling, prediction analysis, survival analysis, and factor analysis with a focus on their applications in health sciences and related fields, including biostatistics, medical statistics, and bioinformatics.

This book is intended for students, researchers, and professionals eager to harness the combined power of biostatistics, data science, and the R programming language while gathering vital statistical knowledge needed for cutting-edge scientists in all fields. Whether you are initiating your journey into the world of data science and statistical analysis or seeking to enhance your skills in handling any simple to complex data, including biological, health, and medical data, this book serves as your comprehensive and ready-to-use programming-based handy guide.

Azizur Rahman
Faruq Abdulla
Md. Moyazzem Hossain
April 2024
Wagga Wagga (Australia), Dhaka (Bangladesh),
and Newcastle upon Tyne (UK).

About the Authors/Editors

Azizur Rahman is an associate professor at the School of Computing, Mathematics and Engineering and the 'Data Mining Research Group' leader at Charles Sturt University, Australia. He earned a BSc (Honours) in Statistical Science, an MSc (Thesis) in Biostatistics, and a PhD in Economics and Statistics from the University of Canberra under the supervision of Professor Ann Harding, AO FASSA. He worked as a biostatistical research fellow in the Faculty of Health and Medical Sciences at the University of Adelaide. Professor Rahman is a statistician and data scientist with expertise in developing and applying novel methodologies, models, and technologies. He designs projects to understand multidisciplinary research issues within various fields with the interaction or adaptation of statistics, data science, AI, and ML. Professor Rahman develops data-centric 'alternative computational methods in microsimulation modelling technologies,' which are handy tools for decision-making processes in government and nongovernmental organizations, precision estimation, policy analysis, and evaluation. He founded and runs the 'Data Analytics Lab' at Charles Sturt. Professor Rahman has accrued more than $4.03 million of external research funding and over 203 scholarly publications and received several awards, including the 2023 Charles Sturt Excellence Awards and the ANZRSAI's 2023 Outstanding Service Award.

Faruq Abdulla is an outstanding graduate researcher, statistician, and data scientist, adeptly practicing in academia and industry. His expertise includes applying and developing sophisticated statistical, data science, and machine learning methodologies, models, and techniques in biological and medical sciences. With a keen focus on high-dimensional simulation and real-world data, he tackles pressing public health challenges, thereby contributing to evidence-based policy formulation. He has completed an MSc (Thesis) and a BSc (Honors) in Statistics from the Islamic University, Kushtia, Bangladesh. His academic excellence is evident through his first place in his class in order of merit at both the BSc and MSc levels, earning him the prestigious Presidential Gold Medal for achieving the highest marks in the Faculty of Applied Science & Technology in the MSc final examination. Moreover, Abdulla actively contributes to the scientific community by advancing scientific knowledge through his research findings published in renowned international peer-reviewed and high-impact journals indexed in SCOPUS and SCI. Additionally, he serves as a discerning reviewer for esteemed peer-reviewed journals published by world-class publishers.

Md. Moyazzem Hossain is an applied statistician and data scientist specializing in developing and applying contemporary statistical and data science methodologies, models, and techniques and currently holding the position of Professor in the Department of Statistics and Data Science at Jahangirnagar University, Bangladesh. Hossain earned his PhD from the School of Mathematics, Statistics, and Physics at Newcastle University, UK. He also obtained his BSc (Honors), MSc (Thesis), and MPhil from the Department of Statistics, Jahangirnagar University, Bangladesh. Hossain's outstanding contributions have been recognized through accolades such as the 'Best Conference Paper' award at the Australia and New Zealand Regional Science Association International 45th Annual Conference, held at Charles Sturt University, Wagga Wagga, Australia, on 1–2 December 2022. His research findings have been disseminated through numerous peer-reviewed publications in esteemed journals. Additionally, Hossain has served as an academic editor for *PloS ONE* and contributed as a reviewer for various international journals.

1

Introductory Data Sciences

1.1 Data, Statistics, and Science

Nowadays, data is everywhere, and the security of personal data is a significant concern. This 21st century's modern artificial intelligence (AI) and metaverse-centric world are heavily data-dependent. As a result, the definition of data is a broad concept. Simply put, 'data are collected facts on subjects utilised to analyze relevant interesting question for making an informative decision' (Rahman & Harding, 2017). For example, each student's body mass index (BMI) score is one piece of data. You can collect a group of students' BMI scores to create a dataset and then use this to calculate the mean score of a class by comparing it with other classes or the national mean value.

Moreover, in computing aspects, data is collected from observation of the quantities, characters, and symbols of the object(s) or anything in an electronic form that can be stored and used by a computer. Scientific processes perform data operations to form a basis for reasoning, planning, or inference (Rahman, 2020). For example, AI-based medical modeling or biostatistical intervention is solely rooted in comprehensive data, algorithms, and programming.

Statistics is a highly interdisciplinary field of knowledge that bridges the science (that teaches us to discover or know) and the art (that teaches us to do traditionally or innovatively). It concerns developing and applying knowledge by studying procedures for collecting, analyzing, interpreting, and presenting simple to high dimensional complex data from the population(s) (Oxford Reference, 2008). 'How do you deal with variation in data?' and 'How do you measure uncertainty of outcomes?' are two significant questions in statistics (Fisher, 1925). Notably, in real-life data, there are many circumstances in which the outcome is uncertain due to high variability in data. However, in some situations, the uncertainty is purely linked to the outcome of interest is not ascertained yet (e.g., a nurse may not know whether a critical patient will survive one week more) or if the outcome is already determined but it is not revealed to someone yet (e.g., a patient may not know whether s/he has a particular disease just after a clinical test).

Furthermore, according to the Science Council (2009), 'science is the pursuit and application of knowledge and understanding of the natural and social world following a systematic methodology based on evidence.' This definition covers only the empirical science part. There are three additional parts in science: theoretical, computational, and data science. Table 1.1 presents a general outline of the science with its classification, foundation, and application.

It is noted that the foundation of theoretical science is logic and pure mathematics, and it covers some main application domains such as computer science, applied math, and statistics. Besides, computational science is based on computing algebra, numerical analysis,

TABLE 1.1

An Overall Outline of Science

Classification	Theoretical Science	Empirical Science		Computational Science	Data Science
		Natural Science	Social Science		
Foundation	Pure math, logic, including fuzzy logic	Physics, soil/ earth science, astronomy, chemistry, biology	Economics, psychology, sociology, psychology, political science	Computing algebra, numerical analysis, simulation algorithms, computational methods	Data, big-data, computer, algorithms, data mining, programming, AI/ML
Application	Computer science, applied math, statistics	Engineering, agricultural science, medicine, pharmacy	Business admin, archaeology, anthropology, law	Complex systems (e.g. weather, finance, biology, airflow around a plane, motion of stars in a galaxy), predictive analytics	Almost all domains of knowledge, including statistical data science.

simulation algorithms, computational methods, and models (Harjule et al., 2023). The application domains cover complex systems, including weather, finance, complex biology, airflow around a plane, motion of stars in a galaxy, predictive analytics, and modeling of climate change impacts. Moreover, the 'fourth paradigm' of science is data science, which enriches the developments of other parts, especially around the data-driven objectives. It asserts that 'everything about science is changing because of the impact of information technology' and the data deluge, such as big data concepts. Although further details on data science will be covered in later sections, a quick definition of the scientific method with a biostatistical example follows.

1.1.1 Scientific Method

The scientific method systematically investigates problems in a reasoned, orderly, logical, and defensible manner. The scientific method distinguishes any conclusions that someone makes from the mere speculations or opinions that someone else not using the method may offer. The exemplar demonstration of the key components of the scientific method is evident in the exploration of DNA, a pivotal achievement that unfolded over a decade from 1944 to 1953 (Wikipedia, 2023a–c). The sequence of events in this instance encompasses:

- *Question*: Preliminary investigations had unveiled DNA's chemical composition (comprising four nucleotides), the structure of individual nucleotides, and other attributes. While the Avery–MacLeod–McCarty experiment in 1944 identified DNA as the carrier of genetic information, the mechanism for how this information was stored within DNA remained unclear.
- *Hypothesis*: Scientists formulated a hypothesis proposing that DNA possessed a helical structure.
- *Prediction*: The prediction was made that if DNA indeed had a helical structure, its X-ray diffraction pattern would exhibit an X-shaped form. This prediction was

based on the helix transform mathematics derived by other researchers, establishing a purely mathematical construct independent of the biological issue.

- *Experiment*: Another researcher conducted X-ray diffraction using pure DNA, resulting in the production of photo 51. The obtained results displayed the anticipated X-shaped pattern.
- *Analysis*: Upon scrutinizing the detailed diffraction pattern, a scientist promptly identified it as a helix. Subsequently, the scientist and a colleague developed their model, incorporating this newfound information alongside previously known details about DNA's composition, particularly Chargaff's base pairing rules.

This pivotal discovery served as a catalyst for subsequent investigations into genetic material, paving the way for further studies in fields such as molecular genetics. This pioneering modeling work was awarded the Nobel Prize in 1962. More detailed stepwise guidelines for proceeding with a scientific method with relevant examples will be discussed in a later chapter (e.g., Crawford & Stucki, 1990).

1.2 What Is Data Science?

As previously mentioned, data science is conceptualized as the 'fourth paradigm' of science, adding to the traditional trio of theoretical, empirical, and computational sciences. This paradigmatic shift underscores the transformative impact of data-driven approaches on the scientific landscape, driven by the dynamic nature of science and the rapid pace of technological innovations leading to an influx of data. Data science, positioned as a multidisciplinary field, employs scientific methodologies, processes, algorithms, and systems to glean knowledge and insights from data (Wikipedia, 2023a).

The unifying concept of data science integrates statistics, data analysis, machine learning, and related methodologies to comprehend and analyze real-world phenomena through data. Drawing on applied mathematics, statistics, computer science, and information and communication technologies, data science represents the empirical synthesis of actionable knowledge throughout the entire data lifecycle process.

The American Statistical Association (ASA) has identified three emerging foundational professional communities in data science: database management, statistics, and machine learning, encompassing distributed and parallel systems (ASA, 2015). The pivotal roles of statistics, artificial intelligence (AI), and machine learning within data science are emphasized. Framing questions statistically enables the leverage of data resources to extract knowledge and generate improved solutions (Rahman et al., 2020). Statistical inference, grounded in the acknowledgment of randomness in data, empowers researchers to pose questions about underlying processes and quantify uncertainty in their responses.

The statistical framework within data science facilitates the differentiation between causation and correlation, enabling the identification of interventions that can bring about changes in outcomes. Moreover, it enables scientists to devise methods for prediction and estimation, quantifying the level of certainty, and performing these tasks using algorithms characterized by predictable and reproducible behavior. In essence, statistical methods within data science serve the fundamental purpose of accumulating knowledge from raw data, a core function integral to the broader field of data science.

1.2.1 Microdata vs Big Data

Microdata refers to the detailed information collected on the interested attributes of sampling units (i.e., observable objects) of a micro-population, including a group of individuals, households, or establishments. Microdata can be composed by a census, survey, or experimental study. As an illustration, in a national census, various pieces of information such as age, employment status, residential address, educational attainment, and numerous other variables may be gathered, each documented individually for every individual who provides a response. This is microdata. Typically, microdata is not readily available for analysis and/or modeling and needs to be simulated (Rahman & Harding, 2017). These sorts of datasets are beneficial for business organizations, policymakers, and researchers. Three specific examples are the following:

1. Enterprises utilize microdata approximations to construct customer (e.g., patient) profiles, identify market segments, and decide on optimal business site placements.
2. State and local authorities leverage microdata for delineating political boundaries, assessing the effects of public policies, and gauging the demand for healthcare facilities, educational institutions, road infrastructure, recreational spaces, and fire protection.
3. Scholars employ microdata to investigate urban expansion, public health matters, environmental circumstances, and societal patterns.

Furthermore, the term 'big data' encompasses any collection of data, or datasets, of such considerable size or complexity that conventional data management methods become challenging to apply, necessitating the modification of traditional data processing applications to address them. An example of big data is biomarker or genetic data, which can be used to detect specific diseases as early as possible, the risk of developing a disease, progression of a disease, response, and toxicity to a given treatment. Further details about big data are discussed in the next section.

1.2.2 Big Data – Definition and Sources

Big data is also data, but in addition, it possesses the characteristics that are often known to as the four Vs:

- *Volume* – How much data is there? A collection of data in huge size and yet growing exponentially with time.
- *Variety* – How diverse are different forms of data? Sources, types, and resolutions of big data are widely varied.
- *Velocity* – At what speed is data in motion? The speed of new data generation and handling is significant in the big data environments.
- *Veracity* – How accurate is the data? Varying levels of noise and processing errors are quite common in big data, which leaves the data in doubt.

These four attributes set big data apart from the data encountered in conventional data management and analysis tools. The difficulties posed by these characteristics are evident in nearly every facet, including data capture, curation, storage, search, sharing, transfer, and visualization. Moreover, handling big data necessitates the application of specialized

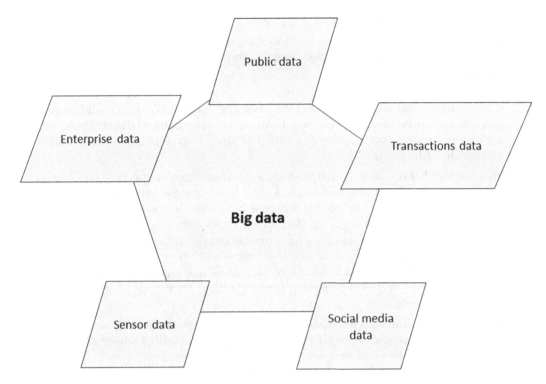

FIGURE 1.1
A diagram of sources of big data.

techniques to extract meaningful insights. The statistic shows that 500+ terabytes of new data get ingested into Facebook's social media databases daily. This data is mainly generated regarding photo and video uploads, message exchanges, and comments. If scientists wish to investigate the pattern of use of Facebook for a particular age group of clients or if there is any relationship between Facebook activities and mental health, they need to deal with big data and data science techniques.

A depiction of big data sources is presented in Figure 1.1. Big data sources are widespread, including various forms of information gathered from transactions, enterprise activities, citizen science or public data, sensors, social media, and many more. For example, sensor data are increasingly being gathered by cheap and numerous information-sensing mobile devices, aerial means (remote sensing), software logs, cameras, microphones, radio-frequency identification (RFID) readers, and wireless sensor networks.

1.2.3 Types of Data

With the progression of technology and evolving terminology, various expressions have been employed to characterize data and its utilization and analysis. As a result, the types of data can be described as follows:

Structured data: Structured data refers to any data located in a consistent field within a record or file. This type of data exhibits a high level of organization, facilitating easy ordering, and processing through data analysis tools, and is readily searchable through simple search operations. It is predominantly housed in relational databases and spreadsheets. Examples of structured data include tabulated data-sets with one or more variables, such as a patient's response to a drug, and five measurements (i.e., age, sex, weight, height, and BMI).

Unstructured data: Unstructured data has no predefined format or organization, i.e., information that does not reside in a traditional row-column database. This makes it much more challenging to collect, process, and analyze. As you might expect, this is the opposite of structured data. Examples of unstructured data include doctor's notes, emails, word processing files, PDF files, digital images, video, audio, and social media posts.

Semi-structured data: Semi-structured data deviates from a specific data model but retains some level of structure, lacking a fixed or rigid schema. This type of data does not reside in a relational database but possesses certain organizational properties that facilitate analysis. We can store this data in a relational database with some processing. Examples of semi-structured data include information on emails – where sender, recipient, date, time and other fixed fields might be added to the unstructured contents of the email message, Extensible Markup Language (XML) and other markup language files, JavaScript Object Notation (JSON) documents.

Raw data: Information collected from a source but not formatted or analyzed. Suppose an agricultural scientist sets up a greenhouse experiment with a sensor device that records the humidity and temperature in the greenhouse each hour. In that case, the list of humidity and temperature readings for every hour, as recorded within a computer or research lab server, is an example of 'raw data'.

Cleaned/processed data: Data that has been transformed from the raw data to add value by preprocessing and/or formatting. The above raw data example is an example of cleaned data if the scientist processed/formatted the data to identify and/or fix any incorrect or corrupt data before conducting research analysis.

Open data: Data that anyone can access, use, or share – may be internal or external. A range of open datasets is available at https://data.gov.au/. This is also called public domain data.

Personal data: Personal data encompasses any details about a living individual that can be identified or is identifiable. The aggregation of various pieces of information, when combined, can lead to the identification of a specific person and constitutes personal data. Patients' data is held by a hospital or doctor, which could be an example of personal data that uniquely identifies a person with his/her unique ID number or name and date of birth.

Univariate data: This data contains information from only one variable. For example, public health students' final exam score data from a biostatistics subject is univariate.

Bivariate data: This data contains information from only two relevant variables. For example, public health students' final exam scores and the overall mark data from a biostatistics subject are bivariate.

Multivariate data: This data contains information from more than two relevant variables. For example, the gradebook data for a cohort of medical students, which contains scores for all assessment items (e.g., online quizzes, assignments, practical exam, final exam, and the overall grade) from medical statistics class.

Furthermore, the detailed classification of structured data will be discussed in a later chapter.

1.2.4 Big Data Wheel and Uncertainty

A big data wheel can represent the overall cycle of massive data generation, making it handy for specific purposes. There are typically three challenges in a big data wheel, these being:

- Technical (i.e., how to generate, store, and manipulate big data);
- Statistical or computational (i.e., how to clean/preprocess, model and analyze, learn, decide, and report from big data); and
- Measuring efficacy (i.e., how to implement knowledge from big data and how to evaluate their effectiveness).

Figure 1.2 depicts a big data wheel focusing on the big data statistical challenge. However, the validity of big data could be called into question – since big data has considerable uncertainty. Typically, the proportion of uncertainty in big data increases

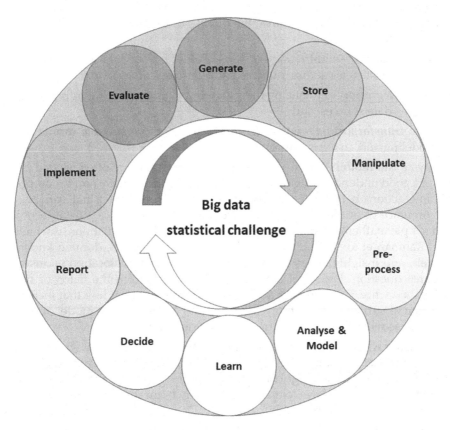

FIGURE 1.2
Statistical challenge in a big data wheel.

exponentially with a rapid increase in volume and variety of data sources. Data integrity can be compromised in several ways, such as human data entry errors, sensor data processing errors, or errors that occur during data transmission in the big data process. Given that the primary objective of this book is the data science perspective of biostatistics with real-world biostatistical data analysis, we will limit and focus our discussion on the next section to the statistical challenge.

1.2.5 Statistical Data Science Approaches

Data science is an evolutionary extension of statistics, aiding us in dealing with the vast volume of data generated in the present era. It adds computational approaches from computing science to the repertoire of statistical methods. As a result, to handle and analyze data, you will need computing tools such as R. Although various computing tools are available to data scientists, this subject will focus on R since it is an open and free software attainable for all users.

The statistical challenge in the data science process typically consists of five steps, as seen in the big data wheel chart (Figure 1.2). We will briefly introduce these steps and handle them in more detail from Chapter 2.

1. **Preprocess**: Typically, data collection is an error-prone process. So, after the data manipulation phase, you enhance the quality of the data by preprocessing techniques (Rahman, 2019):
   - *Data cleansing* – eliminating erroneous values from a data source and addressing inconsistencies across multiple data sources;
   - *Data integration* – enhancing data sources by amalgamating information from various data sources; and
   - *Data transformation* – ensuring that the data is in a suitable format for your modeling and analysis purposes.

2. **Analysis and model**: Various data exploration methods are often needed in order to get a good understanding of a dataset. As an illustration, we aim to comprehend the interactions between variables, the distribution of individual variables, and the presence of outliers. We mainly use descriptive statistics, visual techniques, and simple mathematical modeling such as probability calculations to achieve this (e.g. Rahman et al., 2024). We then use inferential analysis, domain knowledge, models, and insights about the data we found in data exploration to answer the research question. We select appropriate techniques from the fields of statistics and data science. Constructing a model is an iterative procedure that includes the selection of variables, executing the model, and examining the model diagnostics. All of these will be gradually introduced to you in later chapters.

3. **Learn**: Once we perform an analysis or undertake data modeling, we have the output or results at hand. In this step, we need to read and interpret the results. This requires technical knowledge. In many real-world cases, we may need only some part or parts of an analysis to answer the research question. However, we need to be careful to interpret the results accurately. Do not worry if you feel intimidated by this step; it will be taught step by step in this subject.

4. **Decide**: We decide based on the learning from our analysis findings. A decision is always objective and an evidence-based, data-driven outcome. The statistical

reliability of each decision is to be checked before reporting them to the relevant industries.

5. **Report**: Finally, we communicate the appropriate decision(s) to our intended audience (e.g. a healthcare business) through a suitable reporting format (e.g., presentation or formal industry report). The communication of decisions can be in many forms, from written and/or verbal presentations to research reports. We need valuable and specific skills to communicate our findings and conclusions to the audience confidently and professionally.

Frequently, there is a necessity to progress and automate the execution of these steps because the business intends to apply the insights acquired from the data in another project or enable an operational process to utilize the results derived from our model. For example, health policymakers can use the outcome of a health model to introduce changed health policies and processes if it produces reliable estimates of disease occurrence and prevalence (say, estimates of COVID-19 prevalence, smoking and cancer or cardiovascular diseases across remote or regional Australia).

1.2.6 The Scope of Data Science

Artificial intelligence (AI) is the most popular word worldwide in 2023. We require data science techniques and algorithms to design AI/machine learning (ML) assisted models for any social, economic, environmental, or complex system problems. The US Bureau of Labor Statistics (2023) reports that data science and computer information research is expected to grow by 23% from 2022 to 2032, much faster than the average for all occupations. With a sharp and exponential flow of data coming from various mechanisms, including IoT, AI, sensors, and digital transformation of industries, data science is stimulating about 80% of the total market in healthcare.

Thus, in this data-driven era, big data is our big reality, and the scope of data science is significantly enormous. Data science could bring benefits to many areas of our society. A few areas are listed below:

- *Health care*: Management of chronic disease, delivery of personalized medicine, finding correlations between lifestyle variables and illness. More specific examples of current data science applications to the biostatistics domain could be to model 'adverse drug interactions'. The US Food and Drug Administration (FDA) has updated its policies, and they require medical practitioners/researchers to report any adverse drug impacts, along with patient's other information, including disease status and health outcomes. Since the system data is vast, dynamic, and messy, researchers can determine drug interactions associated with good and bad health outcomes using natural language processing.

- *Security*: Detection of crime, predictions of patterns, protecting the country against external and internal attacks, forensics.

- *Citizen science*: Personalized experience of citizens, direct marketing, infrastructure need, intelligent traffic management, understanding what social media says about something important to the community.

- *Tax compliance*: Fraud identification and avoidance checking.

- *Energy*: The usage of energy and optimization of its use can be modeled using data science such as AI.

- *Environment*: Data science can be used in air pollution monitoring, protecting coral reefs, predicting weather patterns and extremes, and earthquake monitoring.

- *Agriculture*: Management of biosecurity, sustainable food production, control of feral animals and weeds.

Data science and big data applications have become ubiquitous in both commercial and noncommercial contexts. The range of use cases is extensive, and the examples provided here merely scratch the surface of the vast array of possibilities. Companies across nearly every industry leverage big data and data science techniques to gain insights into various aspects such as customers, processes, staff, achievements, competition, and products. Many organizations utilize data science to enhance customer experiences, employing strategies like cross-selling, up-selling, and personalization of offerings. A notable instance is Google AdSense, which gathers data from internet users to tailor relevant commercial messages to individuals browsing the internet.

Financial institutions employ data science to forecast stock market movements, assess the risk associated with specific loans, and devise strategies to attract new clients for their services. It's noteworthy that a significant portion, at least 50%, of global trades are automatically executed using trading algorithms developed with the assistance of big data and data science techniques.

Government and nongovernment organizations are also taking advantage of big data. For example, the Bureau of Meteorology is using data science techniques and significant investments in data collection to advance our capability in weather forecasting, environment monitoring, and climate study. Data.gov serves as the primary open data repository for the US government. An example illustrating the utilization of data science and big data involves Edward Snowden, who revealed how both the American National Security Agency and the British Government Communications Headquarters employed these technologies to monitor millions of individuals. These organizations amassed a staggering five billion data records from diverse sources such as Google Maps, Angry Birds, email, and text messages. Subsequently, they applied data science techniques to extract the crucial meaning or significant aspects of this vast dataset. Furthermore, nongovernmental organizations leverage big data analysis to both generate funds and champion their causes. As an illustration, the World Wildlife Fund employs data scientists to enhance the efficacy of their fundraising initiatives.

Moreover, educational providers such as universities use data science in their research and enhance their students' study experience. The rise of massive open online courses (MOOC) increasingly produces data on student experience, which allows universities to study how this type of learning can complement traditional classes.

1.3 Introduction to R

R, a freely available statistical programming language, serves as cost-free statistical software. It was developed with a user-friendly graphical interface (GUI), facilitating the teaching and learning of statistics or data science courses. By simplifying software complexity, it eliminates barriers to understanding content (Ihaka & Gentleman, 1996). R boasts numerous packages, such as R Commander, some equipped with drop-down menus for streamlined statistical data analysis. As a leading alternative to commercial statistical packages like SPSS, the R package is particularly useful for beginners, as it reveals the underlying R code for each analysis performed.

1.3.1 Installing R

To install R, a set of instructions need to be followed. Here, we outline the installation steps based on using the Chrome browser. While it is not mandatory, you have the option to freely download and install R from this source: https://www.google.com/chrome/.

The following links provide good instructions for installation of R:

https://cran.r-project.org/doc/manuals/r-release/R-admin.html

Installing R requires a range of systematic steps. An example of these steps is as follows:

a. R can be obtained by downloading it from the Comprehensive R Archive Network (CRAN). Open your browser and search for CRAN.

b. Once you access the CRAN page, choose the version compatible with your operating system – Linux, Mac OS X, or Windows.

c. Upon reaching the CRAN download page, various options will be available. Opt for the base subdirectory, as this installs the fundamental packages required for a basic setup. The process of installing additional useful packages within R, rather than from the webpage, will be covered later.

d. Click on the link for the latest version to initiate the download.

e. If you are using Chrome, the download progress will be displayed at the bottom of your browser. Once the installer file is downloaded, click on that tab to commence the installation process. Other browsers may have different procedures, so locate where they store downloaded files and initiate the procedure by clicking on them.

f. Progress through various choices to complete the installation. It is recommended to stick with the default selections.

g. Note: When choosing the language, it is recommended to opt for English for ease of following this book. Continue selecting all the default options.

h. The appearance may differ on a Mac, but the process involves accepting the defaults as well.

i. After following all the default settings and successfully completing the installations, you will receive a message stating, 'Congratulations! You have installed R.'

1.3.2 Installing RStudio

Installing RStudio involves a few straightforward steps. Here is a general guide on how to install RStudio on a Windows operating system. If you are using a different operating system, the steps might vary slightly.

- Install of R is prerequisite for installation of RStudio.
- Go to the RStudio Download page.

FIGURE 1.3
RStudio interface.

- Under 'Installers,' find the RStudio Desktop section and choose the appropriate installer for your operating system (Windows, macOS, or Linux).
- Run the downloaded RStudio installer.
- Once the installation is complete, you can open RStudio.
- If you are using Windows, you can find RStudio in the Start menu or on your desktop.
- RStudio requires R to be installed on your system.

Remember that these instructions might change slightly depending on the version of R and RStudio available at the time of your installation. After successful completion of the installation process of R and RStudio, you will get the following interface if you open RStudio (Figure 1.3).

1.3.3 Object, Class, Environment, Console, and Script in R

Object and Class

In R, objects are fundamental as the language is object-based. Objects, which encompass anything saved in a variable, serve as the basis for all operations in R. On the other hand, the class of an object dictates the functions that can be applied to it, while the data type or structure determines its class. Some examples are given in Table 1.2.

Environment

The Environment window displays these objects, simplifying management by allowing users to observe random access memory (RAM) occupancy. Objects can be exported, imported via .RData files, and unnecessary ones can be deleted to enhance coding efficiency. Some common formula and functions are listed in Table 1.3.

TABLE 1.2

Examples of Object and Class

Object	Class
`number <- 42`	Numeric
`text <- 'Hello, World!'`	Character
`is_valid <- TRUE`	Logical
`vector <- c(1, 2, 3, 4, 5)`	Numeric vector

`data_frame <- data.frame(`
`  name = c('Alice', 'Bob'),`
`  age = c(25, 30),`
`  score = c(90, 85)`
`)`
	Data frame containing columns for name, age, and score

TABLE 1.3

Some Important Formula and Function Used to Handle Objects Stored in Environment

Formula	Function
`ls()`	Finding all the names of objects in the environment
`save.image('MyBackup.RData')`	Saving all the objects that are there in the environment
`save(Cities, file='cityobj.RData')`	Saving one object from the environment
`rm(Cities)`	Removing an object
`rm(list=ls())`	Removing all the object from the environment
`load('MyBackup.RData')`	Loading a RData file

Console

R Studio is regarded as the central hub of the R programming language, functioning as its IDE (Integrated Development Environment). The console within R Studio allows users to write, execute, and view code output simultaneously. However, code written in the console cannot be saved as a script, and once executed, it cannot be edited.

Script

An R script is a file containing a series of R commands and functions that can be executed in a sequential manner. It allows users to write and save a set of R code for later use, share with others, or automate repetitive tasks. R scripts typically have a '.R' file extension.

1.3.4 R Packages/Libraries

After installing R and RStudio, you can use the R console in RStudio to install additional packages as needed for your analysis. In R, packages are collections of R functions, data, and compiled code designed to enhance the capabilities of the R programming language. As R is a modular language, its packages play an important role in expanding the capabilities of R. In R, the packages can be divided into two parts – system and user where system packages are those that are provided by the R by default whereas user libraries are the third-party libraries that the user downloads from CRAN (Comprehensive R Archive Network).

TABLE 1.4

Functions Used for Installing and Loading Packages Along with Examples

Task	Function	Example
Installing packages	`install.packages()`	*Single package*: `install.packages('ggplot2')`

		Multiple packages:
		`install.packages(c('dplyr',`
		`'tidyr', 'readr'))`

Loading packages	`library()` or `require()` N.B. `require()` *returns a logical value* *indicating whether the package is available.*	`library(ggplot2)` or `require(ggplot2)`
Checking installed packages	`installed.packages()`	`installed_packages <-` `installed.packages()`
Updating packages	`update.packages()`	`update.packages('tidyr')`
Removing packages	`remove.packages()`	`remove.packages('ggplot2')`

Installing and loading packages is a common and important task in R. Some functions used for installing and loading packages are listed in Table 1.4.

Please note that it is good practice to install and load only the packages you need for a particular analysis to keep your R environment clean and efficient. Packages can have dependencies, so installing a package may also install other packages required for its functionality.

1.3.5 R Help

In R, there are several ways to access help and documentation for functions, packages, and general usage. In R script, write and run the commands in the example column of Table 1.5 for getting help about the provided topic in the argument in the function such as mean and data frames.

1.3.6 Operators in R

R operators are symbols or keywords that perform operations on variables and values. They are fundamental components of R programming and are used to manipulate and perform calculations on data. Therefore, understanding how to use R operators effectively is essential for writing efficient and expressive code in the R programming language. Here are some common types of operators in R (Table 1.6).

TABLE 1.5

Functions Used for Getting Help in R Along with Examples

Function	Task	Example
`help()`	Provides information about a specific function or topic	`help(mean)` `help('data frames')`
`?`	It is a shorthand for the help() function	`?mean` `?'data frames'`
`example()`	Shows examples of the usage of a specified function	`example(mean)`
`apropos()`	Helps to find functions related to a keyword	`apropos('mean')`
`vignette()`	Provides detail about a package	`vignette('caret')`

TABLE 1.6

Some Common Types of Operators in R

Type	Operator	Function
Arithmetic	+	Addition
	−	Subtraction
	*	Multiplication
	/	Division
	^	Exponentiation
	%%	Modulus − remainder after division
	%/%	Integer division
Relational	==	Equal to
	!=	Not equal to
	<	Less than
	>	Greater than
	<=	Less than or equal to
	>=	Greater than or equal to
Logical	&	Element-wise AND
	|	Element-wise OR
	!	Logical NOT
	&&	Logical AND
	||	Logical OR
Assignment	<-	Leftward assignment
	=	Alternative assignment
	->	Rightward assignment
Other	%in%	Matches elements in a vector
	%*%	Matrix multiplication

1.3.7 Rules of Writing Code in R

Writing correct syntax in R is crucial for the proper execution of code. Here are some fundamental rules for writing syntax in R:

a. R is case-sensitive. Variable names, function names, and other identifiers must be written with consistent capitalization. Example: `myVariable` and `MyVariable` are treated as different.

b. When creating objects in R, one must make sure that the name of the object doesn't start from a number, doesn't have any symbol (. and _ are acceptable symbols generally used in the object name to denote space) and doesn't coincide with another preexisting function or object.

c. Avoid using reserved words (keywords) as variable or function names. Examples of reserved words include `if`, `else`, `while`, and `function`.

d. Use `<-` or `=` for variable assignment. While both are generally interchangeable, `<-` is the preferred assignment operator in R.

e. Statements are generally terminated by a newline character. However, multiple statements can be written on a single line using a semicolon (`;`).

f. Parentheses `()`, curly brackets `{}`, and square brackets `[ ]` are used for grouping, function calls, and indexing, respectively.

g. Use single (') or double (') quotation marks for character strings. Consistency is key, but it's common to use single quotes.

h. Follow standard mathematical and logical operator precedence. Use parentheses to explicitly define the order of operations when necessary.

i. When calling a function, use parentheses even if there are no arguments. Omitting parentheses may lead to unexpected behavior.

j. All the comments in R are to be preceded by the # symbol.

1.3.8 Data Types and Structure in R

Variable in R is linked to a specific data type. Each R data type requires varying amounts of memory and supports specific operations. The R programming language has several fundamental data types summarized in Table 1.7.

In R, data structures are objects that are used to store and organize data in an efficient manner. They can be differentiated on the basis of their homogeneity, dimensions, and purpose of data analysis. An overview of data structures in R is illustrated in Figure 1.4.

TABLE 1.7

Data Types in R

Basic Data Types	Values	Examples
Numeric	Set of all real numbers	`numeric <- 3.14`
Integer	Set of all integers, Z	`integer <- 42`
Logical	TRUE and FALSE	`logical <- TRUE`
Complex	Set of complex numbers	`complex <- 1 + 2i`
Character	'a', 'b', 'c', ..., '@', '#', '$', ..., '1', '2', etc.	`character <- 'Hello'`

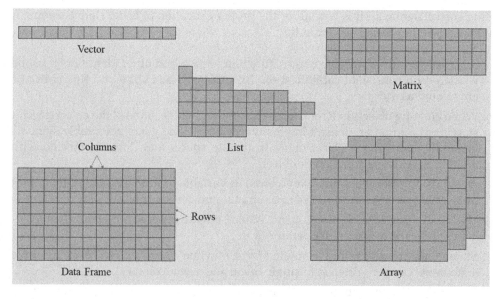

FIGURE 1.4
Overview of data structures in R.

TABLE 1.8

Widely Used Data Structures in R

Structure	Definition	Function	Example
Vector	A vector is the simplest data structure in R, representing a one-dimensional array of elements of the same data type	`c()`	`numeric_vector <- c(1, 2, 3, 4, 5)` `character_vector <- c('apple', 'orange', 'banana')`
Matrices	A matrix is a two-dimensional data structure where elements are arranged in rows and columns	`matrix()`	`matrix_data <- matrix(c(1, 2, 3, 4, 5, 6), nrow = 2, ncol = 3)`
Arrays	Arrays are multi-dimensional extensions of matrices, allowing storage of data in more than two dimensions	`array()`	`array_data <- array(c(1, 2, 3, 4, 5, 6), dim = c(2, 3, 1))`
Lists	A list is a versatile data structure that can contain elements of different types, including other lists	`list()`	`my_list <- list(name = 'John', age = 25, scores = c(90, 85, 92))`

Data frames	A data frame is a two-dimensional table-like data structure where columns can be of different data types	`data.frame()`	`student_data <- data.frame(`
			`name = c('Alice', 'Bob', 'Charlie'),`
			`age = c(22, 25, 24),`
			`grade = c('A', 'B', 'C')`
			`)`

Data tables	The `data.table` package introduces a data table, an extension of data frames with enhanced functionality for large datasets	`data.table()`	`library(data.table)`
			`dt <- data.table(name = c('Alice', 'Bob', 'Charlie'),`
			`age = c(22, 25, 24))`

Table 1.8 demonstrates some commonly used data structures in R.

1.3.9 Functions in R

A function in R refers to a block of code or a sequence of statements designed to execute a particular task. In R, a function is treated as an object, enabling the R interpreter to transfer control to the function, along with any required arguments essential for the function to carry out its actions. Subsequently, the function executes its designated task, relinquishing control and potentially providing results to the interpreter, which may then be stored in other objects. R has two types of function:

a. User-defined function

R language allow us to write our own function, these are specific to what a user wants and once created they can be used like the built-in functions. These functions are called user-defined functions. In R, `function()` keyword is used to create a function. A general function in R can be defined as below:

```
function_name <- function(arg_1, arg_2, …) {
                    Function body
                }
```

The different parts of a function are defined below:

i. Function name – This denotes the specific name assigned to the function and is stored in the R environment as an object under this name.

ii. Arguments – Arguments serve as placeholders within a function. When the function is called, you provide a value for each argument. While arguments are optional, a function can have none. Additionally, arguments may have default values.

iii. Function body – The function body encompasses a set of statements that delineate the operations performed by the function.

iv. Return value – The return value is the outcome produced by a function, specifically the result of evaluating the last expression in the function body.

To call a function, use the function name followed by parenthesis, like `function_name()`. Some examples of user-defined functions along with their assigned tasks are presented in Table 1.9.

b. Built-in function

Built-in functions in R are predefined functions that are available in R programming languages to perform common tasks or operations. Therefore, you do not need to define statement for the built-in function. You can call any built-in function for the respective task when required and you will get your desired return by providing argument to that function. We can call one or more built-in functions while defining your own (user-defined) functions. R has many built-in functions; some are listed in Table 1.10.

TABLE 1.9

Example of User-Defined Functions Along with Their Assigned Tasks

Function	Task

`function_1 <- function(x) {`
`  y <- x^2`
`  return(y)`
`}`
	To compute the square of an arbitrary number x

`function_2 <- function(x) {`
`  y <- x + 5`
`  return(y)`
`}`
	To add 5 with an arbitrary number x

`function_3 <- function(x) {`
`  y <- x * 5`
`  return(y)`
`}`
	To multiply an arbitrary number x by 5

`function_4 <- function(x) {`
`  y <- x / 5`
`  return(y)`
`}`
	To divide an arbitrary number x by 5

`function_5 <- function(x) {`
`  y <- x^2 + x +7`
`  return(y)`
`}`
	To compute the equation, $y = x^2 + x + 7$, where x is an arbitrary number

`function_6 <- function(x, y) {`
`  z <- (x^2 + y^2)/3`
`  return(z)`
`}`
	To compute the equation, $z = (x^2 + y^2)/3$, where x and y are arbitrary numbers

TABLE 1.10

Some Built-in Functions Along with Their Assigned Tasks

Function	Task
Mathematical Functions	
sum()	Calculates the sum of numeric values
min(), max()	Finds the minimum or maximum value in a set of values
abs()	Calculates absolute value of a number
sqrt()	Calculates square root of a number
round()	Rounds a number to the nearest integer
exp()	Calculates exponential of a number
log()	Calculates natural logarithm of a number
cos(), sin(), tan()	Calculates cosine, sine, and tang of a number
Statistical Functions	
mean()	Compute arithmetic mean of a numeric vector
median()	Compute median of a numeric vector
cor()	Calculates the correlation between two numeric vectors
sd()	Calculates the standard deviation of a numeric vector
var()	Calculates the variance of a numeric vector
Data Manipulation Functions	
unique()	Returns the unique values in a vector
subset()	Subsets a data frame based on conditions
merge()	Combines two data frames by common columns
aggregate()	Groups data according to a grouping variable
order()	Uses ascending or descending order to sort a vector
Data Exploration Functions	
head(), tail()	Displays the first or last few rows of a data frame
summary()	Generates summary statistics for a data frame or object
table()	Creates a table of counts for categorical variables
Character/String Functions	
paste()	Concatenates strings
grep()	Searches for patterns in character vectors
toupper(), tolower()	Converts characters to uppercase or lowercase
Data Frame Functions	
dim()	Returns the dimensions of a data frame
names()	Retrieves or sets the column names of a data frame
str()	Displays the structure of an R object
Logical Functions	
ifelse()	Applies conditional statements
any(), all()	Check if any or all elements of a logical vector are TRUE
Date and Time Functions	
Sys.Date(), Sys.time()	Obtain the current date or date-time
as.Date(), as.POSIXct()	Convert character strings to Date or POSIXct objects
File Input/Output Functions	
read.csv()	Reads data from a CSV file
write.csv()	Save data to write a CSV file
read.table()	Reads data from a tabular
write.table()	Save a tabular file with data

These are just a few examples, and R provides a rich set of built-in functions for a wide range of tasks.

1.3.10 Examples of Writing Code Using R Script

Objects under Different Data Structures

Mathematical Operation

Let us start with writing code for some basic mathematical operations in R. The following are some examples:

R-code:

```
# Adding two numbers
addition <- 7 + 2
print(addition)

# Adding two numbers by calling built-in function
addition <- sum(7, 2)
print(addition)

# Subtracting two numbers
subtraction <- 9 - 4
print(subtraction)

# Multiplying two numbers
multiplication <- 6 * 7
print(multiplication)

# Dividing two numbers
division <- 15 / 3
print(division)

# Calculating the power of a number
exponentiation <- 2^3
print(exponentiation)

# Calculating the remainder of the division
modulo <- 10 %% 3
print(modulo)

# Calculating the square root of a number calling built-in function
sqr <- sqrt(25)
print(sqr)

# Calculating the average of two numbers
average <- (3+4)/2
print(average)
```

Output:

```
Console  Terminal ×  Background Jobs ×
R  R 4.3.2 · E:/Book/Chapter 1/
> # Adding two numbers
> addition <- 7 + 2
> print(addition)
[1] 9
>
> # Adding two numbers by calling built-in function
> addition <- sum(7, 2)
> print(addition)
[1] 9
>
> # Subtracting two numbers
> subtraction <- 9 - 4
> print(subtraction)
[1] 5
>
> # Multiplying two numbers
> multiplication <- 6 * 7
> print(multiplication)
[1] 42
>
> # Dividing two numbers
> division <- 15 / 3
> print(division)
[1] 5
>
> # Calculating the power of a number
> exponentiation <- 2^3
> print(exponentiation)
[1] 8
>
> # calculating the remainder of the division
> modulo <- 10 %% 3
> print(modulo)
[1] 1
>
> # calculating the square root of a number calling built-in function
> sqr <- sqrt(25)
> print(sqr)
[1] 5
>
> # calculating the average of two numbers
> average <- (3+4)/2
> print(average)
[1] 3.5
>
```

Vector and Related Operation

We can generate different types of vector using the following R-code:

R-code:

```
# Creating a numeric vector using c() function
num_vector <- c(1, 2, 3, 4, 5)
print(num_vector)

# Creating a character vector using c() function
char_vector <- c('apple', 'banana', 'orange')
print(char_vector)
```

```
# Creating a sequence of numbers using seq() function
seq_vector <- seq(1, 10, by = 2)
print(seq_vector)

# Creating a vector by replicating values using rep() function
rep_vector <- rep(2, times = 5)
print(rep_vector)

# Creating a named vector using c() and functions
name_vector <- c(apple = 3, banana = 5, orange = 2)
print(name_vector)
```

Output:

```
Console   Terminal ×   Background Jobs ×
R  R 4.3.2 · E:/Book/Chapter 1/
> # Creating a numeric vector using c() function
> num_vector <- c(1, 2, 3, 4, 5)
> print(num_vector)
[1] 1 2 3 4 5
>
> # Creating a character vector using c() function
> char_vector <- c("apple", "banana", "orange")
> print(char_vector)
[1] "apple"  "banana" "orange"
>
> # Creating a sequence of numbers using seq() function
> seq_vector <- seq(1, 10, by = 2)
> print(seq_vector)
[1] 1 3 5 7 9
>
> # Creating a vector by replicating values using rep() function
> rep_vector <- rep(2, times = 5)
> print(rep_vector)
[1] 2 2 2 2 2
>
> # Creating a named vector using c() and functions
> name_vector <- c(apple = 3, banana = 5, orange = 2)
> print(name_vector)
 apple banana orange
     3      5      2
> |
```

The following R-code can be used to perform different vector operations:

R-code:

```
# Adding, subtracting, multiplying, dividing, and comparing two numeric
vectors element-wise
x <- c(1, 2, 3); y <- c(4, 5, 6)
# Adding
z1 <- x + y
print(z1)
```

```
# Subtracting
z2 <- x - y
print(z2)
# Multiplying
z3 <- x * y
print(z3)
# Dividing
z4 <- x / y
print(z4)
# Comparing
z5 <- x == y
print(z5)

# Finding length and value at a particular position of a vector
# Length
x <- c(1, 2, 3, 4, 5)
length_x <- length(x)
print(length_x)
# Value at 4th position
x_4 <- x[4]
print(x_4)

# Converting numeric vector to character and factor vector
x <- c(5, 3, 2, 9, 1)
cha_x <- as.character(x)
print(cha_x)
fac_x <- as.factor(x)
print(fac_x)

# Log-transformation of vector
x <- c(5, 4, 8, 2, 3)
log_e_x <- log(x)
print(log_e_x)
log_10_x <- log10(x)
print(log_10_x)
```

Output:

```
Console   Terminal ×   Background Jobs ×                                          ▭ ◻
R  R 4.3.2 · E:/Book/Chapter 1/
> # Adding, subtracting, multiplying, dividing, and comparing two numeric vectors element-wise
> x <- c(1, 2, 3); y <- c(4, 5, 6)
> # Adding
> z1 <- x + y
> print(z1)
[1] 5 7 9
> # Subtracting
> z2 <- x - y
> print(z2)
[1] -3 -3 -3
> # Multiplying
> z3 <- x * y
> print(z3)
[1]  4 10 18
> # Dividing
> z4 <- x / y
> print(z4)
[1] 0.25 0.40 0.50
> # Comparing
> z5 <- x == y
> print(z5)
[1] FALSE FALSE FALSE
>
> # Finding length and value at a particular position of a vector
> # Length
> x <- c(1, 2, 3, 4, 5)
> length_x <- length(x)
> print(length_x)
[1] 5
> # Value at 4th position
> x_4 <- x[4]
> print(x_4)
[1] 4
>
> # Converting numeric vector to character and factor vector
> x <- c(5, 3, 2, 9, 1)
> cha_x <- as.character(x)
> print(cha_x)
[1] "5" "3" "2" "9" "1"
> fac_x <- as.factor(x)
> print(fac_x)
[1] 5 3 2 9 1
Levels: 1 2 3 5 9
>
> # Log-transformation of vector
> x <- c(5, 4, 8, 2, 3)
> log_e_x <- log(x)
> print(log_e_x)
[1] 1.6094379 1.3862944 2.0794415 0.6931472 1.0986123
> log_10_x <- log10(x)
> print(log_10_x)
[1] 0.6989700 0.6020600 0.9030900 0.3010300 0.4771213
> |
```

Matrix and Related Operation

We can create matrix and perform different matrix operations using the following R-code:

R-code:

```
# Creating a matrix
mat1 <- matrix(c(1, 2, 3, 4), nrow = 2)
mat2 <- matrix(c(5, 6, 7, 8), nrow = 2)
print(mat1)

# Adding two matrices
add <- mat1 + mat2
print(add)

# Subtracting two matrices
subt <- mat1 - mat2
print(subt)

# Multiplying two matrices element-wise
mult1 <- mat1 * mat2
print(mult1)

# Multiplying two matrices to get the matrix product
mult2 <- mat1 %*% mat2
print(mult2)

# Transposing a matrix
tran <- t(mat1)
print(tran)

# Finding the inverse of a square matrix
inv <- solve(mat1)
print(inv)
```

Output:

```
Console   Terminal ×   Background Jobs ×                                            ▭ ▢
R  R 4.3.2 · ~/
> # Creating a matrix
> mat1 <- matrix(c(1, 2, 3, 4), nrow = 2)
> mat2 <- matrix(c(5, 6, 7, 8), nrow = 2)
> print(mat1)
     [,1] [,2]
[1,]    1    3
[2,]    2    4
>
> # Adding two matrices
> add <- mat1 + mat2
> print(add)
     [,1] [,2]
[1,]    6   10
[2,]    8   12
>
> # Subtracting two matrices
> subt <- mat1 - mat2
> print(subt)
     [,1] [,2]
[1,]   -4   -4
[2,]   -4   -4
>
> # Multiplying two matrices element-wise
> mult1 <- mat1 * mat2
> print(mult1)
     [,1] [,2]
[1,]    5   21
[2,]   12   32
>
> # Multiplying two matrices to get the matrix product
> mult2 <- mat1 %*% mat2
> print(mult2)
     [,1] [,2]
[1,]   23   31
[2,]   34   46
>
> # Transposing a matrix
> tran <- t(mat1)
> print(tran)
     [,1] [,2]
[1,]    1    2
[2,]    3    4
>
> # Finding the inverse of a square matrix
> inv <- solve(mat1)
> print(inv)
     [,1] [,2]
[1,]   -2  1.5
[2,]    1 -0.5
> |
```

Array and Related Operation

The following R-code can be used to define array and naming row and columns of array:

R-code:

```
# Creating array
array1 <- array(seq(1:12), dim = c(2, 3, 2))
# The first and second number in 'dim' specifies the number of rows and
columns
# The last number in 'dim' specifies the number of matrices
print(array1)
```

```
# Naming rows and columns
row_names <- c('Row1','Row2')
col_names <- c('Col1','Col2','Col3')
matrix_names <- c('Matrix1','Matrix2')
array1 <- array(seq(1:12), dim = c(2,3,2), dimnames = list(row_names,
col_names, matrix_names))
print(array1)
```

Output:

```
Console   Terminal ×   Background Jobs ×

R  R 4.3.2 · ~/
> # Creating array
> array1 <- array(seq(1:12), dim = c(2, 3, 2))
> # The first and second number in 'dim' specifies the number of rows and columns
> # The last number in 'dim' specifies the number of matrices
> print(array1)
, , 1

     [,1] [,2] [,3]
[1,]    1    3    5
[2,]    2    4    6

, , 2

     [,1] [,2] [,3]
[1,]    7    9   11
[2,]    8   10   12

>
> # Naming rows and columns
> row_names <- c("Row1","Row2")
> col_names <- c("Col1","Col2","Col3")
> matrix_names <- c("Matrix1","Matrix2")
> array1 <- array(seq(1:12), dim = c(2,3,2), dimnames = list(row_names, col_names, matrix_names))
> print(array1)
, , Matrix1

      Col1 Col2 Col3
Row1     1    3    5
Row2     2    4    6

, , Matrix2

      Col1 Col2 Col3
Row1     7    9   11
Row2     8   10   12

> |
```

We can apply different functions to the previously defined array using the following R-code:

R-code:

```
# Accessing elements of the array
element <- array1[1, 2, 1]
cat('\nDesired Element:', element)

# Access Entire Row or Column
# Access entire elements at 2nd column of 1st matrix
column_2_mat1 <- array1[,c(2),1]
cat('\n2nd Column Elements of 1st matrix:', column_2_mat1)
# Access entire elements at 1st row of 2nd matrix
row_1_mat2 <- array1[c(1), ,2]
cat('\n2nd Column Elements of 1st matrix:', row_1_mat2)

# Check if Element Exists using %in% operator
11 %in% array1
13 %in% array1

# Find total elements (length) in array1
len <- length(array1)
cat('Total Elements:', len)

# Applying a function to each element of the array
squared <- apply(array1, c(1, 2), function(x) x^2)
print(squared)

# Finding the sum across dimensions of the array
sum_array <- apply(array1, 3, sum)
cat('\nThe sum across dimensions:', sum_array)
```

Output:

```
Console   Terminal ×   Background Jobs ×                                    ─ ♭
 R  R 4.3.2 · ~/
> # Accessing elements of the array
> element <- array1[1, 2, 1]
> cat("\nDesired Element:", element)

Desired Element: 3>
> # Access Entire Row or Column
> # Access entire elements at 2nd column of 1st matrix
> column_2_mat1 <- array1[,c(2),1]
> cat("\n2nd Column Elements of 1st matrix:", column_2_mat1)

2nd Column Elements of 1st matrix: 3 4> # Access entire elements at 1st row of 2nd matrix
> row_1_mat2 <- array1[c(1), ,2]
> cat("\n2nd Column Elements of 1st matrix:", row_1_mat2)

2nd Column Elements of 1st matrix: 7 9 11>
> # Check if Element Exists using %in% operator
> 11 %in% array1
[1] TRUE
> 13 %in% array1
[1] FALSE
>
> # Find total elements (length) in array1
> len <- length(array1)
> cat("Total Elements:", len)
Total Elements: 12>
> # Applying a function to each element of the array
> squared <- apply(array1, c(1, 2), function(x) x^2)
> print(squared)
, , Col1

        Row1 Row2
Matrix1    1    4
Matrix2   49   64

, , Col2

        Row1 Row2
Matrix1    9   16
Matrix2   81  100

, , Col3

        Row1 Row2
Matrix1   25   36
Matrix2  121  144

>
> # Finding the sum across dimensions of the array
> sum_array <- apply(array1, 3, sum)
> cat("\nThe sum across dimensions:", sum_array)

The sum across dimensions: 21 57
> |
```

List and Related Operation

The following R-code can be used to define list and access elements of list:

R-code:

```
# Creating a list with different types of elements
list1 <- list(
  name = 'Rocky',
  age = 27,
```

```
  subj = c('English', 'Mathematics', 'Science'),
  grades = c(91, 94, 83),
  is_student = TRUE
)
print(list1)

# Accessing elements of the list
name <- list1$name
age <- list1$age
subj <- list1$subj
grades <- list1$grades
cat('\nName:', name)
cat('\nAge:', age)
cat('\nSubject:', subj)
cat('\nGrade:', grades)
```

Output:

```
Console   Terminal   Background Jobs

R  R 4.3.2 · ~/
> # Creating a list with different types of elements
> list1 <- list(
+    name = "Rocky",
+    age = 27,
+    subj = c("English", "Mathematics", "Science"),
+    grades = c(91, 94, 83),
+    is_student = TRUE
+ )
> print(list1)
$name
[1] "Rocky"

$age
[1] 27

$subj
[1] "English"      "Mathematics" "Science"

$grades
[1] 91 94 83

$is_student
[1] TRUE

>
> # Accessing elements of the list
> name <- list1$name
> age <- list1$age
> subj <- list1$subj
> grades <- list1$grades
> cat("\nName:", name)

Name: Rocky> cat("\nAge:", age)

Age: 27> cat("\nSubject:", subj)

Subject: English Mathematics Science> cat("\nGrade:", grades)

Grade: 91 94 83
> |
```

The following R-code can be used to add new elements to the predefined list and remove element from the predefined list:

R-code:

```
# Adding new elements to the list
list1$city <- 'New York'
list1$hobbies <- c('Travelling', 'Reading', 'Gaming')
print(list1)

# Removing elements from the list
list1$subj <- NULL
print(list1)
```

Output:

```
Console   Terminal ×   Background Jobs ×
R  R 4.3.2 · ~/
> # Adding new elements to the list
> list1$city <- "New York"
> list1$hobbies <- c("Travelling", "Reading", "Gaming")
> print(list1)
$name
[1] "Rocky"

$age
[1] 27

$grades
[1] 91 94 83

$is_student
[1] TRUE

$city
[1] "New York"

$hobbies
[1] "Travelling" "Reading"    "Gaming"

>
> # Removing elements from the list
> list1$subj <- NULL
> print(list1)
$name
[1] "Rocky"

$age
[1] 27

$grades
[1] 91 94 83

$is_student
[1] TRUE

$city
[1] "New York"

$hobbies
[1] "Travelling" "Reading"    "Gaming"

> |
```

We can apply different functions to the predefined list using the following R-code:

R-code:

```
# Applying a function to each element of the list
# Sum of grades
sum_grade <- sum(list1$grades)
cat('\nThe sum of grades is:', sum_grade)
# Squared of grades
grade_square <- lapply(grades, function(x) x^2)
print(grade_square)
```

Output:

```
Console   Terminal ×   Background Jobs ×

R  R 4.3.2 · ~/
> # Applying a function to each element of the list
> # Sum of grades
> sum_grade <- sum(list1$grades)
> cat("\nThe sum of grades is:", sum_grade)

The sum of grades is: 268> # Squared of grades
> grade_square <- lapply(grades, function(x) x^2)
> print(grade_square)
[[1]]
[1] 8281

[[2]]
[1] 8836

[[3]]
[1] 6889

> |
```

Data Frame and Related Operation

The R-code in the next page can be used to define data frame, accessing column of the data frame, adding new column to the column, removing column from data frame, sub-setting from data frame with different conditions, and modifying values in the data frame.

R-code:

```r
# Creating data frame
df <- data.frame(
  id = c('N54', 'N23', 'N41'),
  sex = c('F', 'F', 'M'),
  age = c(27, 20, 24)
)
print(df)

# Accessing columns of the data frame
col.age <- df$age
cat('\nAge:', col.age)

# Adding new column to the data frame
df$subj <- c('Phisics', 'Mathematics', 'Geography')
df$grade <- c(83, 92, 97)
print(df)

# Removing a column from the data frame
df$subj <- NULL
print(df)

# Sub-setting rows based on a condition
df_subset <- df[df$age > 20, ]
print(df_subset)

# Modifying values in the data frame
df$grade[1] <- 88
print(df)
```

Output:

```
Console  Terminal ×  Background Jobs ×                                                    ▭ ☐
R  R 4.3.2 · ~/
> # Creating data frame
> df <- data.frame(
+     id = c("N54", "N23", "N41"),
+     sex = c("F", "F", "M"),
+     age = c(27, 20, 24)
+ )
> print(df)
   id sex age
1 N54   F  27
2 N23   F  20
3 N41   M  24
>
> # Accessing columns of the data frame
> col.age <- df$age
> cat("\nAge:", col.age)

Age: 27 20 24>
> # Adding new column to the data frame
> df$subj <- c("Phisics", "Mathematics", "Geography")
> df$grade <- c(83, 92, 97)
> print(df)
   id sex age        subj grade
1 N54   F  27     Phisics    83
2 N23   F  20 Mathematics    92
3 N41   M  24   Geography    97
>
> # Removing a column from the data frame
> df$subj <- NULL
> print(df)
   id sex age grade
1 N54   F  27    83
2 N23   F  20    92
3 N41   M  24    97
>
> # Sub-setting rows based on a condition
> df_subset <- df[df$age > 20, ]
> print(df_subset)
   id sex age grade
1 N54   F  27    83
3 N41   M  24    97
>
> # Modifying values in the data frame
> df$grade[1] <- 88
> print(df)
   id sex age grade
1 N54   F  27    88
2 N23   F  20    92
3 N41   M  24    97
> |
```

The following R-code can be used to check the internal structure of an R object (data frame) as well as to see some observations in the head and tail of the data frame:

R-code:

```
# displaying the internal structure of an R object (Data Frame)
str(df)

# Head and Tail of an R object (Data Frame)
head(df, 2)
tail(df, 2)
```

```
# Calculating summary statistics for the data frame
summary.stats <- summary(df)
print(summary.stats)
```

Output:

```
Console   Terminal ×   Background Jobs ×
R 4.3.2 · ~/
> # displaying the internal structure of an R object (Data Frame)
> str(df)
'data.frame':   3 obs. of  4 variables:
 $ id   : chr  "N54" "N23" "N41"
 $ sex  : chr  "F" "F" "M"
 $ age  : num  27 20 24
 $ grade: num  88 92 97
>
> # Head and Tail of an R object (Data Frame)
> head(df, 2)
   id sex age grade
1 N54   F  27    88
2 N23   F  20    92
> tail(df, 2)
   id sex age grade
2 N23   F  20    92
3 N41   M  24    97
>
> # Calculating summary statistics for the data frame
> summary.stats <- summary(df)
> print(summary.stats)
      id                 sex                  age            grade
 Length:3           Length:3           Min.   :20.00   Min.   :88.00
 Class :character   Class :character   1st Qu.:22.00   1st Qu.:90.00
 Mode  :character   Mode  :character   Median :24.00   Median :92.00
                                       Mean   :23.67   Mean   :92.33
                                       3rd Qu.:25.50   3rd Qu.:94.50
                                       Max.   :27.00   Max.   :97.00
> |
```

***Importing and Exporting Data with Different File Format
from External Memory (Hard Drive)***

CSV Data File

R-code:

```
# Reading data from csv data file
data.csv <- read.csv('E:/Book/Chapter 1/Dataset.csv')
head(data.csv, 5)

# Writing data frame as csv data file
write.csv(data.csv, 'E:/Book/Chapter 1/Data.csv', row.names=FALSE)
```

Output:

```
Console   Terminal ×   Background Jobs ×                                    ▬▢
R  R 4.3.2 · ~/
> # Reading data from csv data file
> data.csv <- read.csv("E:/Book/Chapter 1/Dataset.csv")
> head(data.csv, 5)
  id sex age
1  1   M  23
2  2   F  56
3  3   M  14
4  4   M  27
5  5   F  45
>
> # Writing data frame as csv data file
> write.csv(data.csv, "E:/Book/Chapter 1/Data.csv", row.names=FALSE)
> |
```

Excel (xlsx) Data File

R-code:

```
# Reading data from excel (xlsx) data file
# install.packages('readxl')
library(readxl)
data.xlsx <- read_excel('E:/Book/Chapter 1/Dataset.xlsx')
head(data.xlsx, 5)

# Writing data frame as excel (xlsx) data file
# install.packages('writexl')
library('writexl')
write_xlsx(data.xlsx, 'E:/Book/Chapter 1/Data.xlsx')
```

Output:

```
Console   Terminal ×   Background Jobs ×                                    ▬▢
R  R 4.3.2 · ~/
> # Reading data from excel (xlsx) data file
> # install.packages("readxl")
> library(readxl)
> data.xlsx <- read_excel("E:/Book/Chapter 1/Dataset.xlsx")
> head(data.xlsx, 5)
# A tibble: 5 × 3
     id sex     age
  <dbl> <chr> <dbl>
1     1 M        23
2     2 F        56
3     3 M        14
4     4 M        27
5     5 F        45
>
> # Writing data frame as excel (xlsx) data file
> # install.packages("writexl")
> library("writexl")
> write_xlsx(data.xlsx, "E:/Book/Chapter 1/Data.xlsx")
> |
```

Tab-delimited Text (txt) Data File

R-code:

```
# Reading data from tab-delimited text data file
data.text <- read.table('E:/Book/Chapter 1/Dataset.txt', header = TRUE)
# Or data.text <- read.delim('E:/Book/Chapter 1/Dataset.txt', header =
TRUE)
head(data.text, 5)

# Writing data frame as tab-delimited text data file
write.table(data.text, 'E:/Book/Chapter 1/Data.txt', row.names=FALSE,
quote=FALSE)
```

Output:

```
Console   Terminal ×   Background Jobs ×                                              ▬ ☐
R  R 4.3.2 · ~/
> # Reading data from tab-delimited text data file
> data.text <- read.table("E:/Book/Chapter 1/Dataset.txt", header = TRUE)
> # Or data.text <- read.delim("E:/Book/Chapter 1/Dataset.txt", header = TRUE)
> head(data.text, 5)
  id sex age
1  1   M  23
2  2   F  56
3  3   M  14
4  4   M  27
5  5   F  45
>
> # Writing data frame as tab-delimited text data file
> write.table(data.text, "E:/Book/Chapter 1/Data.txt", row.names=FALSE, quote=FALSE)
> |
```

STATA (dta) Data File

R-code:

```
# Reading data from stata i.e. dta data file
# install.packages('haven')
library(haven)
data.dta <- read_dta('E:/Book/Chapter 1/Dataset.dta')
head(data.dta, 5)

# Writing data frame as stata i.e. dta data file
library(haven)
write_dta(data.dta, 'E:/Book/Chapter 1/Data.dta')
```

Output:

```
Console   Terminal ×   Background Jobs ×                                    ─ ▢
 R  R 4.3.2 · ~/
> # Reading data from stata i.e. dta data file
> # install.packages("haven")
> library(haven)
> data.dta <- read_dta('E:/Book/Chapter 1/Dataset.dta')
> head(data.dta, 5)
# A tibble: 5 × 3
     id   age sex
  <dbl> <dbl> <dbl+lbl>
1     1    23 1 [Male]
2     2    56 2 [Female]
3     3    14 1 [Male]
4     4    27 1 [Male]
5     5    45 2 [Female]
>
> # writing data frame as stata i.e. dta data file
> library(haven)
> write_dta(data.dta, "E:/Book/Chapter 1/Data.dta")
> |
```

SPSS (sav) Data File

R-code:

```
# Reading data from spss i.e. sav data file
library(haven)
data.sav <- read_sav('E:/Book/Chapter 1/Dataset.sav')
head(data.sav, 5)

# Writing data frame as spss i.e. sav data file
library(haven)
write_sav(data.sav, 'E:/Book/Chapter 1/Data.sav')
```

Output:

```
Console   Terminal ×   Background Jobs ×                                    ─ ▢
 R  R 4.3.2 · ~/
> # Reading data from spss i.e. sav data file
> library(haven)
> data.sav <- read_sav('E:/Book/Chapter 1/Dataset.sav')
> head(data.sav, 5)
# A tibble: 5 × 3
     id sex          age
  <dbl> <chr+lbl>  <dbl>
1     1 M [Male]      23
2     2 F [Female]    56
3     3 M [Male]      14
4     4 M [Male]      27
5     5 F [Female]    45
>
> # writing data frame as spss i.e. sav data file
> library(haven)
> write_sav(data.sav, "E:/Book/Chapter 1/Data.sav")
> |
```

R (rds) Data File

R-code:

```
# Reading data from R i.e. rds data file
data.rds <- readRDS('E:/Book/Chapter 1/Dataset.rds')
head(data.rds, 5)

# Writing data frame as R i.e. rds data file
saveRDS(data.rds, file = 'E:/Book/Chapter 1/Data.rds')
```

Output:

```
Console   Terminal ×   Background Jobs ×
R  R 4.3.2 · ~/
> # Reading data from R i.e. rds data file
> data.rds <- readRDS("E:/Book/Chapter 1/Dataset.rds")
> head(data.rds, 5)
  id sex age
1  1   M  23
2  2   F  56
3  3   M  14
4  4   M  27
5  5   F  45
>
> # Writing data frame as R i.e. rds data file
> saveRDS(data.rds, file = "E:/Book/Chapter 1/Data.rds")
> |
```

JSON Data File

R-code:

```
# Reading data from json data file
# install.packages('jsonlite')
library(jsonlite)
data.json <- fromJSON('E:/Book/Chapter 1/Dataset.json')
head(data.json, 5)

# Writing data frame as json data file
data.json = toJSON(data.json)
write(data.json, 'E:/Book/Chapter 1/Data.json')
```

Output:

```
Console   Terminal ×   Background Jobs ×                                    ▬ ☐
R  R 4.3.2 · ~/
> # Reading data from json data file
> # install.packages("jsonlite")
> library(jsonlite)
> data.json <- fromJSON("E:/Book/Chapter 1/Dataset.json")
> head(data.json, 5)
  id sex age
1  1   M  23
2  2   F  56
3  3   M  14
4  4   M  27
5  5   F  45
>
> # Writing data frame as json data file
> data.json = toJSON(data.json)
> write(data.json, "E:/Book/Chapter 1/Data.json")
> |
```

Basic Statistical Analysis over Variable

The variable can be vector, data of list, column of matrix, column of array, column of data frame, and column of data table. The following R-code can be used to compute measures of central value:

R-code:

```
# Define two vectors
x <- c(5, 8, 10, 3, 6, 3)
y <- c(4, 9, 0, 1, 2, 4)
# Sum
sum_x <- sum(x)
print(sum_x)
# Arithmetic mean
a.mean_x <- mean(x)
print(a.mean_x)
# Geometric mean
g.mean_x <- exp(mean(log(x)))
print(g.mean_x)
# Harmonic mean
# We need to install and call the package named 'psych'
# install.packages('psych')
library(psych)
h.mean_x <- harmonic.mean(x)
print(h.mean_x)
# Median
median_x <- median(x)
print(median_x)
# Mode
# We can define here a user-defined function for getting mode
```

```
mode_x <- function(x) {
  uniqv <- unique(x)
  uniqv[which.max(tabulate(match(x, uniqv)))]
}
print(mode_x(x))
print(mode_x(y))
```

Output:

```
Console   Terminal    Background Jobs

R  R 4.3.2 · E:/icddrb/TB/
> # Define two vectors
> x <- c(5, 8, 10, 3, 6, 3)
> y <- c(4, 9, 0, 1, 2, 4)
> # Sum
> sum_x <- sum(x)
> print(sum_x)
[1] 35
> # Arithmetic mean
> a.mean_x <- mean(x)
> print(a.mean_x)
[1] 5.833333
> # Geometric mean
> g.mean_x <- exp(mean(log(x)))
> print(g.mean_x)
[1] 5.277266
> # Harmonic mean
> # We need to install and call the package named "psych"
> # install.packages("psych")
> library(psych)
> h.mean_x <- harmonic.mean(x)
> print(h.mean_x)
[1] 4.768212
> # Median
> median_x <- median(x)
> print(median_x)
[1] 5.5
> # Mode
> # We can define here a user-defined function for getting mode
> mode_x <- function(x) {
+     uniqv <- unique(x)
+     uniqv[which.max(tabulate(match(x, uniqv)))]
+ }
> print(mode_x(x))
[1] 3
> print(mode_x(y))
[1] 4
>
```

Some common measures of dispersion such as range, variance, and standard deviation as well as correlation between two variables can be computed using the following R-code:

R-code:

```
# Define two vectors
x <- c(2, 5, 2, 1, 5, 7, 8, 9, 1, 0, 3, 4)
y <- c(6, 2, 7, 9, 1, 3, 1, 5, 5, 10, 13, 11)
# Range
# We can define here a user-defined function for getting range
range_x <- function(x) {
```

```
  ran <- max(x) - min(x)
  return(ran)
}
print(range_x(x))
print(range_x(y))
# Variance
var_x <- var(x)
print(var_x)
# Standard deviation
sd_x <- sd(x)
print(sd_x)
# Correlation
cor_xy <- cor(x, y)
print(cor_xy)
```

Output:

```
Console   Terminal ×   Background Jobs ×                                                    ▭ ☐
R  R 4.3.2 · E:/icddrb/TB/
> # Define two vectors
> x <- c(2, 5, 2, 1, 5, 7, 8, 9, 1, 0, 3, 4)
> y <- c(6, 2, 7, 9, 1, 3, 1, 5, 5, 10, 13, 11)
> # Range
> # We can define here a user-defined function for getting range
> range_x <- function(x) {
+     ran <- max(x) - min(x)
+     return(ran)
+ }
> print(range_x(x))
[1] 9
> print(range_x(y))
[1] 12
> # Variance
> var_x <- var(x)
> print(var_x)
[1] 8.628788
> # Standard deviation
> sd_x <- sd(x)
> print(sd_x)
[1] 2.93748
> # Correlation
> cor_xy <- cor(x, y)
> print(cor_xy)
[1] -0.5704095
>
```

Plotting Data Using `plot()` *Function*

R possesses the capability to generate various types of plots, and notably, these can be customized extensively. Achieving the desired appearance of a plot involves leveraging the numerous arguments provided in plotting functions. R has a generic function `plot()` which can help to make many plots of two argument vectors. If we provide one vector of continuous data, it plots that on the y-axis against the index on the x-axis. The `plot()` function has many arguments as input listed in Table 1.11.

TABLE 1.11

Arguments of `plot()` Function in R

Argument	Description	Example
bg	The color to be used for the background	bg = 'red'
cex	Character size and expansion	cex = 1.5, cex = 0.8
cex.axis	The magnification to be used for axis annotation	cex.axis = 1.2
cex.lab	The magnification to be used for x and y label	cex.lab = 0.8
cex.main	The magnification to be used for main titles	cex.main = 1.3
cex.sub	The magnification to be used for sub-titles	cex.sub = 0.9
col	Color	col = 'black', col = '#ff0000'
family	Font on the plot	family = 'Arial'
fg	The color to be used for the foreground of plots	fg = 'yellow'
font	An integer which specifies which font to use for text. Italic, bold, etc.	font = 3
font.axis	The font to be used for axis annotation	font.axis = 2
font.lab	The font to be used for x and y labels	font.lab = 3
font.main	The font to be used for plot main titles	font.main = 2
font.sub	The font to be used for plot sub-titles	font.sub = 2
lty	Line type	lty = 2
lwd	Line width	lwd = 3
main	Plot primary title	main = 'Iris'
pch	Use the following R-code to see the scatter plot symbol for points:	pch = 1, pch= 'p'

```
par(mar = c(0, 0, 3, 3))
plot(1, type = "n", xlab = "", ylab = "", xlim =
    c(0, 5), ylim = c(0, 5), axes = FALSE)
points(rep(1:5, 5), rep(5:1, each = 5), pch =
    1:25, col = "blue", cex = 2)
text(rep(1:5, 5), rep(5:1, each = 5), labels =
    1:25, pos = 2)
```

Use the following R-code to see the scatter plot symbol for points:
Output of different scatter plot symbols as below:

TABLE 1.11   (Continued)

Argument	Description		Example
srt	The string rotation in degrees		`srt = 90`
sub	Subtitle of plot		`sub = 'All data'`
xlab, ylab	Label of the *x* or *y* axis		`xlab = 'Distance (Miles)'`
xlim, ylim	Min/max *x* or *y* axis values		`xlim = c(0, 10)`
xpd	If true, allows plotting outside the plot		`xpd = TRUE`
Type	Declare types of plot as below:		`type = 'l'`
	Value	**Description**	
	'p'	Points plot (default)	
	'l'	Line plot	
	'b'	Both line and points	
	'c'	Empty points joined by lines	
	'o'	Over-plotted points and lines	
	'h'	Histogram-like plot	
	's'	Step plot (horizontal first)	
	'S'	Step plot (vertical first)	
	'n'	No plotting	

Please note that,

i. In R, it is possible to display two or more graphs atop each other rather than opening a new window for each graph. Instead of reusing `plot()`, this can be achieved by employing the `lines()` and `points()` functions for the second line and scatter graph, respectively, however, the arguments in these functions are same as of `plot()` function. We do not need to provide x-axis in these functions.

ii. When dealing with multiple graphs within a single window, it is possible to add a legend into each graph, serving as a guide for the viewer to interpret the graphical elements. The `legend()` function can be used to add legend. It has many arguments, however, some important arguments are listed in Table 1.12.

iii. R also offers the capability to merge multiple graphs into a single image for ease of viewing using the `par()` function. It is necessary to set the layout parameters prior to calling the `plot()` function for our graph.

iv. To add straight lines to an existing plot in R, the straightforward `abline()` function can be utilized. This function accepts four arguments: a, b, h, and v. Here, a and b denote the slope and intercept, respectively; h signifies the *y*-coordinate points for horizontal lines, while v signifies the *x*-coordinate points for vertical lines.

v. We can also add any text to the plot using the `text()` function. It has many arguments, some important arguments are listed in Table 1.13.

TABLE 1.12

Some Important Arguments of `legend()` Function in R

Arguments	Function
x, y	The *x* and *y* coordinates determine the position of the legend and can also be specified using keywords such as `topleft, top, topright, left, center, right, bottomleft, bottom,` and `bottomright`.
legend	A vector of characters or expressions with a length of at least one, to be displayed in the legend.
col	The color of points or lines displayed in the legend.
lty, lwd	The line types and widths for lines featured in the legend. Either of these must be specified for drawing lines.
pch	The plotting symbols displayed in the legend, specified as a numeric vector or a vector of 1-character strings. It must be specified for drawing symbols.

TABLE 1.13

Some Important Arguments of `text()` Function in R

Arguments	Function
x, y	Numeric vectors of coordinates where the text labels should be placed. If the length of the *x* and *y* vectors differs, the shorter one is recycled.
labels	A character vector or expression that defines the text to be displayed. If the length of labels exceeds the length of the x and y vectors, the coordinates are recycled to match the length of labels.
cex	A numeric character expansion factor; when multiplied by `par('cex')`, it determines the final character size. The values `NULL` and `NA` are interchangeable and equivalent to `1.0`.
col, font	The color and font to be employed, potentially as vectors. These values default to the global graphical parameters specified in `par()`, unless `vfont` is set to `NULL`.

Saving and Exporting Plots Generated by `plot()` *Function*

   i. We have the option to store our plots in various file formats using the respective functions such as pdf (`pdf()`), jpeg (`jpeg()`), png (`png()`), or tiff (`tiff()`). These functions, with their arguments, need to be called before the plot is created.

  ii. The initial argument for these functions is the path and/or filename where the plot should be saved.

 iii. Additionally, when utilizing these functions, one can specify the image size by using arguments such as `width`, `height`, `units`, background color (`bg`), and resolution (`res`).

  iv. If the file does not exist, it will be generated; however, if it already exists, it will be overwritten without notification.

   v. Subsequently, the connection should be closed by executing `dev.off()` after creating the plot.

Please note that all these functions, except for `pdf()`, save a single image (last plotted image before `dev.off()`) though it could be a multi-panel plot). Conversely, `pdf()` is designed to save multiple images as distinct pages.

Examples of Plotting Data Using `plot()` *Function*

We can overlay more than one plot into a single window using the following R-code:

R-code:

```
tiff('E:/Book/Chapter 1/Figure 1.5.tiff', width = 300, height = 300,
units = 'mm', bg = 'white', res = 300)
x <- 1:10
y1 <- x^2
y2 <- 2 * x
y3 <- 3 * x
# Plotting the first graph
plot(x, y1, type = 'b', pch = 18, col = 'blue', lwd = 2, ylim = c(0,
max(y1, y2, y3)), xlab = ", ylab = ")
# Adding lines for the second and third graphs
lines(x, y2, type = 'b', pch = 19, col = 'red', lwd = 2)
lines(x, y3, type = 'b', pch = 8, col = 'orange', lwd = 2)
# Adding legend
legend('topleft', legend = c('y = x^2', 'y = 2x', 'y = 3x'),
       col = c('blue', 'red', 'orange'), pch = c(18, 19, 8), lwd = 2)
# Adding straight line
abline(h = 50, lty = 2, lwd = 2, col = 'black')
abline(v = 5, lty = 2, lwd = 2, col = 'black')
abline(a = 10, b = 4.4, lty = 3, lwd = 3, col = 'darkgreen')
# Adding a title and labels to the axes
title(main = 'Plot of X versus f(X)', col.main = 'blue', font.main = 4)
title(xlab = 'X', col.lab = 'red')
title(ylab = 'Y = f(X)', col.lab = 'red')
# Adding text
text(3.5, 100, 'This is an example of plotting in R', col = 'purple', cex
= 1.2)
dev.off()
```

The output plot is illustrated in Figure 1.5.

The following R-code can be used to visualize multiple plots in separate panel into a single window:

R-code:

```
tiff('E:/Book/Chapter 1/Figure 1.6.tiff', width = 300, height = 300,
units = 'mm', bg = 'white', res = 300)
x <- 1:10
y1 <- x^2
y2 <- log(x)
```

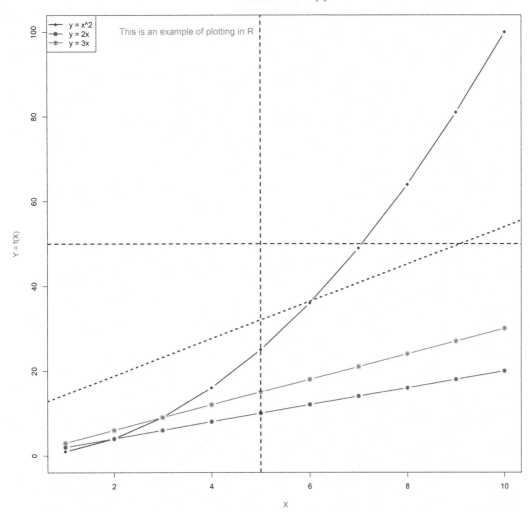

FIGURE 1.5
Overlaying multiple plots into a single window.

```
y3 <- exp(x)
y4 <- sqrt(x)
# Creating 2 by 2 panel window
par(mfrow = c(2,2))
# Plotting the graphs
plot(x, y1, type = 'b', pch = 18, col = 'blue', lwd = 2, xlab = 'X', ylab
= expression(y = x^2))
plot(x, y2, type = 'b', pch = 19, col = 'red', lwd = 2, xlab = 'X', ylab
= expression(y = log(x)))
plot(x, y3, type = 'b', pch = 8, col = 'orange', lwd = 2, xlab = 'X',
ylab = expression(y = e^x))
```

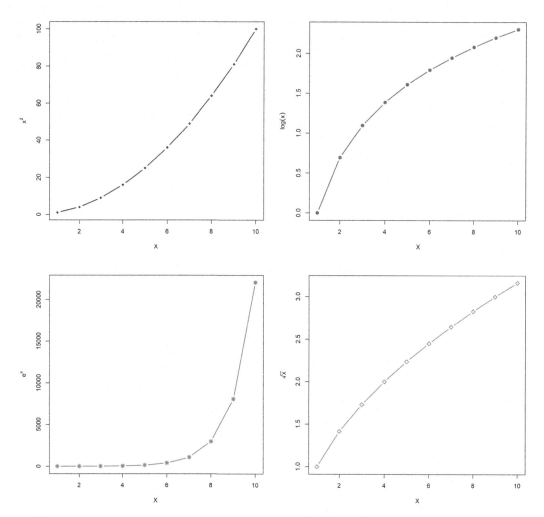

FIGURE 1.6
Multiple plots in separate panel into a single window.

```
plot(x, y4, type = 'b', pch = 5, col = 'purple', lwd = 2, xlab = 'X',
ylab = expression(y = sqrt(x)))
dev.off()
```

The output plot is shown in Figure 1.6.

Plotting Data Using `ggplot()` *Function*

'ggplot2' is a powerful and widely used data visualization package in the R programming language. Developed by Hadley Wickham, ggplot2 is part of the tidyverse collection of packages, which emphasizes a consistent and coherent approach to data manipulation, exploration, and visualization. The philosophy behind ggplot2 is based on the Grammar of Graphics, which defines a set of rules for constructing graphics by combining different elements. This grammar consists of data, aesthetics, and geometric objects (geoms) that represent the visual elements of the plot. The basic structure of the ggplot is given below:

```
ggplot(data = <DATA>, mapping = aes(<MAPPINGS>)) +
  <GEOM_FUNCTION>()
```

Or, we can specify aesthetics for each given GEOM_FUNCTION() independently instead of the aesthetics defined globally in the ggplot() function as like below:

```
ggplot(data = <DATA>) +
  <GEOM_FUNCTION>(mapping = aes(<MAPPINGS>))
```

Here are some key components of ggplot2:

- *Data*: The first step in creating a plot with ggplot2 is to specify the dataset. This is done by passing a data frame to the ggplot() function.
- *Aesthetics*: Aesthetics define how variables in the dataset map to visual elements in the plot. Common aesthetics include x and y for the axes, color for color-coding points, fill color, size and shape for point, labels, transparency (alpha), line width, and line type.
- *Geometrics*: It defines how data being displayed. GEOMS are the graphical representations of data points which include points, lines, and bars.
- *Facets*: Faceting allows you to create multiple plots based on one or more categorical variables, providing a way to visualize subsets of the data.
- *Statistics*: ggplot2 can seamlessly integrate statistical summaries and transformations into data visualizations. Examples include binning, smoothing, descriptive, and intermediate.
- *Coordinates*: These coordinate functions give you flexibility in how you present your data visually and allow you to adapt the coordinate system to suit the requirements of your analysis or the nature of your data. Examples include Cartesian, fixed, polar, flip, map, and limits.
- *Themes and customization*: ggplot2 provides flexibility for customizing the appearance of plots. You can modify axis labels, titles, legends, and overall themes.

Saving and Exporting Plots Generated by ggplot() Function

The output of ggplot() function can be saved using ggsave() function as like below:

- ggsave('Path/plot name.jpg', generated plot)
- ggsave('Path/plot name.png', generated plot)
- ggsave('Path/plot name.tiff', generated plot)
- ggsave('Path/plot name.pdf', generated plot)

To gain further insight into the theoretical foundations of ggplot2, I suggest exploring 'The Layered Grammar of Graphics' (http://vita.had.co.nz/papers/layered-grammar.pdf) and 'The Cheat sheet of Data Visualization with ggplot2' (https://github.com/rstudio/cheatsheets/blob/main/data-visualization.pdf).

Example of Plotting Data Using `ggplot()` *Function*

An illustration of plotting data using the `ggplot()` function is presented below, employing hypothetical data. The process is outlined step by step, with each step introducing a different layer to enhance the plot.

R-code:

```
# Generating a data frame
library(MASS)
set.seed(01234)
data <- as.data.frame(mvrnorm(n=200, mu=c(0, 0), Sigma=matrix(c(5, 3, 4,
4), ncol=2)))
colnames(data) <- c('x', 'y')
data$z1 <- as.factor(sample(1:2, 200, replace=TRUE))
data$z2 <- as.factor(sample(1:3, 200, replace=TRUE))
data$z3 <- as.factor(sample(1:4, 200, replace=TRUE))
# Installing the package ggplot2
install.packages('ggplot2')
# Loading the package ggplot2
library(ggplot2)
# Data and aesthetic layer
plot <- ggplot(data = data, aes(x = x, y = y))
# Geometric layer
plot1 <- plot + geom_point(size = 2, col = 'red', shape = 20)
# Save the plot1 as tiff format
ggsave('E:/Book/Chapter 1/Figure 1.7.tiff', plot1)
```

The output plot is illustrated in Figure 1.7.

We can change the color, size, and shape of the points according to the factor variables included in the dataset using the following R-code:

R-code:

```
# Changing color, size, and shape according to the factor variables
plot2 <- plot + geom_point(aes(col = z1, size = z2, shape = z3))
ggsave('E:/Book/Chapter 1/Figure 1.8.tiff', plot2)
```

The output plot is illustrated in Figure 1.8.

We can add facet layer to the plot in order to partition the plot according to a factor variable such as z3-variable using the following R-code:

R-code:

```
# Facet Layer
plot3 <- plot + geom_point(aes(col = z1))
```

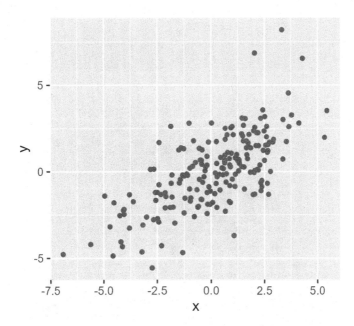

FIGURE 1.7
Scatter plot of *x*-variable versus *y*-variable.

```
# Partition the plot into rows and columns according to z3
plot4 <- plot3 + facet_wrap(facets = vars(size = z3))
ggsave('E:/Book/Chapter 1/Figure 1.9.tiff', plot4, width=5, height=5,
units = 'in', dpi = 300)
# Partition the plot into rows according to z3
plot5 <- plot3 + facet_grid(z3 ~ .)
ggsave('E:/Book/Chapter 1/Figure 1.10.tiff', plot5, width=6, height=6,
units = 'in', dpi = 300)
# Partition the plot into columns according to z3
plot6 <- plot3 + facet_grid(. ~ z3)
ggsave('E:/Book/Chapter 1/Figure 1.11.tiff', plot6, width=8, height=4,
units = 'in', dpi = 300)
```

The output plot is illustrated in Figures 1.9–1.11.

The following R-code can be used to add statistics layer to the plot illustrated in Figure 1.9.

R-code:

```
# Statistics layer
plot7 <- plot4 + geom_smooth(method = lm, col = 'blue')
ggsave('E:/Book/Chapter 1/Figure 1.12.tiff', plot7, width=5, height=5,
units = 'in', dpi = 300)
```

The output plot is illustrated in Figure 1.12.

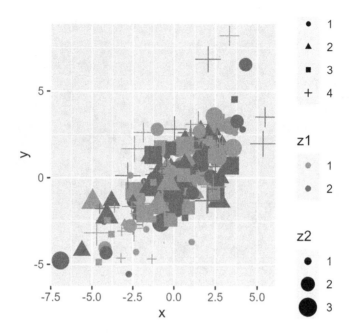

FIGURE 1.8
Scatter plot of x-variable versus y-variable with different color, size, and shape of the points according to the factor variables z_1, z_2, and z_3, respectively.

We can add color factor, statistics layer, coordinates layer, theme layer, and customization layer simultaneously to the plots illustrated in Figure 1.7 using the following R-code:

R-code:

```
# Adding color factor
plot8 <- plot + geom_point(aes(col = z2))
# Statistics layer
plot9 <- plot8 + geom_smooth(method = lm, col = 'blue')
# Coordinates layer
plot10 <- plot9 + coord_cartesian(xlim = c(min(data$x), max(data$x)),
ylim = c(min(data$y), max(data$y)))
# Themes and customization layer
plot11 <- plot10 + theme(plot.background = element_rect(fill =
'lightblue', colour = 'gray')) +
labs (title = 'Scatter plot of x-variable versus y-variable',
     x = 'x-variable',
     y = 'y-variable')
ggsave('E:/Book/Chapter 1/Figure 1.13.tiff', plot11, width=8, height=5,
units = 'in', dpi = 300)
```

The output plot is illustrated in Figure 1.13.

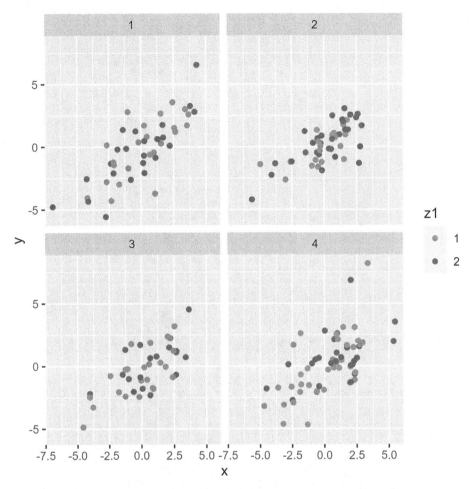

FIGURE 1.9
Scatter plot of x-variable versus y-variable (partitioned into rows and columns according to the factor variable z_3).

1.4 Conclusion and Exercises

This section presents the summary of this chapter and includes problem-solving exercises.

1.4.1 Concluding Remarks

Data science is becoming part of our everyday life. This introductory chapter has provided a range of highlights from data science, including data and microdata, big data, types of data, big data challenge and uncertainty, statistics to data science, and the scope of data science. In addition, this introductory chapter provides a comprehensive overview of the fundamental concepts and principles that form the bedrock of R, a powerful and versatile programming language for statistical computing and data analysis. It covers object, class, environment, console, R-script, help functions, operators, rules of writing code, data types and structure, built-in and user-defined functions, and R packages. This chapter also encompasses instructions for the installation of R, RStudio, and relevant packages, guiding

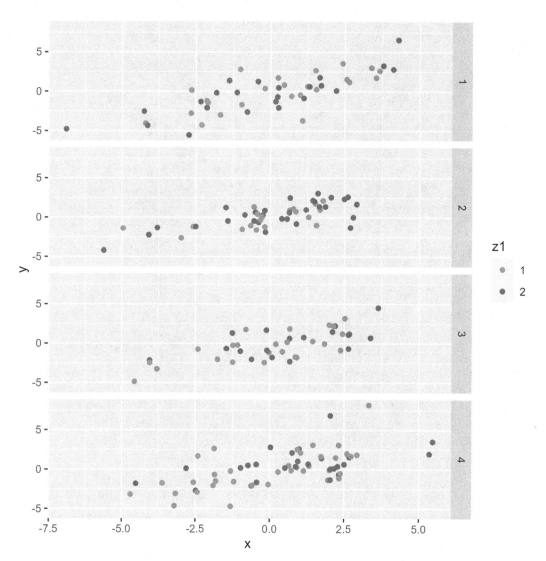

FIGURE 1.10
Scatter plot of x-variable versus y-variable (partitioned into rows according to the factor variable z_3).

users through the process of coding using R-scripts. Moreover, the provided example code caters to a range of tasks, including basic operations, importing data, basic statistical analysis, and data plotting, making it valuable for both beginners and individuals seeking to reinforce their comprehension. As we move forward in our journey with R, the proficiency gained in these fundamental coding principles will undoubtedly serve as a springboard for more advanced analyses and complex programming tasks. R software will be utilized to analyze all examples and exercises in this book; therefore, you must download and install R/RStudio on your computer by following the detailed instructions in this chapter, and you should complete this chapter's reading and exercises before jumping to the next chapters.

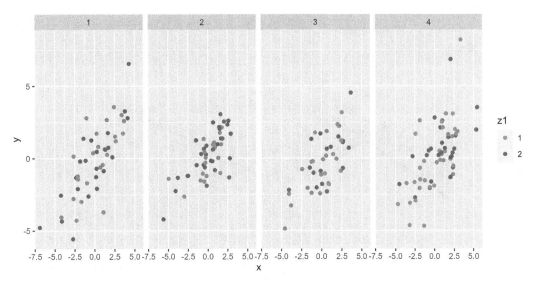

FIGURE 1.11
Scatter plot of x-variable versus y-variable (partitioned into columns according to the factor variable z_3).

1.4.2 Practice Questions

Exercise 1.1 Suppose that a public health practitioner/researcher collects data on the age (in years), gender, weight (in kg), body temperature, severity of illness (out of the five scores) and the overall risk condition levels (say, from H (high), M (moderate) and N (no) risk condition) of 27 patients in a local hospital, and then enter the data into R software for performing some analyses using data science techniques.

    a. Which statement about this data is correct?

        i. It is an unstructured dataset.

        ii. It is a semi-structured dataset.

        iii. It is a univariate structured dataset.

        iv. It is a multivariate semi-structured dataset.

        v. It is a multivariate structured dataset.

    b. What will happen if the health practitioner/researcher types m (in small letters) instead of M (in capital letters) for males for the last observation of the data while entering them into R?

Exercise 1.2

    a. Download and install R on your computer and provide a screenshot of RGui interface.

    b. List some basic classes of objects used in R functions or expressions.

Exercise 1.3 Using R-script in RStudio,

    a. Generate two vectors, 'odd_numbers' and 'even_numbers', containing odd and even integers from 1 to 100, respectively.

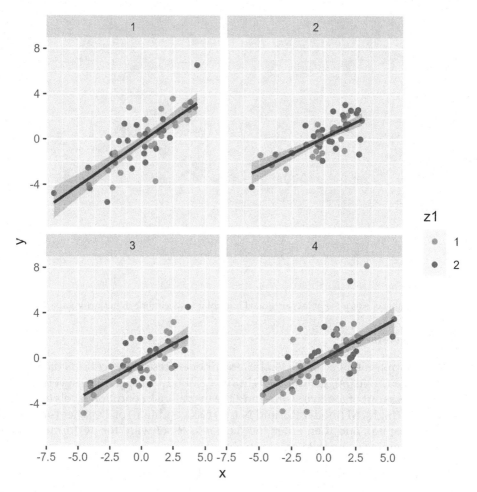

FIGURE 1.12
Scatter plot of x-variable versus y-variable with linear fitted line   (partitioned into rows and columns according to the factor variable z_3).

b. Determine the length of the vectors 'odd_numbers' and 'even_numbers'.

c. Create a vector 'result_vector' by adding the vectors 'odd_numbers' and 'even_numbers'.

d. Construct a data frame named 'numbers_df' with the vectors 'odd_numbers', 'even_numbers', and 'result_vector'.

e. Remove the 'odd_numbers' vector from the data frame.

f. Add another vector 'subtraction_vector' to the data frame by subtracting 'even_numbers' from 'result_vector'.

g. Reduce the data frame under the condition that values in 'subtraction_vector' are greater than 20.

h. Define a function 'calculate_z_sum' to compute the sum of values for $z = 3x^3 + 2y^2 + 5$.

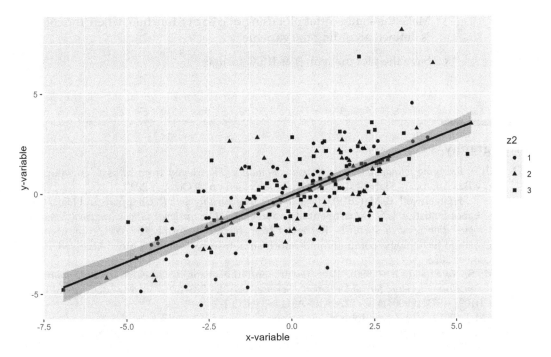

FIGURE 1.13
Scatter plot of *x*-variable versus *y*-variable with linear fitted line, colored with z_3-factor variable, and customized coordinates, axes, title, and theme.

    i. Use the 'calculate_z_sum' function to check the result considering the vectors 'result_vector' and 'subtraction_vector' from the reduced data frame as x and y, respectively.

Exercise 1.4 Imagine you have a dataset named 'My_Data' stored in CSV format within the Chapter 1 folder, which is located under the 'Book' directory on the 'E' drive of your computer's hard drive.

    a. Import the 'My_Data' data file as a data frame named 'my_data' into R from your hard drive.

    b. Check the dimensions of the data frame 'my_data'

    c. Check the variable types of the data frame 'my_data'

    d. Change the *z*-variable as a factor variable.

    e. Calculate sum, mean, and standard deviation of the variable '*x*'.

    f. Make a summary of all variables in the data frame 'my_data'

    g. Calculate correlation between '*x*' and '*y*'

    h. Make a scatter plot of '*x*' versus '*y*' using the plot() function with blue-colored small square points, and title the plot 'Scatter plot of *x* versus *y*' with axes labeled '*X*' and '*Y*'.

    i. Make the same scatter plot using ggplot() function.

j. Make the same scatter plot using `ggplot()` function, where the color is defined according the variable 'z'.

k. Save the plot made in j) as JPEG format.

Bibliography

ASA. (2015). *Emerging foundational professional communities*. Retrieved from https://en.wikipedia. org/wiki/American_Statistical_Association. Accessed on 09 October 2023.

Boehmke, B., & Greenwell, B. M. (2019). *Hands-on machine learning with R*. Chapman and Hall/CRC.

Bureau of Labor Statistics, U.S. Department of Labor. (2023). *Occupational outlook handbook, computer and information research scientists*. Retrieved from https://www.bls.gov/ooh/computer-and-information-technology/computer-and-information-research-scientists.htm. Accessed on 04 December 2023.

Crawford, S., & Stucki, L. (1990). Peer review and the changing research record. *Journal of the American Society for Information Science*, 41(3), 223–228. https://doi.org/10.1002/(SICI)1097-4571(199004)41:3<223::AID-ASI14>3.0.CO;2-3

Crawley, M. J. (2012). *The R book*. John Wiley & Sons.

Davies, T. M. (2016). *The book of R: a first course in programming and statistics*. No Starch Press.

De Vries, A., & Meys, J. (2015). *R for dummies*. John Wiley & Sons.

Fisher, A. R. (1925). *Statistical methods for research workers*. Oliver and Boyd.

Grolemund, G. (2014). *Hands-on programming with R: Write your own functions and simulations*. O'Reilly Media, Inc.

Harjule, P., Rahman, A., Agarwal, B., & Tiwari, V. (2023). A review of computational statistics and artificial intelligence methodologies. In Harjule et al. (Eds), *Computational statistical methodologies and modelling for Artificial Intelligence* (1st ed., pp. 3–22). CRC Press. https://doi.org/10.1201/9781003253051-2

Hossain, M. M. (2021). Statistics for data science and policy analysis. *Journal of the Royal Statistical Society Series A: Statistics in Society*, 184(4), 1612.

Ihaka, R., & Gentleman, R. (1996). R: a language for data analysis and graphics. *Journal of computational and graphical statistics*, 5(3), 299–314.

Kabacoff, R. (2022). *R in action: Data analysis and graphics with R and Tidyverse*. Simon and Schuster.

Lander, J. P. (2014). *R for everyone: Advanced analytics and graphics*. Pearson Education.

Long, J. D., & Teetor, P. (2019). *R cookbook: proven recipes for data analysis, statistics, and graphics*. O'Reilly Media.

Matloff, N. (2011). *The art of R programming: A tour of statistical software design*. No Starch Press.

Mount, J., & Zumel, N. (2019). *Practical data science with R*. Simon and Schuster.

Oxford Reference. (2008). *A dictionary of statistics* (2nd ed.). Oxford University Press.

R Core Team, R. (2013). *R: A language and environment for statistical computing*. R Foundation for Statistical Computing. http://www.R-project.org/

Rahman, A. (2019). Statistics-based data preprocessing methods and machine learning algorithms for big data analysis. *International Journal of Artificial Intelligence*, 17(2), 44–65.

Rahman, A. (2020). *Statistics for data science and policy analysis*. Springer.

Rahman, A. & Harding, A. (2017). *Small area estimation and microsimulation modeling*. CRC Press. https://doi.org/10.1201/9781315372143

Rahman, A., Russell, K., Kemp, M. & Ip, R. (2020). *Scientific data analysis – Study guide*. Charles Sturt University.

Rahman, A., Othman, N., Kuddus, M. A., & Hasan, M. Z. (2024). Impact of the COVID-19 pandemic on child malnutrition in Selangor, Malaysia: A pilot study. *Journal of Infection and Public Health*, 17(5), 833–842. https://doi.org/10.1016/j.jiph.2024.02.019

Science Council. (2009). *A definition of science*. Retrieved from https://sciencecouncil.org/about-science/our-definition-of-science/. Accessed on 07 September 2023.

Wickham, H., Çetinkaya-Rundel, M., & Grolemund, G. (2023). *R for data science*. O'Reilly Media, Inc.

Wikipedia. (2023a). *Data science*. Retrieved from https://en.wikipedia.org/wiki/Data_science. Accessed on 08 October 2023.

Wikipedia. (2023b). *Scientific method*. Retrieved from https://en.wikipedia.org/wiki/Scientific_method. Accessed on 08 October 2023.

Wikipedia. (2023c). *Scientific method*. Retrieved from https://en.wikipedia.org/wiki/Scientific_method#DNA_example. Accessed on 08 October 2023.

2

Contemporary Concepts of Biostatistics

2.1 Biostatistics

A definition of statistics is provided in Chapter 1. But what is biostatistics? How is it different than statistics? Although these sorts of questions would be addressed throughout the book, a data-centric definition of biostatistics follows.

Biostatistics can be considered as a specialized discipline in statistical science whose applications are focused on biological sciences, including medical sciences and public health. In terms of data, the application of biostatistics comprises various interdisciplinary knowledge domains, including biology, medicine, health, epidemiology, genetics, diseases, agriculture, nutrition, bioinformatics, and counting. The basic components of biostatistics entail designing a biostatistical hypothesis, collecting relevant data from a fully new experiment or survey or from existing data sources such as survey data or big data, preprocessing and presentation of data, and utilizing statistical data science techniques to model or find appropriate results for making evidence-based decisions and valid conclusions. Therefore, typically the term 'biostatistics' is used to distinguish the specific application of various statistical concepts, methods, tools, and techniques in the overall field of biological sciences, including agriculture and health sciences.

As undergraduate or postgraduate learners, educators, researchers, scholars, and practitioners, we involve situations or studies that are involved in generating data. Once the data are collected, we employ descriptive biostatistical methods to preprocess, organize, summarize, and effectively present the data for further explorations in relation to our analysis plan to address the hypothesis. Some biostatistical situations are mentioned below:

- **Clinical trials**: A pharmaceutical company is testing a new drug for hypertension. In this circumstance, biostatisticians design the clinical trial, randomize participants into treatment and control groups, and analyze the data to determine the drug's effectiveness and safety.

- **Epidemiological studies**: Biostatisticians investigate the association between smoking and lung cancer. Hence, biostatistical methods are used to analyze data from large populations to validate the association.

- **Public health surveys**: A government health agency conducts a survey to assess the prevalence of a specific disease in a population. Biostatisticians analyze the survey data to estimate the disease burden, identify trends, and inform public health interventions.

DOI: 10.1201/9781003426189-2

- **Genetic studies**: Scientists study the genetic basis of a rare disease by conducting genome-wide association studies (GWAS). Biostatisticians analyze genetic data to identify genetic variants associated with the disease.

- **Nutritional research**: Nutritionists investigate the relationship between dietary habits and the risk of developing diabetes. Biostatistics are used to analyze data from food diaries, clinical measurements, and health outcomes to draw conclusions about the impact of diet on health.

  A researcher is interested in classifying a study population according to the nutritional health outcomes (Rahman et al., 2024), e.g., children in a particular city or a country pertaining to the proportion of children who are malnourished.

- **Survival analysis**: Demographers and biostatisticians may analyze data collected from patients such as cancer patient records to estimate the survival rates after a certain treatment. Survival analysis techniques help in understanding the time until an event of interest (e.g., death) occurs.

- **Environmental health studies**: Environmental scientists and biostatisticians may be involved in studying the impact of climate change and environmental factors (such as air pollution or water contamination) on health outcomes. Statistical methods help assess associations and risks.

- **Quality control in healthcare**: Hospitals may use biostatistical methods to monitor and improve the quality of healthcare services. This includes analyzing patient outcomes, infection rates, and other performance metrics.

- **Vaccine efficacy studies**: Biostatisticians play a crucial role in evaluating the effectiveness of vaccines in preventing diseases. They analyze data from vaccine trials to assess the level of protection conferred by the vaccine.

- **Bioinformatics**: Advances in biotechnology allow for the collection of genomic, proteomic, and other molecular data. Biostatisticians analyze these high-dimensional datasets to identify genetic markers, pathways, and potential therapeutic targets for diseases.

- **Behavioral observations**: Researchers may observe and record behaviors related to health outcomes, such as addiction to smart devices and social media, dietary habits, physical activity, or smoking, to assess their impact on disease risk.

- **Electronic health records (EHR)**: Health systems and hospitals maintain electronic health records that contain a wealth of patient information. Biostatisticians can analyze EHR data to study disease patterns, treatment outcomes, and healthcare utilization.

2.2 Scope of This Book beyond Biostatistics

Nevertheless, statistical methods are widely used in various fields; therefore, this book aids learners, educators, researchers, scholars, and practitioners in diverse fields beyond biostatistics, as it presents statistical theories in a generalized manner, even though the examples primarily focus on biostatistics. The following examples highlight the versatility of statistical methods across various disciplines:

- **Environmental science**: Environmental scientists might use statistical methods to analyze air or water quality measurements over time, identifying trends and correlations with industrial emissions, population density, or other factors.

- **Education**: In education, a common statistical application is the analysis of test scores to assess the effectiveness of a teaching method. For instance, researchers might use statistical techniques such as *t*-tests or analysis of variance (ANOVA) to compare the average test scores of students exposed to different teaching approaches.

- **Social science**: Social scientists may use statistical techniques like descriptive statistics and inferential statistics to analyze survey responses on topics such as public opinion, social attitudes, or demographic trends.

- **Psychology**: Psychologists use statistical methods for designing experiments to test psychological hypotheses in order to identify underlying factors influencing psychological traits. They might use *t*-tests or analysis of variance (ANOVA) to analyze data from experiments comparing the effects of different treatments or interventions on a group of participants.

- **Economics**: Economists might use regression to analyze how changes in income (dependent variable) are related to changes in consumer spending, interest rates, or other factors (independent variables).

- **Finance**: In finance, statistical applications are widespread, and one notable example is the calculation of risk and return for investment portfolios. Modern portfolio theory, developed by Harry Markowitz, uses statistical methods to analyze the historical performance of different assets and their correlations. By applying concepts such as variance and covariance, investors can construct diversified portfolios that aim to maximize returns for a given level of risk. Additionally, Value at Risk (VaR), a statistical measure, is often used to estimate the potential loss in value of a portfolio under adverse market conditions.

- **Marketing**: Market analysts might use statistical methods to analyze customer demographics, purchasing patterns, and preferences. Through techniques such as regression analysis or clustering, marketers can identify target audiences, tailor marketing campaigns to specific customer segments, and optimize resource allocation for maximum return on investment.

- **Manufacturing and quality control**: Statistical process control (SPC) charts monitor key parameters in the manufacturing process, such as dimensions, weights, or defect rates, over time. By employing control charts and statistical techniques, manufacturers can identify variations that may indicate potential issues in the production process.

- **Sports analytics**: In sports analytics, a common statistical application is player performance analysis using advanced metrics. For example, in baseball, the use of statistics like on-base percentage (OBP), slugging percentage (SLG), and wins above replacement (WAR) allows teams to evaluate a player's overall contribution to the team. Moreover, predictive modeling is used to forecast sports match outcomes.

2.3 Basic Statistical Terms and Concepts

A range of conceptual terms that are used in statistics, including biostatistics, are presented in Table 2.1. It covers simple definitions with biostatistical examples for each of the concepts. These basic terms and concepts are essential for scientific investigation and statistical data analysis. As Table 2.1 is self-explanatory and most of the concepts will be treated further in relevant places within later chapters, we will not describe them further here in this chapter.

TABLE 2.1

A Presentation of the Basic Concepts of Biostatistics with Definitions and Examples

Terms	Definition	Example
Population	A statistical population is the entire set of subjects having one or more interested attribute/s for a study. In statistics, a population could be finite or infinite.	The total number of Australians who smoke daily (an example of a finite population), and the blood pressure measurements of all smokers in Australia (an example of an infinite population).
Parameter	A parameter is the statistical presentation of an interested characteristic of the population. In most cases, the Greek symbols are used to denote parameters.	The normal distribution has two parameters such as μ (i.e., population mean) and σ (i.e., population standard deviation).
Sample	A sample is a set of collected observations from the study population. A sample is also called data.	A collection of data from 2,500 smokers in New South Wales, Australia.
Random sample	When a sample is collected randomly (e.g., each population unit has an equal and independent chance of being selected in the sample) from the population, it is referred to as a random sample.	The weights of 150 patients were selected randomly from a hospital record of 2,500 patients.
Statistic	A statistic is the presentation of an interested characteristic of the sample.	The mean or average of the sample is represented by $\bar{x}$, and called a statistic.
Data	A set of collected information is called data.	Birth measurements (e.g., weight, length, sex) of 25 newborn babies in Sydney constitute a dataset.
Variable	A variable is a characteristic that has different measured values in the dataset. Variables could be classified into two major categories: categorical and numeric variables.	Weight, length, and sex are all examples of variables.
Nominal scale	Variables whose possible values are just names or labels that are arbitrarily assigned to attributes for the classification of the data.	Nominal variables are sex (female or male) and blood type (A, B, AB, O).
Ordinal scale	Variables whose possible values could be numerical or alphabetical. These values can be put in an order (e.g., from best to worst, or least to greatest), but we cannot do other mathematical operations on them.	Different stages of a particular cancer for 74 patients.
Interval scale	Numerical values denote units of equal magnitude and rank-order on a scale without an absolute zero.	Measurements of the body temperature of patients.

(Continued)

TABLE 2.1 (Continued)

Terms	Definition	Example
Ratio scale	Numerical values denote units of equal magnitude and rank-order on a scale with an absolute zero.	The duration of effectiveness time of a particular medicine administrated to patients.
Distribution	The systematic organization of a collection of data	The distribution of the data can be presented by a graph or frequency distribution.
Random variable	When a probability measure is assigned to each value of a variable that comes from a random phenomenon, then the variable is called a random variable.	Survival time of cancer patients in London.
Normal distribution	A grouping of data that is graphically symmetrical and bell-shaped.	Patients' anatomic (e.g., weight) and physiologic (intelligence score) traits are normally distributed.
Sampling distribution	The probability distribution of a statistic is called a sampling distribution. Sampling distributions constitute the test statistic for statistical inference (e.g., statistical hypothesis testing).	The distribution of $\bar{x}$ is a sampling distribution (i.e., a normal or t distribution).

2.4 Process of Biostatistical Data Analysis

Data analysis is a systematic process that involves inspecting, cleaning, transforming, and modeling data to extract useful information, draw conclusions, and support decision-making. Here a general outline of the biostatistical data analysis process is illustrated in Figure 2.1.

2.4.1 Define the Research Problem, Objective, and Questions

Defining the research problem or objective is the foundation of any research process. It provides a roadmap to the researchers for the entire research process, guiding them in formulating hypotheses, designing methodologies, and analyzing results. Clarity, relevance, and

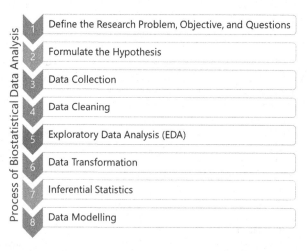

FIGURE 2.1
General outline of the biostatistical data analysis process.

feasibility are key principles to uphold in this crucial stage of research planning. It is necessary to clearly define the key research questions whose answers are sought by the researchers because it helps the data analysts by providing a complete data analysis framework. For instance, a team of researchers undertook a study to address the research problem of investigating the prevalence, determinants, and consequences of problematic smartphone use (PSU) among preschoolers. The study aimed to accomplish three key objectives: first, to estimate the prevalence of problematic smartphone users among preschoolers; second, to identify the determinants of PSU among this age group; and third, to assess the consequences of PSU on the physical and mental health of preschoolers. The study addressed various research questions, including the prevalence of PSU among preschoolers, the reasons for preschoolers' access to smartphones, the determinants influencing PSU in this demographic, the extent to which significant factors (if any) increase or decrease the likelihood of PSU, the potential adverse effects of PSU on the physical and mental health of preschoolers, and the degree to which PSU significantly elevates the likelihood of physical and mental health problems in this population (Abdulla et al., 2023a).

2.4.2 Formulate the Hypothesis

Formulating research hypotheses is a crucial step in the research process, as it guides the investigation and analysis of data. A hypothesis is a declarative statement that predicts the relationship between variables or anticipates the outcome of a research study. A well-constructed hypothesis guides the collection and analysis of data, providing a framework for drawing meaningful conclusions and contributing to the advancement of knowledge in a particular field. For example, Abdulla et al. (2023a) investigated several hypotheses related to problematic smartphone use (PSU) among preschoolers in their study such as:

a. There is no significant relationship between preschoolers' PSU and the duration of time they spend using smartphones per day.

b. There is no significant relationship between preschoolers' PSU and the duration of time their mothers spend using smartphones per day.

c. There is no significant relationship between preschoolers' PSU and the duration of time their fathers spend using smartphones per day.

d. Preschoolers' PSU is not significantly associated with their mothers' education level.

e. Preschoolers' PSU is not significantly associated with their fathers' education level.

f. Preschoolers' PSU is not significantly associated with their mothers' profession.

g. Preschoolers' PSU is not significantly associated with their fathers' profession.

h. There is no association between preschoolers' PSU and their mothers' age.

i. There is no association between preschoolers' PSU and the number of siblings they have.

j. There is no association between preschoolers' PSU and their family income.

k. Preschoolers' PSU does not have an adverse effect on their physical health.

l. Preschoolers' PSU does not have an adverse effect on their mental health.

These hypotheses were investigated to gain a better understanding of the factors potentially contributing to problematic smartphone use among preschoolers in their study population.

2.4.3 Data Collection

Data collection is a fundamental process in the field of research, analytics, and information gathering that will be used to solve the research problem and generate knowledge. It involves systematically measuring the variables of interest to answer research questions, test hypotheses, or make informed decisions. The quality of collected data significantly influences the validity and reliability of any analysis or study. Thoughtful planning, careful execution, and ethical considerations contribute to the reliability and validity of collected data. It is also crucial to ensure data completeness, i.e., whether the dataset includes all the data intended to be collected or not. Because it is a critical aspect of data quality, and incomplete data can hinder the reliability and validity of analyses and conclusions.

2.4.4 Data Cleaning

Data cleaning is the process of identifying and correcting errors, inconsistencies, inaccuracies, addressing missing values, and removing duplications in datasets. It is a necessary step in the data preparation phase before analysis or modeling, as the quality of the data directly impacts the reliability and validity of the results. Therefore, through systematic and careful cleaning processes, researchers can ensure that their data accurately reflects the real-world phenomena they are studying.

2.4.5 Exploratory Data Analysis

Exploratory data analysis (EDA) is an important phase in the data analysis process, serving as the compass that guides researchers and analysts through the intricacies of a raw dataset. This multifaceted approach involves a meticulous examination of the dataset's characteristics, patterns, and outliers to derive meaningful insights. EDA employs a combination of statistical techniques and visualizations to unravel the inherent structure of data, facilitating a deeper understanding of its distribution, central tendencies, and relationships between variables. By meticulously cleaning and transforming the data, EDA lays the foundation for subsequent analyses, hypothesis testing, and model development. In essence, exploratory data analysis is the enlightening preamble to the broader narrative of data-driven discovery and understanding. Different techniques of EDA are covered in Chapters 3, 4, 5, 8, 10, and 11.

2.4.6 Data Transformation

Following exploratory data analysis, data analysts can determine whether data transformation is necessary. If deemed necessary, data transformation becomes a pivotal step in the realm of data analysis, encompassing the modification of the structure or format of raw data to enhance its suitability for subsequent analysis or modeling. Data transformation not only refines the data's usability but also contributes to improving model performance and the overall interpretability of analytical results. It is a fundamental step that bridges raw data with the analytical techniques employed to extract meaningful insights. There are various techniques to transform the data. Some popular techniques are mentioned below:

- **Normalization**: Numeric normalization ensures that variables with different scales are brought to a common scale, preventing one variable from dominating the analysis due to its magnitude. Scale numerical variables to a standard range (e.g., between 0 and 1) to ensure that variables with different scales contribute equally to the analysis.

- **Standardization**: Standardization transforms data to have a mean of zero and a standard deviation of one, aiding algorithms that assume a standard distribution.

- **Log transformation**: Apply the natural logarithm to data, especially useful for data that exhibits exponential growth or has a highly skewed distribution.

- **Box-cox transformation**: A family of power transformations that includes the logarithm as a special case. It is useful for stabilizing variance and making the data more closely approximate a normal distribution.

- **Categorical encoding**: Categorical variables are often transformed into numerical representations through techniques like one-hot encoding, enabling their incorporation into mathematical models.

- **Binning or discretization**: Group continuous data into discrete bins or intervals. This can simplify the analysis and capture nonlinear relationships.

- **Imputation**: Handle missing data by replacing missing values with estimated or calculated values. Imputation methods include mean, median, or mode imputation, as well as more advanced techniques like regression imputation.

- **PCA (principal component analysis)**: Reduce dimensionality by transforming data into a new set of uncorrelated variables (principal components) that retain most of the original information.

- **Interaction terms**: Create new variables by combining two or more existing variables. Interaction terms can capture relationships between variables that may not be apparent when considered individually.

- **Smoothing**: Reduce noise in time series data or other fluctuating datasets by applying smoothing techniques such as moving averages.

The choice of data transformation techniques depends on the nature of the data, the objectives of the analysis, and the requirements of the specific modeling or analytical methods being used.

2.4.7 Inferential Statistics

Inferential statistics are the analytical cornerstone of drawing meaningful conclusions about populations based on information gathered from representative samples. Unlike descriptive statistics, which summarize and describe features of a dataset, inferential statistics extend insights beyond the observed sample to make inferences about the entire population. This branch of statistics utilizes probability theory and various statistical techniques to assess the likelihood of observed results, estimate population parameters, and evaluate the generalizability of findings through hypothesis testing. It allows researchers, analysts, and decision-makers to derive actionable insights, quantify uncertainty, and make informed predictions in situations where examining an entire population is impractical. Different contemporary inferential techniques are covered in a few chapters of this book.

2.4.7.1 Sample Size Calculation

The importance of sample size calculation in inferential statistics cannot be overstated, as it directly influences the reliability and precision of statistical inferences drawn from a sample to make predictions about an entire population. Determining an optimal sample size is crucial for achieving statistical power – the ability of a study to detect meaningful effects when they truly exist. A well-calculated sample size ensures that the study is appropriately powered to detect significant differences or associations, reducing the risk of Type-II errors (failing to detect a real effect). Moreover, it contributes to the precision of estimates, influencing the width of confidence intervals and enhancing the accuracy of hypothesis tests. Inadequate sample sizes can lead to underpowered studies, compromising the validity of statistical analyses and hindering the generalizability of findings. Conversely, an excessively large sample size may incur unnecessary costs and resources. Hence, sample size calculation is a strategic and essential step in the design of studies, providing a balance between statistical power, precision, and resource efficiency in inferential statistics.

The most important features of a sample comprise how the data was collected, the sample size, the distribution of variables, and whether the variables are independent or paired. With these features, the possible sources of bias could be discovered, and the appropriate techniques and statistical testing process might be chosen to prevent such bias in the data. Hence, the calculation of sample size is an important thing in biostatistics, especially when we are collecting data or designing an experiment for data collection.

The calculation of an appropriate sample size depends on several factors, each playing a crucial role in ensuring the reliability and precision of statistical inferences. Here are key factors that influence sample size determination:

- **Effect size**: The magnitude of the effect or difference that researchers aim to detect. A larger effect size often requires a smaller sample size for detection.
- **Significance level (α)**: The chosen level of significance (e.g., 0.05) represents the probability of rejecting a null hypothesis when it is true. A lower significance level requires a larger sample size.
- **Power ($1 - \beta$)**: The probability of correctly rejecting a false null hypothesis. Higher power, often set at 0.80 or 0.90, requires a larger sample size.
- **Population variability (standard deviation)**: The amount of variability or dispersion in the population. A higher standard deviation typically necessitates a larger sample size to achieve precision.
- **Desired precision (margin of error)**: The acceptable level of error or precision in estimating a parameter. A smaller margin of error requires a larger sample size.
- **Study design**: The design of the study, whether it is a cross-sectional study, longitudinal study, or experimental design, can impact the required sample size.
- **Resource constraints**: Practical considerations, including time, budget, and available resources, can influence the feasibility of obtaining a larger sample.

Considering these factors collectively ensures that the sample size is appropriately tailored to the specific characteristics and objectives of the study, striking a balance between statistical power, precision, and practical constraints.

If we would like to calculate the confidence interval (CI), we will see the following form of a formula for estimation:

$$\text{point estimate of parameter} \pm \text{margin of error}.$$

Here, the maximum difference deduced by CI between the point estimate and the true value of the parameter is the margin of error. There are ways to select the value of n (sample size). One simple way is to choose a number that will give us a prearranged estimate for the margin of error of CI. In many situations, this could be fairly challenging though. Thus, we limit our calculation of the sample size to two simple cases in the following way.

a. When the population mean (μ) of a normal distribution can be estimated, and the value of the population standard deviation (σ) is known; and

b. When the population proportion (p) can be estimated, and the data size is large enough to assume that the distribution of a sample proportion ($\hat{p}$) follows an approximately normal distribution with appropriate parameters.

For the first situation a), suppose a $100(1 - \alpha)\%$ CI for the population mean (i.e., μ) of the normal distribution. When we have a desired margin of error at the maximum of m, then the sample size can be calculated by using the following statistical formula:

$$n \geq \left(\frac{Z_{\alpha/2} \times \sigma}{m} \right)^2,$$

where $Z_{\alpha/2}$ is the critical value of the Z score at the probability of $\frac{\alpha}{2}$, and α is a desired cut-off value of the level of significance.

Example 2.1

Consider we aim to calculate a 95% CI for the mean of a normal distribution population which has a standard deviation of 1. We also considered that the margin of error is to be no more than 0.05. What sample size do you need to achieve such a reliability of our estimation?

A 95% CI estimation involves that $\alpha = 0.05$, and hence $Z_{\alpha/2} = Z_{0.025} = 1.96$. From the question we also have $\sigma = 1$, and the margin of error we want is $m \leq 0.05$. By using these values, we obtain the sample size as:

$$n \geq \left(\frac{1.96 \times 1}{0.05} \right)^2 = 1536.64 \cong 1537.$$

In R, we use:

```
n=((((1.96*1)/(0.05))^2)
n
[1] 1536.64
```

Therefore, we need a sample size of at least 1,537 observations. We must always round up the calculation of sample size to ensure that the margin of error must not go over the predefined value of m. When the margin of error is important in a biostatistical investigation, it is required to determine the appropriate sample size before we collect the sample observations through a sample survey or an experiment.

Now, for the first situation b), consider a $100(1 - \alpha)\%$ CI for a population proportion (p). When a predefined margin of error is $\leq m$, the sample size can be calculated by using the following statistical formula:

$$n \geq \left(\frac{Z_{\alpha/2}}{m} \right)^2 p(1-p).$$

It could be noted that this calculation includes p (i.e., typically unknown). When we have an estimated value, i.e., $\hat{p}$ (maybe from an earlier study or experiment) of the population proportion p, then we replace p by the estimated value $\hat{p}$. But, if we do not have any knowledge of p, we could utilize the greatest possible value of $p(1 - p)$.

Example 2.2

We would like to obtain a 95% CI for the proportion of US voters who support a particular health policy measure introduced at the recent federal budget. When we consider the margin of error is to be $m \leq 0.05$, and an estimated value of $\hat{p} = 0.45$ (e.g., this estimate is from an opinion survey conducted six weeks earlier on the health policy that introduced in the budget).

By using the given values, we obtain the sample size as:

$$n \geq \left(\frac{1.96}{0.05}\right)^2 \times 0.45(1 - 0.45) = 380.3184 \cong 381.$$

Using R:

```
n=((1.96)/(0.05))^2)*(0.45*(1-0.45))
n
[1] 380.3184
```

Therefore, we need a sample size of at least 381 observations.

Moreover, if there is no earlier survey estimate available for p, we could use $p = 0.5$ since it does not support either 'yes' or 'no' to the health policy opinion. It also provides the greatest possible value of the margin of error. In this case, the calculated sample size is $n \geq 385$.

It is Important to note that the required sample size is not dependent to the size of the population from which sample has been taken. This is because typically the population is adequately big enough for collecting a random sample of size n observations from that population. Ultimately, the sample matters, but not the fraction of the overall population in which the sample belongs. Given that a random sampling procedure is utilized, a sample of 385 could give us a 95% CI estimation for the true population proportion with a margin of error ≤0.05.

A suitable sample size is central to any well-organized research study or designed experiment. Typically, the level of significance, power, effect size, and alternative hypothesis (i.e., one-sided or two-sided test) all influence the determination of an appropriate sample size for a scientific and data-centric investigation. For example, a stimulating relationship exists between the sample size (n) and power ($1 - \beta$), i.e., when n increases, so does the power to detect a real effect. Besides this, while n increases, a designed study can detect smaller and smaller effects with greater and greater power.

Therefore, the role of n in the power of a statistical test must be considered before a student or researcher goes on to sophisticated statistical analyses, including the analysis of variance, analysis of covariance, and regression modeling. An investigator can choose power and estimate the necessary sample size in advance or conduct power analysis afterward. In this book, we edge the scope of power analysis here with an example below and consider predetermining sample size estimation is best. However, interested readers are recommended to explore relevant studies in the literature (e.g., Cohen, 1988; Serdar et al., 2020; Champely et al., 2022; Zhang et al., 2022; Qiu et al., 2022). A comprehensive review of sample size, power, and effect size with their interrelationship is provided by Serdar et al. (2020). The study also describes simplified and practical estimation procedures using various analytic software, including R packages with clinical and public health data.

In research investigation, when one plans to conduct a hypothesis testing of the population mean (μ) against a known value of the mean parameter (μ_0), then hypotheses can be defined as $H_0 : \mu = \mu_0$ and $H_1 : \mu \neq \mu_0$. Now, we can use the following formulas to estimate our desired **effect size** (δ) and the necessary minimum **sample size** (n) to ensure that our test has a stipulated level of **power** ($1 - \beta$),

$$\delta = \frac{|\mu_a - \mu_0|}{\sigma}$$

$$n = \left(\frac{z_{(1-\alpha/2)} + z_{(1-\beta)}}{\delta}\right)^2 \text{ with } z = \frac{\mu_a - \mu_0}{\sigma / \sqrt{n}}$$

$$1 - \beta = \phi\left(z - \phi^{-1}\left(z_{(1-\alpha/2)}\right)\right) + \phi\left(-z - \phi^{-1}\left(z_{(1-\alpha/2)}\right)\right)$$

where
  μ_a is a specific value of the mean parameter under H_1 (i.e., other than μ_0),
  σ is standard deviation of the population,
  ϕ is the standard normal distribution function,
  ϕ^{-1} is the standard normal quantile function,
  α is the level of significance (type I error), and
  β is type II error.

It is to be noted that the inputs in $n = \left(\dfrac{z_{(1-\alpha/2)} + z_{(1-\beta)}}{\delta}\right)^2$ include the desired power, the level of significance and the effect size. Indeed, the effect size is chosen judiciously to represent a clinically meaningful or practically significant difference in the parameter of interest. Finally, the equation $1 - \beta = \phi(z - z_{(1-\alpha/2)}) + \phi(-z - z_{(1-\alpha/2)})$ result validates that for an estimated fixed sample size, our designed power of the test can be achieved with the chance of detecting a real difference (i.e., effect size), if there is one, between the estimate of sample mean and the population mean.

To demonstrate the use of the formulas mentioned above, let us consider an example concerning a study of osteoporosis in post-menopausal women (Chow et al., 2008). Osteoporosis and osteopenia (or decreased bone mass) most commonly develop in post-menopausal women. The consequences of osteoporosis are vertebral crush fractures and hip fractures. Osteoporosis is diagnosed when vertebral bone density is more than 10% below what is expected for sex, age, height, weight, and race. Usually, bone density is reported as standard deviation (SD) from mean values. The World Health Organization (WHO) defines osteopenia as a bone density value greater than one SD below peak bone mass levels in young women and osteoporosis as a value greater than 2.5 SD below the same measurement scale. In medical practice, most clinicians suggest therapeutic intervention should be begun in patients with osteopenia to prevent progression to osteoporosis.

Example 2.3

Suppose that the mean bone density before the treatment is 1.5 SD (μ_0= 1.5 SD) and after treatment is expected to be 2.0 SD (i.e., μ_a= 2.0 SD). We have the difference of $\mu_a - \mu_0$ = (2.0 − 1.5) SD = 0.5 SD. Now, by the formula of δ, n, and $1 - \beta$, at $\alpha = 0.05$, the required sample size for having a 90% power (i.e., $\beta = 0.10$) for correctly detecting an effect size of δ change from pretreatment to post-treatment can be obtained, and then the power of the test can be verified with the following R codes.

Using R:

```
mu_a=2
mu_0=1.5
sigma=1
(delta=(abs(mu_a- mu_0)/sigma))
alpha=0.05
beta=0.10
(n=(sigma*(qnorm(1-alpha/2)+qnorm(1-beta))/(mu_a-mu_0)) 2)
ceiling(n)#
z=(mu_a-mu_0)/sigma*sqrt(n)
(Power=pnorm(z-qnorm(1-alpha/2))+pnorm(-z-qnorm(1-alpha/2)))
```

Outputs:

```
> mu_a=2
> mu_0=1.5
> sigma=1
> (delta=(abs(mu_a- mu_0)/sigma))
[1] 0.5
> alpha=0.05
> beta=0.10
> (n=(sigma*(qnorm(1-alpha/2)+qnorm(1-beta))/(mu_a-mu_0))^2)
[1] 42.02969
> ceiling(n)#
[1] 43
> z=(mu_a-mu_0)/sigma*sqrt(n)
> (Power=pnorm(z-qnorm(1-alpha/2))+pnorm(-z-qnorm(1-alpha/2)))
[1] 0.9000001
```

Similarly, when research aims to carry out a hypothesis testing of the population proportion (p) against a known value of the parameter (p_0), then hypotheses can be written as: $H_0 : p = p_0$ and $H_1 : p \neq p_0$. Now, the following formulas follow:

$$\delta_2 = \frac{|p_a - p_0|}{\sqrt{p_a(1-p_a)}}$$

$$n_2 = \left(\frac{z_{(1-\alpha/2)} + z_{(1-\beta)}}{\delta_2}\right)^2 \text{ with } z = \frac{p_a - p_0}{\sqrt{p_a(1-p_a)}}$$

$$1 - \beta_2 = \phi\left(z - \phi^{-1}\left(z_{(1-\alpha/2)}\right)\right) + \phi\left(-z - \phi^{-1}\left(z_{(1-\alpha/2)}\right)\right)$$

$$\delta = \frac{|p_a = p_0|}{\sqrt{p_a(1-p_a)}} \tag{2.4}$$

$$n = \left(\frac{z_{(1-\alpha/2)} + z_{(1-\beta)}}{\delta}\right)^2, \text{ with } z = \frac{p_a - p_0}{\sqrt{\dfrac{p_a(1-p_a)}{n}}} \tag{2.5}$$

$$1 - \beta = \Phi\left(z - z_{(1-\alpha/2)}\right) + \Phi\left(-z - z_{(1-\alpha/2)}\right) \tag{2.6}$$

where

p_a is a specific value of the mean parameter under H_1 (i.e., other than p_0),

ϕ is the standard normal distribution function,

ϕ^{-1} is the standard normal quantile function,

α is the level of significance (type I error), and

β_2 is type II error.

Consider the same example study by Chow et al. (2008) to demonstrate the application of the above formulas. Assume the investigation is also interested in examining the treatment effect in terms of the response rate at the end of the study. Sample size calculation can then be carried out based on the response rate to achieve the desired power. The definition of a responder should be given in the study protocol prospectively. For example, a subject may be defined as a responder if there is an improvement in bone density by more than one standard deviation (SD) or 30% of the measurements of bone density.

Example 2.4

Suppose that the response rate of the patient population under study after treatment is expected to be around 50% (i.e., $p = 0.50$). By δ_2, n_2 and $1 - \beta_2$, at $\alpha = 0.05$, the required sample size for having a 90% power (i.e., $\beta_2 = 0.10$) for correctly detecting effect size of δ_2 change between the post-treatment response rate and the reference value of 30% (i.e., $p_0 = 0.30$) can be obtained, and then the power of the test can be verified with the following R codes.

Using R:

```
p_a=0.5
p_0=0.3
(delta=(abs(p_a- p_0))/sqrt(p_a*(1-p_a)))
alpha=0.05
beta=0.10
(n=p_a*(1-p_a)*((qnorm(1-alpha/2)+qnorm(1-beta))/(p_a-p_0))^2)
ceiling(n) #
z=(p_a-p_0)/sqrt(p_a*(1-p_a)/n)
(Power=pnorm(z-qnorm(1-alpha/2))+pnorm(-z-qnorm(1-alpha/2)))
```

Outputs:

```
> p_a=0.5
> p_0=0.3
> (delta=(abs(p_a- p_0))/sqrt(p_a*(1-p_a)))
[1] 0.4
> alpha=0.05
> beta=0.10
> (n=p_a*(1-p_a)*((qnorm(1-alpha/2)+qnorm(1-beta))/(p_a-p_0))^2)
[1] 65.67139
> ceiling(n) #
[1] 66
> z=(p_a-p_0)/sqrt(p_a*(1-p_a)/n)
> (Power=pnorm(z-qnorm(1-alpha/2))+pnorm(-z-qnorm(1-alpha/2)))
[1] 0.9000001
```

Moreover, the process of determining the power of the statistical test for two independent sample cases is fairly similar to that of the one-sample situation we just described. Similar factors apply to the estimation process (Hickey et al., 2018). Additionally, some typical attributes of the power of a test are:

1. One-sided tests usually have more power.
2. An increasing level of significance (α) generally increases power.

3. A statistical test typically has more power against a bigger effect size.
4. An increasing sample size increases power.

2.4.7.2 Sample Size for RCTs, Cohort, Case-Control or Cross-Sectional Studies

Sample size calculation for continuous and binary variables in designed study or randomized controlled trials (RCTs) experiment does not differ from sample size calculation in other fields (i.e., approaches demonstrated earlier). However, time-to-event analysis (TEA), on the other hand, poses hurdles peculiar to RCTs (Wittes, 2002). The author recommends trying at least two techniques to determine the sample size for TEA. Suppose you get similar results that indicate correct estimations. However, getting very dissimilar results indicates an early warning that something tricky is happening, which requires further investigation. The various techniques need diverse parameterizations, and the researcher must deliberately consider how to translate data from the literature to the sample size calculation formulas. These sorts of translations will differ from one technique to another in any case.

There are four steps to calculate sample size in designing an RCT (Zhong, 2009):

1. The null and alternative hypotheses should be specified correctly, along with the α and β values.

2. The facts of relevant parameters of interest to be collected (e.g., from the literature and/or by conducting a pilot study).

3. The sample size must be estimated based on several reasonable parameters. For example, the choice of null and alternative hypothesis should be adjusted according to the study objective. Also, some investigators may find challenges in choosing a noninferiority/equivalence/superiority margin (in both statistical and clinical aspects). This parameter has clinical magnitude, which must be thoughtfully finalized and practical for the study.

4. The δ estimate should be chosen based on the explicit discussion of clinical experts and statisticians (i.e., not simply relying on only biostatisticians' or clinicians' advice).

There are specific risks in determining impractical effect size. If the δ estimate is too big, some inefficacious medicines could arrive at partitioners that could be considered as noninferiority/equivalence. In contrast, if the δ value is too small, some potential effective medicines could be neglected in practice. Besides, another point to keep in mind is that when δ is robustly obtained, it must not be altered throughout the study. Other factors affecting the sample size calculation for RCT include the dropout rate and underlying event rate in the population (Gupta et al., 2016).

Moreover, there are real-world challenges in obtaining the appropriate number of subjects for other studies such as 'cohort,' 'case-control,' and cross-sectional studies. A cohort study starts with defining a disease-free population or outcome by the exposure of interest and follows it until the disease/outcome of interest occurs (Charan & Biswas, 2013). Given that exposure is identified before the outcome in a cohort study, it has a temporal context to appraise causality. Hence, cohort study has the potential to offer robust and data-centric scientific results. Two specific advantages of cohort study are: (i) it can examine rare exposures since participants are chosen by their exposure status, and (ii) it can concurrently analyze multiple outcomes. Although researchers can get reliable results from a well-designed cohort study, it requires a big sample size and a potentially long follow-up period, resulting in attrition bias and a costly venture.

Additionally, case-control research determines observations by outcome status at the start of the study project. Outcomes of interest could be whether the individual has experienced a specific type of surgery, complication, or disease. When the outcome status is known, and individuals are grouped as cases, we choose individuals from the same population for the controls. Then, by conducting interviews, searching available source records, or using survey tools, data are collected on interested variables, including the exposure to a risk factor or several risk factors. A case-control study is suitable for investigating rare outcomes or events with a lengthy latency time since individuals are designated by the outcome status from the beginning. Therefore, case-control research is relatively quick and inexpensive to implement, compared to cohort studies. Typically, it needs a smaller sample size and can assess a single outcome of interest with multiple exposures or risk factors. However, in the sampling process, three crucial points must be carefully considered that follow (Song & Chung, 2010):

1. Clearly define the inclusion and exclusion criteria for the study beforehand to select homogenous cases.
2. Cases should be chosen from various sources, including hospital patients, clinic patients, or community individuals (e.g., patients from community registries could be a useful source of cases).
3. Selected cases should be representative of instances of interest in the target population.

While the case-control study approach looks convenient in many situations, the validity of the findings could be questionable. For example, external validity issues would arise if the selected cases do not represent the study population. Also, if cases are picked from only one source (e.g., a particular hospital), perceived risk factors could mainly be distinctive to that single source. So, the researcher could not generalize the findings to the wider community and strongly advocate for their implementation.

Cross-sectional studies examine the data on disease and exposure at one-particular-time point. Because the temporal relationship between disease occurrence and exposure cannot be established, cross-sectional studies cannot assess the cause-and-effect relationship (Song & Chung, 2010; Rahman et al., 2020; Abdulla et al., 2023b). Although appropriate recruitment of participants, sample size, and number of variables are important for any cross-sectional studies, commonly, they are at risk of low response rates or participation bias from subjects. Given that a larger sample size provides better power, a sample size of at least 60 subjects is recommended for a cross-sectional study. However, this would depend on the suitability of the research question and the variables being measured.

Furthermore, in observational research settings where data are already collected (e.g., electronic health records or insurance claims studies), the utility of these calculations is less apparent. Hence, it is important to note that in secondary analysis, the decision to proceed should not be based on sample size alone. As one of the objectives of this book is to minimize most of the technical contents and focus on applied aspects of data analysis with R, providing all formulae for sample size calculations is beyond this book's scope. Interested readers may explore relevant literature (e.g., Wittes, 2002; Song & Chung, 2010; Charan & Biswas, 2013; Gupta et al., 2016).

There are also various online options available that can assist in sample size calculations. It is to be noted that those online calculators may not have been validated, though. So, it is strongly recommended that a well-validated software platform be utilized. Besides, interested readers may check websites: 'OpenEpi' (http://www.openepi.com), 'Power and sample size' (http://powerandsamplesize.com), 'Biomath' (http://www.biomath.info), 'Java applets for power and sample size' (https://homepage.stat.uiowa.edu/~rlenth/Power/), and 'Sample for clinical trials/scientific experiments' (http://hedwig.mgh. harvard.edu/sample_size/size.html) for a few potentially usable calculators to estimate sample size (all accessed on 27 March 2024).

2.4.7.3 *Importance of Effect Size and Precision*

The previous section discussed effect size, sample size, and power calculations. When testing the null hypothesis in a designed study, we try to ensure a high degree of certainty that the study will be able to provide adequate power for the test. However, we can have another aim for the designed study. We could use the study to estimate the magnitude of the effect – to report (i.e., the treatment improves the recovery rate by 2, 5, or 7 units). With this specific goal in mind, study designing should focus not only on the power of the test but also on the precision with which it will allow us to estimate the magnitude of the effect to be detected. The precision with which we can report the rate difference is a function of the required confidence level, the sample size, and the variance of the outcome index.

Effect size is a primary factor in calculating the power of the tests and has little (if any) impact on determining precision. Although effect size does not directly affect precision, it may indirectly relate to precision. In particular, for techniques such as *t*-tests, ANOVA and ANCOVA that utilize mean differences of groups, the effect size is a function of the mean difference and the standard deviation within groups. The mean difference does not impact precision; however, the standard deviation affects both effect size and precision. That means a smaller standard deviation value provides higher power and better precision in the results. Additionally, for techniques that use proportions, the absolute value of the population proportion or correlation affects the variance estimation, which could impact precision. For example, variance estimation is a function of population proportions in Z-tests for proportions. Typically, variance is highest for proportions near 0.5, lower for proportions near 0 or 1, and relatively stable for other values between 0.1 and 0.9.

2.4.7.4 *Probability Theory*

Probability theory are fundamental concepts in statistical inference and data modeling, playing a crucial role in capturing and quantifying uncertainty within datasets for making informed inferences about populations based on sample data. Probability is employed in estimating parameters and hypothesis testing, where it assesses the likelihood of observing sample results under different hypotheses. Probability distributions, such as the normal distribution, underpin key statistical methods like constructing confidence intervals and forming the basis for sampling distributions. These distributions offer a framework for understanding the behavior of sample statistics, essential for making accurate predictions about population parameters. It enables the data modelers to incorporate randomness and variability into their models for characterizing the behavior of variables and making predictions about future observations. Overall, a solid understanding of probability and probability distributions enhances the robustness and reliability of data models, facilitating more accurate and meaningful insights from complex data. Probability theories are discussed along with examples in Chapter 6.

2.4.7.5 *Statistical Estimation and Hypothesis Testing*

In many situations, we want to comprehend something about the real value of a parameter. For a biostatistical example, a pharmacist dispensing an over-the-counter painkiller could be interested to know in how long it takes for the medicine to take effect. In this example, the variable is time: a quantitative continuous variable. The population might be the collection of times that it takes for this medicine to work for all adults in a community such as all adults in Florida. Also, in this case, the parameter of interest is the mean time, which can be defined as μ (typically unknown). However, if we consider that a representative sample mean (i.e., not is the same as μ) is $\bar{x}$, which is the average time obtained from a small sample of adults from the community. Here, the value of $\bar{x}$ is just a measure of the real value of unknown parameter μ.

Alternatively, we might be interested in whether a person supports a particular health policy of the US government. In this case, the variable is qualitative ('1 = Yes' or '0 = No'). The population might be the responses of all individuals on the US electoral registration, and the parameter of interest could be p, i.e., the percentage of US voters in favor of the health policy. This is different from a sample proportion $\hat{p}$, i.e., the percentage of people in support if we took a small sample of voters and asked their opinions.

In both examples illustrated above, we are concerned with two important statistical concepts:

Statistical estimation: Using sample data, we find our best estimate for the real value of an unknown parameter of the population. From the above examples, it looks naturally realistic to estimate the population mean, μ, of a quantitative variable X by the sample mean, $\bar{x}$, of a randomly collected sample from that population. The statistical theory supports this estimation of unknown parameter μ of the population. Additionally, the theory supports estimating the proportion, p, of a population that has a qualitative characteristic by the sample proportion, $\hat{p}$, of a randomly collected sample from that specific population.

It is significant that the sample observations must be random, which implies that every possible sampling unit of the population has the equal likelihood of being chosen in the sample under the condition that the simple random sampling is employed. If we sample the health policy opinions of only one part of the US people (e.g., female capital city voters, or young regional city voters), we may not get a random and representative sample.

Moreover, to assess the statistical reliability of an estimated value of a parameter we calculate a range of measures such as the sample standard deviation, standard error of the mean and confidence intervals. Besides, a confidence interval measure depends on the level of significance which is also known as the type-I errors. Typically, a smaller value of these measures indicates a more reliable estimated value of our parameter. Further details are discussed at relevant places in later chapters.

Hypothesis testing: Using sample data, we check statistical assumptions and test a claim about the real value of an unknown parameter of the population. The previous section discussed the case where we did not know the value of a population parameter (μ or p), and we wanted to estimate its value. However, in other circumstances, we know a value that the parameter is claimed to have, and we want to check whether this claim is true.

There are many examples in real life. Consider three of them:

Example 2.5

A medical research paper on that painkiller states that, the mean of the time for the medicine to take effect on patients is 30 seconds. Certainly, there will be variability in sample observations since not every patient's circumstance (such as body composition, and demographics) will exactly be the same which could impact to the effectiveness time of that painkiller.

Example 2.6

A hospital claims that at least 85% (a proportion, p) of its patients are satisfied with their services.

How can we test if these claims are true? It will not be enough to measure the effectiveness time of the medicine from one patient, or survey one outgoing patient in the hospital – because there is always variability in real-life data.

Example 2.7

The pharmaceutical company which produces the painkiller might claim that, with an average of 25 seconds the medicine will take effect. However, it is very likely that we will get a patient who reveals that it takes less than 25 seconds, and just as possible that we will get a patient who reveals that it takes more than 25 seconds. Therefore, we could take a random sample of patients administrated with that painkiller (or survey of outgoing patients) and determine if the data in our collected sample is consistent with the actual claim.

It could be noted here that the claim is not about the sample data; i.e., we are not saying that the mean effectiveness time of the medicine for a sample of 30 (say) patients that we have observed is 25 seconds. Instead, the claim is about the population parameter; i.e., the mean effectiveness time of the medicine is 25 seconds for all patients who took that painkiller. Hence, the claim is about the fact that the population mean μ, not the sample mean $\bar{x}$, is equal to 25 seconds. In this situation, we will conduct an appropriate statistical hypothesis testing. Further details about the statistical hypothesis testing with systematic analysis of data using R programming are provided in later chapters.

Statistical significance vs clinical significance: In medical statistics, in order to validate any difference identified by a clinical test is considered to be statistically significant – we should obtain the p-value (i.e., the lowest probability of error to conclude that the difference is statistically significant) for the analysis, which is somewhat linked to the power of the test. Then we compare it with a threshold cut-off value, i.e., an acceptable probability for the type-I error (which is usually 0.10, 0.05 or 0.01. If the estimated p-value is less than the predefined cut-off value, then we determine that the detected difference value is statistically significant.

Moreover, a statistical significance does not always guarantee clinical significance though. This is mainly because the p-value in the statistical analysis could be influenced by the size of data. Typically, a smaller sample size tends to generate a larger p-value, and the analysis outputs may have less practical implication; i.e., it may not be statistically significant. In such a situation, even if the results could have clinical significance, their findings would be misinterpreted because of the insufficient sample observations (Altman & Bland, 1995). Hence, we may consider that it is not always the statistical p-value, but the area under the curve (AUC) measure that validates the precision of a diagnostic test, especially when we analyze clinical, medical, or biological data.

2.4.7.6 *Approaches of Statistical Estimation and Hypothesis Testing*

Biostatistical data analysis methods can be classified as parametric and nonparametric techniques. To select an appropriate analysis method for data analysis, we need a decent understanding about the nature of the data since each method has its own statistical assumptions.

Parametric methods: Parametric approaches utilize the numerical data that is randomly collected from a population that is normally distributed. Parametric methods could be applied to many ratio-scale and interval-scale datasets when those observations are drawn from a population that follows the normal distribution with the appropriate mean and variance parameters. Parametric techniques can be applied to obtain measures of central tendency and variability of the data. We will discuss further about various parametric methods and their applications with real data in the next few chapters of the book.

Nonparametric methods: Nonparametric approaches are utilized to analysis data that is collected from a nonnormally distributed population or datasets which do not meet the criteria of parametric approaches. For example, to analyze the ordinal scale data (i.e., severity of COVID-19, Trauma score, or Injury severity score) we could use nonparametric techniques. We will discuss further about various nonparametric approaches and their applications with real-world data in Chapter 9 of this book.

Moreover, to determine if a parametric method or nonparametric method can be used for data analysis, we must check various criteria about the collected data and the statistical assumptions. It is also significant to carefully assess the statistical validity of those analysis methods assumptions. Besides, in order to choose the best analysis method, the number of data points and/or groups should also be taken into consideration. Figure 2.2 summarizes the key parametric and nonparametric analysis approaches for each situation of our contemporary data.

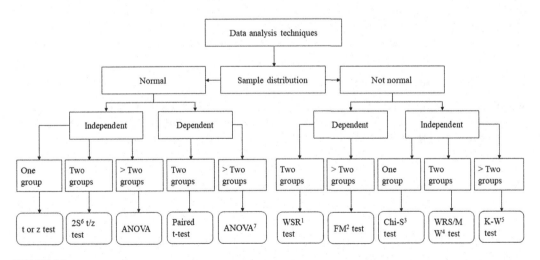

FIGURE 2.2
A summary of parametric and nonparametric methods for biostatistical data analysis.

Notes: [1]Willcoxon signed rank (WSR) test; [2]Friedman (FM) test; [3]Chi-squired (Chi-S) test; [4]Willcoxon rank sum (WRS)/Mann-Whitney (MW) test; [5]Kruskal-Wallis (KS) test; [6]Two samples (2S) test; [7]Analysis of variance (ANOVA).

2.4.8 Data Modeling

In biostatistics, data modeling is an essential component of the research process, contributing to the understanding of complex biological phenomena through statistical modeling in order to make inferences and facilitate effective decision-making. This process includes selecting appropriate statistical models, specifying relationships between variables, and estimating model parameters. The data modeling process in biostatistics often involves the application of regression models, survival analysis, and other specialized statistical methods tailored to the unique characteristics of biomedical data. These models help to investigate associations between variables and make comparisons between groups, facilitating the identification of factors influencing health outcomes or disease progression and predict outcomes. Precision and accuracy are paramount in biostatistical modeling, as the results directly impact healthcare decisions, clinical trials, and public health policies. Rigorous interpretation of biostatistical models is crucial for drawing meaningful conclusions and ensuring that research findings contribute to advancements in medical knowledge and inform public health policy for improving patient outcomes.

Data modeling serves as a versatile tool for both quantitative prediction and classification tasks, offering distinct types tailored to each objective. For quantitative prediction, linear regression is a fundamental choice, enabling the estimation of a continuous outcome variable based on one or more predictor variables. In contrast, for classification purposes where the goal is to assign observations to predefined categories, logistic regression is a go-to method for binary outcomes, providing probabilities that an event will occur. The choice between regression models for quantitative prediction or classification depends on the nature of the outcome variable and the specific goals of the analysis, showcasing the adaptability of regression techniques across diverse data-driven scenarios. The authors discussed different modeling techniques along with real-life examples in Chapters 11 and 12.

Model performance indicators: Model performance indicators are metrics used to assess the effectiveness and efficiency of a model. These indicators help quantify how well a model is performing in terms of its ability to make accurate predictions or classifications. The choice of performance indicators depends on the nature of the problem and the goals of the model. Here are some common model performance indicators:

2.4.8.1 Mean Squared Error (MSE)

Mean squared error (MSE) is a commonly used metric for evaluating the performance of regression models. It measures the average of the squared differences between the predicted values and the actual values. The goal is to minimize the MSE, as a lower value indicates that the model's predictions are closer to the actual values. The formula of MSE is:

$$\text{MSE} = \frac{1}{n} \sum_{i=1}^{n} \left(y_i - \hat{y}_i \right)^2.$$

2.4.8.2 Root Mean Squared Error (RMSE)

Root mean squared error (RMSE) is a metric commonly used to evaluate the performance of a regression model. It is derived from the mean squared error by taking its square root and provides a measure of the average magnitude of errors in the same units as the target variable. The formula of RMSE is:

$$\text{RMSE} = \sqrt{\frac{1}{n}\sum_{i=1}^{n}\left(y_i - \hat{y}_i\right)^2}.$$

2.4.8.3 R-squared (Coefficient of Determination)

R-squared (coefficient of determination) is a statistical metric that measures the proportion of the variance in the dependent variable that is explained by the independent variables in a regression model. In other words, it provides an indication of how well the independent variables explain the variability in the dependent variable. The R-squared value ranges from 0 to 1, where '0' indicates that the model does not explain any of the variability in the dependent variable and '1' indicates that the model explains all the variability in the dependent variable. The formula of R-squared is:

$$R^2 = 1 - \frac{\sum_{i=1}^{n}\left(y_i - \hat{y}_i\right)^2}{\sum_{i=1}^{n}\left(y_i - \bar{y}\right)^2}.$$

2.4.8.4 Adjusted R²

As a penalty for adding regressors to increase the R^2 value, Henry Theil developed the adjusted R^2, denoted by $\bar{R}^2$, can be written as

$$\bar{R}^2 = 1 - \frac{\sum_{i=1}^{n}\left(y_i - \hat{y}_i\right)^2 \Big/ (n-k)}{\sum_{i=1}^{n}\left(y_i - \bar{y}\right)^2 \Big/ (n-1)} = 1 - \left(1 - R^2\right)\frac{n-1}{n-k}.$$

The adjusted R^2 will increase only if the absolute t value of the added variable is greater than 1. For comparative purposes, therefore, $\bar{R}^2$ is a better measure that R^2.

2.4.8.5 Akaike Information Criterion (AIC)

The Akaike Information Criterion (AIC) is a metric used for model selection, especially in the context of statistical modeling and regression analysis. It provides a balance between the goodness of fit of a model and the complexity of the model (the number of parameters). AIC is used to compare different models and choose the one that best balances accuracy and simplicity. The AIC is based on the principle that a good model should fit the data well, but not at the expense of being too complex. It penalizes models with more parameters to prevent overfitting, which occurs when a model captures noise in the training data rather than the underlying patterns. The formula of AIC is:

$$\text{AIC} = -2\ln\left(\hat{L}\right) + 2k$$

where $\hat{L}$ is the maximum likelihood of the model (the likelihood of the data given the model parameters) and k is the number of parameters in the model.

In cases where the sample size is small relative to the number of parameters, a correction factor can be applied to AIC, resulting in the corrected AIC (AICc). The formula of AICc is:

$$\text{AICc} = \text{AIC} + \frac{2k(k+1)}{(n-k-1)},$$

where n is the sample size.

2.4.8.6 Bayesian Information Criterion (BIC)

The Bayesian Information Criterion (BIC), also known as the Schwarz criterion, is a metric used for model selection, similar to the Akaike Information Criterion (AIC). BIC is particularly useful in the context of statistical modeling and regression analysis. Like AIC, BIC aims to balance model goodness of fit with model complexity. The BIC penalizes models with more parameters more strongly than AIC does. This stronger penalty for complexity makes BIC particularly effective in situations where there is a concern about overfitting. The formula of BIC is:

$$\text{BIC} = -2\ln\left(\hat{L}\right) + k\ln(n),$$

where $\hat{L}$ is the maximized likelihood of the model (the likelihood of the data given the model parameters), k is the number of parameters in the model, and n is the sample size.

2.4.8.7 Accuracy

Accuracy is a commonly used performance metric for classification problems. It measures the overall correctness of predictions made by a model, indicating the proportion of correctly classified instances out of the total instances. The formula for accuracy is:

$$\text{Accuracy} = \frac{TP + TN}{TP + TN + FP + FN},$$

where TP, TN, FP, FN stands for true positive, true negative, false positive, and false negative, respectively.

2.4.8.8 Precision

Precision is a performance metric of a classifier that measures the accuracy of positive predictions made by a model. It is particularly relevant in binary classification problems, where there are two classes: positive and negative. Precision answers the question: 'Of all the instances predicted as positive, how many are truly positive?'. The formula is:

$$\text{Precision} = \frac{TP}{TP + FP}.$$

2.4.8.9 Sensitivity or Recall or True Positive Rate

The measure of sensitivity of a test is its likelihood of signifying a positive outcome in a person affected by a certain illness. Sensitivity is estimated utilizing only disease affected people as the ratio between the number of sick individuals that had a positive test outcome (i.e., the true positives (TP)) and the total number of sick persons that also contains the false negatives (FN) result. The sensitivity measure can be defined as:

$$\text{Sensitivity} = \frac{\text{TP}}{\text{TP} + \text{FN}}.$$

2.4.8.10 F1 Score

The F1 score is a metric in machine learning that combines both precision and recall into a single value. It is particularly useful in binary classification problems, where there are two classes: positive and negative. The F1 score is the harmonic mean of precision and recall, providing a balance between the two metrics. The formula of F1 score is:

$$\text{F1 score} = 2 \times \frac{\text{Precision} \times \text{Recall}}{\text{Precision} + \text{Recall}}.$$

2.4.8.11 Specificity or True Negative Rate

Moreover, the measure of specificity of a test is its likelihood of identifying a negative outcome in a healthy person. Specificity is computed by utilizing only healthy people as the ratio between the number of healthy persons that had a negative test outcome (i.e., the true negatives (TN)) and the total number of healthy persons, which also comprises the false positives (FP) result. The specificity measure can be defined as:

$$\text{Specificity} = \frac{\text{TN}}{\text{TN} + \text{FP}}.$$

2.4.8.12 Receiver Operating Characteristic (ROC) Curve and Area under the Curve (AUC)

The sensitivity and specificity measures could be applied to statistically determine the effective performance of a disease's diagnostic test by creating a relevant ROC curve. It could be noted that the ROC curve is the graph of sensitivity measure against (1-specificity measure). Area under the curve is a handy measure of the test results accuracy which has an important practical interpretation (Hanley & McNeil, 1982; Hajian-Tilaki, 2013). A literature review shows that the ROC curve analysis tool has been applied widely in clinical or epidemiological studies for the appraisal of the diagnostic capability of biomarkers as well as the imaging results classification for sick persons from healthy individuals (Daubin et al., 2011).

Also, the ROC study is concerned with some other technical and robust features, including the appropriateness of selection between the parametric and nonparametric tools, existence of errors in gold standard evaluation or selection of reliable model to generate a ROC curve, and the methods for adjusting confounding in medical or biostatistical data analysis (Lu et al., 2010; Hajian-Tilaki et al., 2011).

2.5 Bayesian Analysis

Although this book deals with data science perspective with specific focus on frequentist norms, a quick overview of Bayesian analysis is considered in this section. The main objective here is how to calculate Bayesian probability and understand comparison of frequentist and Bayesian inferences.

Bayesian analysis depends on the philosophy of Bayes' theorem (also known as Bayes' rule) which is a way to figure out conditional probability. However, in typical conditional probability (discusses in Chapter 6) we find the probability of an event given that some event has already occurred. While in Bayesian probability, we find just the opposite; i.e., we find the cause of some events that have already occurred. For example, your probability of getting infection with COVID-19 is connected to various variables (Rahman et al., 2021), including if you have recently traveled to an affected area, visited an infected person, showed some symptoms of the disease or your age is at risk group (e.g., more than 45 years).

Bayes' theorem gives you the actual probability of an event, based on prior knowledge of conditions/tests that might be related to the event. For example, if the probability that someone has cancer is related to their age or smoking status, using Bayes' theorem the information of age or smoking status can be used to assess the probability of cancer more accurately than can be done without knowledge of the age or smoking status.

Bayes' rule is a deceptively simple formula used to calculate conditional probability (Rahman, 2008; Stone, 2016). The theorem was named after English mathematician Thomas Bayes (1701–1761). The formal definition for the rule is:

$$P(A \mid B) = \frac{P(A \cap B)}{P(B)} = \frac{P(B \mid A)P(A)}{P(B)},$$

where

A is the event of which we are primarily interested in;

B is the event of information connected to the event A;

P(A) is the probability of event A which called 'prior' probability;

P(B) is the probability of event B;

P(B | A) is the conditional probability of event B given that event A has occurred or if A is true; and

P(A | B) is the conditional probability of event A, based on prior knowledge of conditions/tests that might be related to the event with the context of the evidence of event B (i.e., if B is true). This conditional probability is called Bayesian 'posterior' probability.

Example 2.8

In a particular pain clinic, 10% of patients are prescribed narcotic pain killers. Overall, 5% of the clinic's patients are addicted to narcotics (including pain killers and illegal substances). Among the addicts in the clinic, 8% have been prescribed narcotics pain pills. If a patient is prescribed pain pills, what is the probability that they will become an addict?

Here, $P(A$=patient prescribed pain pills$) = 0.10$, $P(B$ = patient an addict$) = 0.05$, and $P(B \mid A) = 0.08$.

Using these values into the Bayesian formula and solve:

$$P(A|B) = P(B|A) * P(A) / P(B) = (0.08^*0.1) / 0.05 = 0.16.$$

Using R:

```
> P_B_given_A=0.08
> P_A= 0.10
> P_B=0.05
> P_B_given_A=0.08
> (P_A_given_B=((0.08*0.10)/(0.05)))
[1] 0.16
```

Hence, the probability of an addict being prescribed pain pills is 0.16 (16%).

The Bayesian rule can be extended to sophisticated forms as well as to be used in many statistical procedures such as in Bayesian regression and prediction analysis (Rahman, 2008; Rahman, 2011; Gelman et al., 2013; McManus et al., 2020; Clyde et al., 2015). In Bayesian statistics, researchers commonly argue that prior probabilities are intrinsically subjective (i.e., your prior information is different from mine), and many data scientists see this as a fundamental drawback to Bayesian analysis (Rahman & Upadhyay, 2015; Stone, 2016). Advocates of the Bayesian approach argue that this is inescapable, and that frequentist methods also entail subjective choices, but this has been a basic source of contention between the 'fundamentalist' supporters of the two statistical paradigms for at least the last 50 years. In contrast, powerful computational tools allow Bayesian methods to tackle large and complex modeling problems with relative ease (Gelman et al., 2013; Rahman & Harding, 2016), where frequentist methods can only approximate. Bayesian modeling methods provide natural ways for people in many disciplines to structure their data and knowledge, and they yield direct and intuitive answers to the practitioner's questions.

There are many varieties of Bayesian analysis (Rahman, 2008; Rahman & Upadhyay, 2015; Clyde et al., 2015). The fullest version of the Bayesian paradigm casts data science problems in the framework of decision-making. It entails formulating subjective prior probabilities to express preexisting information, careful modeling of the data structure, checking and allowing for uncertainty in model assumptions, formulating a set of possible decisions and a utility function to express how the value of each alternative decision is affected by the unknown model parameters (Rahman, 2008). But each of these components can be omitted. Many users of Bayesian methods do not employ genuine prior information, either because it is insubstantial or because they are uncomfortable with subjectivity. The decision-theoretic framework is also widely omitted, with many feeling that subjective prior based inferential results should not really be formulated as a decision. So there are varieties of Bayesian analysis and varieties of Bayesian analysts. But the common strand that underlies this variation is the basic principle of using Bayes' theorem and expressing uncertainty about unknown parameters probabilistically (Rahman, 2009).

Also, in most experiments, the prior probabilities on hypotheses are not known. In this case, our recourse is the art of statistical inference: we either make up a prior (Bayesian) or do our best using only the likelihood (frequentist).

The Bayesian approach models uncertainty by a probability distribution over hypotheses (Rahman, 2017). One's ability to make inferences depends on one's degree of confidence in the chosen prior, and the robustness of the findings to alternate prior distributions may be relevant and important (Rahman et al., 2016).

On the other hand, the frequentist method uses conditional distributions of data given specific hypotheses. The presumption is that some hypothesis (parameter specifying the conditional distribution of the data) is true and that the observed data is sampled from that distribution. In particular, the frequentist approach does not depend on a subjective prior that may vary from one investigator to another. This attribute makes it more robust approach than Bayesian.

These two methods may be further contrasted as follows:

Frequentist	Bayesian
• Never uses or gives the probability of a hypothesis (no prior or posterior)	• Uses probabilities for both hypotheses and data
• Depends on the likelihood for both observed and unobserved data	• Depends on the prior and likelihood of observed data
• Does not require a prior	• Requires one to know or construct a 'subjective prior'
• Dominated statistical practice during the 20th century to date, and still growing due to the invention in data science and analytic software	• Dominated statistical practice before the 20th century and then in recent years due to the invention of powerful computing facilities
• Tends to be less computationally intensive	• May be computationally intensive due to integration over many parameters

Finally, the frequentist decision support measures like *p*-values and confidence intervals continue to dominate research and societal outcomes, especially in the biomedical, public health and life sciences. In the current era of powerful computers and big data, Bayesian methods seem to be undergoing a renaissance in fields like machine learning and genetics, though.

2.6 Conclusion and Exercises

A summary of this chapter is provided in here with some problem-solving exercises in biomedical, public health and biostatistics.

2.6.1 Concluding Remarks

This chapter served as an exploration of modern statistical terms and concepts. It has provided with a contemporary definition and introduction of biostatistics along with scope of this book. This chapter also provides a solid background of the procedure of data analysis including defining research problem, objective, and question; formulating hypothesis; data collection and cleaning; exploratory data analysis; data transformation; inferential statistics; and data modeling. Moreover, a precise overview of calculation of sample size, probability theory, estimation and hypothesis testing, parametric and nonparametric techniques, Bayesian analysis, and some model performance indicators are briefly discussed. It is vital, however, that the readers should go through the contents carefully to gain valuable conceptual knowledge for the later chapters where these terms and concepts would be gradually accounted in detail with practical data analysis using R software throughout the book.

2.6.2 Practice Questions

Exercise 2.1 Write the main difference between biostatistics and statistics.

Exercise 2.2 Identify the population and sample from your field of study or research and provide example of a parameter and a statistic.

Exercise 2.3 What are the different steps in a statistical hypothesis test?

Exercise 2.4 Calculate the required sample size when we aim to calculate a 90% CI for the mean of a normal population which has a standard deviation of 1.5, and the margin of error is provided as $m \le 0.15$.

Bibliography

Abdulla, F., Hossain, M. M., Huq, M. N., Hai, A., Rahman, A., Kabir, R., Peya, F.J., Islam, S., & Khan, H. T. (2023a). Prevalence, determinants and consequences of problematic smartphone use among preschoolers (3–5 years) from Dhaka, Bangladesh: A cross-sectional investigation. *Journal of Affective Disorders*, 329, 413–427.

Abdulla, F., Hossain, M. M., & Rahman, A. (2023b). Determinants of early sexual initiation among female adolescents in Bangladesh: Evidence from a countrywide cross-sectional survey. *Public Health*, 223, 102–109. https://doi.org/10.1016/j.puhe.2023.07.019

Altman, D. G., & Bland, J. M. (1995). Statistics notes: Absence of evidence is not evidence of absence. *Bmj*, 311(7003), 485.

Champely, S., Ekstrom, C., Dalgaard, P., Gill, J. et al. (2022). Basic functions for power analysis, CRAN. Retrieved from https://cran.r-project.org/web/packages/pwr/. Accessed on 25 March 2024.

Charan, J., & Biswas, T. (2013). How to calculate sample size for different study designs in medical research? *Indian Journal of Psychological Medicine*, 35(2), 121–126.

Chow, S., Shao, J., & Wang, H. (2008). *Sample size calculations in clinical research* (2nd ed.). Chapman & Hall/CRC Press.

Clyde, M., Çetinkaya-Rundel, M., Rundel, C., Banks, D., Chai, C., & Huang, L. (2015). An introduction to Bayesian thinking: A companion to the statistics with R course. Retrieved from https://statswithr.github.io/book/. Accessed on 27 March 2024.

Cohen, J. (1988). *Statistical power analysis for the behavioral sciences* (2nd ed.). Lawrence Erlbaum.

Daubin, C., Quentin, C., Allouche, S., et al. (2011). Serum neuron-specific enolase as predictor of outcome in comatose cardiac-arrest survivors: a prospective cohort study. *BMC Cardiovascular Disorders*, 11, 1–13.

Gelman, A., Vehtari, A., Carlin, J., Stern, H.S., Dunson, D., & Rubin, D. (2013). *Bayesian data analysis*. CRC Press.

Glantz, S. A. (1987). *Primer of biostatistics* (2nd ed.). McGraw-Hill Book Co.

Gupta, K. K., Attri, J. P., Singh, A., Kaur, H., & Kaur, G. (2016). Basic concepts for sample size calculation: Critical step for any clinical trials! *Saudi Journal of Anaesthesia*, 10(3), 328–331. https://doi.org/10.4103/1658-354X.174918

Hajian-Tilaki, K. (2013). Receiver operating characteristic (ROC) curve analysis for medical diagnostic test evaluation. *Caspian Journal of Internal Medicine*, 4(2), 627–635.

Hajian-Tilaki, K., Hanley, J. A., & Nassiri, V. (2011). An extension of parametric ROC analysis for calculating diagnostic accuracy when underlying distributions are mixture of Gaussian. *Journal of Applied Statistics*, 38(9), 2009–2022.

Hanley, J. A., & McNeil, B. J. (1982). The meaning and use of the area under a receiver operating characteristic (ROC) curve. *Radiology*, 143(1), 29–36.

Harjule, P., Rahman, A., Agarwal, B., & Tiwari, V. (2023). *Computational statistical methodologies and modeling for artificial intelligence*. CRC Press.

Hickey, J. L., Grant, S. W., Dunning, J., & Siepe, M. (2018). Statistical primer: Sample size and power calculations – Why, when and how? *European Journal of Cardio-Thoracic Surgery*, 54(1), 4–9. https://doi.org/10.1093/ejcts/ezy169

Lopes, B., Ramos, I. C. D. O., Ribeiro, G., Correa, R., Valbon, B. D. F., Luz, A. C. D., ... & Ambrósio Junior, R. (2014). Biostatistics: Fundamental concepts and practical applications. *Revista Brasileira de Oftalmologia*, 73, 16–22.

Lu, Y., Dendukuri, N., Schiller, I., & Joseph, L. (2010). A Bayesian approach to simultaneously adjusting for verification and reference standard bias in diagnostic test studies. *Statistics in Medicine*, 29(24), 2532–2543.

McManus, S., Rahman, A., Horta, A., Coombes, J. (2020). Applied Bayesian modeling for assessment of interpretation uncertainty in spatial domains. In A. Rahman (Ed.), *Statistics for data science and policy analysis* (1st ed., pp. 3–13). Springer. https://doi.org/10.1007/978-981-15-1735-8_1

Qiu, W., Chavarro, J., Lazarus, R., Rosner, B., & Ma, J. (2022). Power and sample size calculation for survival analysis of epidemiological studies, CRAN. Retrieved from https://cran.r-project.org/web/packages/powerSurvEpi/index.html. Accessed on 25 March 2024.

Rahman, A. (2008). *Bayesian predictive inference for some linear models under student-t errors* (1st ed.). VDM Verlag.

Rahman, A. (2009). Objective Bayesian prediction for the matrix-T error regression model. In *2009 International Workshop on Objective Bayes Methodology (O-Bayes09)* (pp. 1–13).

Rahman, A. (2011). Bayesian predictive inference for multivariate simple regression model with matrix-T error. *Pioneer Journal of Theoretical and Applied Statistics*, 1(2), 99–112.

Rahman, A. (2017). Estimating small area health-related characteristics of populations: A methodological review. *Geospatial Health*, 12(1), 3–14. Article 495. https://doi.org/10.4081/gh.2017.495

Rahman, A. (2020). *Malnutrition: Prevalence, risk factors and outcomes*. (Nutrition and diet research progress). Nova Science Publishers.

Rahman, A., Gao, J., D'Este, C., & Ahmed, S. E. (2016). An assessment of the effects of prior distributions on the Bayesian predictive inference. *International Journal of Statistics and Probability*, 5(5), 31–42. https://doi.org/10.5539/ijsp.v5n5p31

Rahman, A., & Harding, A. (2016). *Small area estimation and microsimulation modeling*. CRC Press. https://doi.org/10.1201/9781315372143

Rahman, A., Kuddus, M. A., Ip, H. L., Bewong, M. (2021). A review of COVID-19 modelling strategies in three countries to develop a research framework for regional areas. *Viruses*, 13(11), 2185. https://doi.org/10.3390/v13112185

Rahman, A., Upadhyay, S. K. (2015). A Bayesian reweighting technique for small area estimation. In U. Singh, A. Loganathan, S. K. Upadhyay, & D. K. Dey (Eds.), *Current trends in Bayesian methodology with applications* (1st ed., pp. 503–519). CRC Press.

Rahman, M. A., Rahman, M. S., & Rahman, A. (2020). Associated risk factors for acute malnutrition among rural Bangladeshi children: A cross-sectional analysis. In A. Rahman (Ed.), *Malnutrition: Prevalence, risk factors and outcomes*. (Nutrition and diet research progress) (pp. 73–93). Nova Science Publishers.

Serdar, C. C., Cihan, M., Yücel, D., & Serdar, M.A. (2020). Sample size, power and effect size revisited: simplified and practical approaches in pre-clinical, clinical and laboratory studies. *Biochemia Medica*, 31(1), 010502. https://doi.org/10.11613/bm.2021.010502

Song, J. W., & Chung, K. C. (2010). Observational studies: cohort and case-control studies. *Plastic and Reconstructive Surgery*, 126(6), 2234–2242. https://doi.org/10.1097/PRS.0b013e3181f44abc

Stone, J. V. (2016). *Bayes' rule with R: A tutorial introduction to Bayesian analysis*. Sebtel Press.

Wittes, J. (2002). Sample size calculations for randomized controlled trials. *Epidemiologic Reviews*, 24(1), 39–53.

Zhang, E., Wu, V. Q., Chow, S. C., & Zhang, H. G. (2022). R functions for chapter 3, 4, 6, 7, 9, 10, 11, 12, 14, 15 of sample size calculation in clinical research, CRAN. Retrieved from https://cran.r-project.org/web/packages/TrialSize/. Accessed on 25 March 2024.

Zhong, B. (2009). How to calculate sample size in randomized controlled trial? *Journal of Thoracic Disease*, 1(1), 51–54.

Rahman, A., Othman, N., Kuddus, M. A., & Hasan, M. Z. (2024). Impact of the COVID-19 pandemic on child malnutrition in Selangor, Malaysia: A pilot study. *Journal of Infection and Public Health*, 17(5), 833–842. https://doi.org/10.1016/j.jiph.2024.02.019

3

Summary Statistics and Presentation of Data

3.1 Summarization of Data

Data are a set of measurements or observations that are collected as a source of information and provide a partial picture of reality. It is important to constantly consider what information the data are conveying, how to use them, and what needs to be done to include more helpful information, regardless of whether the data are being gathered for a specific reason or are being used already. Typically, data are gathered in an unprocessed manner, making it challenging to decipher the underlying information for the intended purpose. Therefore, raw data need to be summarized, processed, and analyzed. In public health, data summarization helps in efficient advance analysis, facilitates effective communication, aids in issue identification, supports monitoring and evaluation, and assists in resource allocation. By condensing complex datasets into meaningful summaries, public health professionals can make informed decisions, design evidence-based interventions, and address health challenges more effectively.

Data summarization refers to the process of condensing and extracting key information from a large dataset to provide a concise and meaningful overview. Data summarization aims to simplify complex datasets while retaining the important features and insights. By summarizing the data, analysts and researchers can quickly understand the dataset's main properties, distributions, and relationships, without getting overwhelmed by the detailed individual data points. It involves calculating and presenting of summary statistics, which capture the essential characteristics and patterns within the data.

Example 3.1

Suppose a public health researcher has generated hypothetical outcome data of a tuberculosis (TB) test with a basic demographic history of a particular day. The collected dataset consists of data from 30 individuals and the data looks like in Table 3.1.

The researcher is interested in gathering primary knowledge about how many samples are TB-positive and the factors associated with TB-positive. The problem is how the researcher will explain this dataset to the audience. The dataset can be described in one way: the researcher will present the outcome of the TB test, whether positive or negative, along with the demographic history of the age and gender of each individual. However, this way is quite challenging in the case of a large dataset, and the main problem is that the audience will need help understanding or guessing the meaningful insight of the dataset. Therefore, the appropriate way is to use summary statistics to explain the dataset meaningfully. The researcher can explain the dataset as 70% (21) of the sample showing positive in the TB test and the remaining 30% (9) showing negative. Moreover,

DOI: 10.1201/9781003426189-3

TABLE 3.1

Hypothetical Outcome Data of a TB Test from 30 Patients

Sample #	Gender	Age	TB Result	Sample #	Gender	Age	TB Result	Sample #	Gender	Age	TB Result
1	Male	15	Positive	11	Female	49	Positive	21	Male	35	Positive
2	Male	63	Positive	12	Male	54	Positive	22	Male	44	Positive
3	Female	23	Negative	13	Male	57	Positive	23	Female	24	Negative
4	Male	45	Positive	14	Male	60	Positive	24	Female	40	Positive
5	Female	45	Positive	15	Male	23	Positive	25	Male	51	Negative
6	Female	48	Positive	16	Female	35	Negative	26	Female	15	Negative
7	Female	36	Positive	17	Male	32	Negative	27	Male	43	Positive
8	Male	43	Negative	18	Female	15	Positive	28	Male	38	Positive
9	Male	34	Positive	19	Male	25	Positive	29	Female	35	Negative
10	Male	47	Positive	20	Female	33	Negative	30	Male	36	Positive

among the TB-positive individuals, around 71% (15) are male; on the other hand, 67% (6) of the TB-negative individuals are female. So, we can conclude that the TB positive is more prevalent among males.

Example 3.2

Suppose you have data from an outbreak of gastroenteritis affecting 41 persons who had recently attended a wedding. Your supervisor asked you to describe the ages of the affected persons. Now think about the following questions:

- Why your supervisor asked about the ages of the affected persons?
- How would you describe the ages of the affected persons to your supervisor?

Your supervisor inquired about the age of the affected individuals to determine if there were any specific age groups that were highly affected. So, you could simply list the ages of each person in response to your supervisor. Alternatively, your supervisor might prefer one summary number – a measure of central value and the measure of dispersion. Therefore, saying the mean or median age rather than reciting 41 ages is certainly more efficient and meaningful for the objective of your supervisor. You can also provide a frequency distribution to your supervisor after categorizing the age of the affected persons.

3.2 Statistics Used for Summarizing Data

Statistics provide a set of tools and techniques for summarizing data effectively. These statistical measures and methods help condense complex datasets into meaningful and concise summaries. Here are some commonly used statistics for summarizing data:

Rate: A rate refers to the measurement of one quantity relative to another quantity, typically with respect to time. It represents the frequency or occurrence of an event or phenomenon over a specific period (Megabiaw & Rahman, 2013). Rates are often expressed as a ratio or proportion per unit of time. The incidence rate of a specific infectious disease in a population over a given period, expressed as cases per 1,000 person-years. For instance, if there are 20 new cases of the disease in a population of 10,000 people over one year, the incidence rate would be 2 per 1,000 person-year.

Ratio: A ratio compares two quantities by dividing one quantity by the other. Ratios are used to understand the relative sizes or magnitudes of two variables. A ratio can be expressed in different forms, such as a fraction, decimal, or percentage. For example, the ratio of the number of individuals with a particular genetic mutation to the total number of individuals in a study population. If there are 50 individuals with the mutation and 250 individuals in total, the ratio is 1:5.

Proportion: A proportion represents the relationship of a part to the whole. It refers to the fraction or percentage of a sample or population with a particular characteristic or attribute. Proportions are frequently used to estimate population parameters or make inferences about a larger group based on a representative sample (Rahman & Harding, 2013). The proportion of patients in a clinical trial who experienced a specific side effect after taking a medication. If 15 out of 100 patients developed the side effect, the proportion is 15/100 or 0.15 (15%).

Frequency distributions: Frequency distributions organize data into categories or intervals and record the frequency or count of data points falling into each category. To gain more insight, it is recommended to read the Section 3.4.1 and Example 3.4.

Measures of central tendency: These statistics indicate a dataset's typical or central value. The commonly used measures of central tendency include the mean (average), median (middle value), and mode (most frequently occurring value). For a better understanding, readers can see the relevant examples discussed in Chapter 5.

Measures of dispersion: These statistics quantify the variability or spread of data points. Measures of dispersion include the range (difference between the maximum and minimum values), variance (average squared deviation from the mean), and standard deviation (square root of the variance). Readers are recommended to see the relevant examples discussed in Chapter 5 for clear understanding.

Percentiles: Percentiles divide a dataset into equal parts, indicating the relative position of a specific value within the dataset. The median is the 50th percentile, while quartiles (25th and 75th percentiles) and deciles (10th and 90th percentiles) divide the data into four and ten equal parts, respectively. Readers are encouraged to read the relevant examples covered in Chapter 5 to enhance their understanding of these concepts.

These are just a few examples of the statistics used for summarizing data, but beyond these, there are many other statistics used for summarizing data. The choice of statistical measures depends on the specific data type, research objectives, and the questions being addressed. These statistical measures will be accounted for in detail with relevant biostatistics examples in later chapters. The remainder of this chapter mainly focuses on various tools for graphical summary presentation of data.

3.3 Presentation of Data

Researchers may have uncovered many complex truths that need long explanations while writing. This is where the importance of the presentation of data comes in. Researchers have to present their findings so that the readers can go through them quickly and understand every point that researchers want to showcase. Data presentation can greatly influence audiences by increasing their interest; therefore, it is a vital part of data summarization to identify the main features of the data at a glance using a suitable presentation method. Data presentation tools are useful instruments for communication that may efficiently exhibit vast amounts of complex data in a simplified form, make the data easily legible and understood at the same time, and draw and hold readers' attention.

Inappropriate data presentation makes it difficult for readers and reviewers to understand the material. It takes art to present data in an effective and efficient manner. Presentation techniques need to be chosen based on the audience's comprehension level, the information to be emphasized, the data format, the analysis method to be utilized, and the goal to be covered. Generally speaking, there are three ways that statistics data can be presented: textually, tabularly, and graphically. The benefits and drawbacks of various presentation techniques must be carefully considered before selecting one. Depending on which particular information is going to be emphasized, multiple presenting techniques must be used, even when the same information is being presented. Text or written language should be used when comparing or introducing two values at a specific point in time. A table, on the other hand, works best when all the material needs to be read equally and lets the readers focus on the parts that interest them. Compared to conventional text and tables, graphs are frequently easier to read and comprehend. They enable readers to quickly grasp the general data trend and the conclusions drawn from a comparison of the two groups' results. Researchers can more effectively deliver raw data to audiences by using graphical form. Regardless of the approach taken, there is one thing to constantly keep in mind: presentation should always be kept simple.

Example 3.3

Recall **Example 3.1**. Suppose the researcher wants to present the TB-positive proportion, then the text form is suitable for presenting this information to the audience as 70% (21) of the sample are TB-positive. If the researcher wants to summarize the whole dataset, then the tabular form is suitable to present the summary of the dataset as shown in Table 3.2.

However, if the researcher wants to compare TB's positive and negative proportions among male and female individuals, then the graphical form will be the best method. The visual comparison of TB's positive and negative proportions among male and female individuals is illustrated in Figure 3.1.

TABLE 3.2

Summary of the TB Test Result Dataset.

TB Test	Male [% (*n*)]	Female [% (*n*)]	Total [% (*n*)]	Mean (SD)	Median (IQR)
Positive	83.3 (15)	50.0 (6)	70.0 (21)	40.6 (13.3)	43.0 (14.0)
Negative	16.7 (3)	50.0 (6)	30.0 (9)	32.3 (10.8)	33.0 (15.5)
Total	100.0 (18)	100.0 (12)	100.0 (30)	38.1 (13.0)	37.0 (17.0)

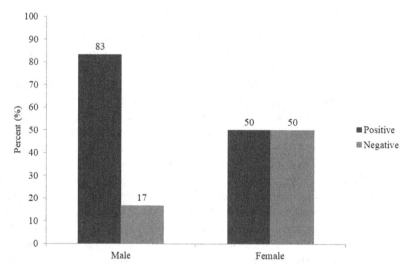

FIGURE 3.1

Comparison of the TB positive and negative rates among male and female individuals.

3.4 Presentation of a Continuous Variable

This section will focus on the summarization and presentation of the continuous data. Continuous data refers to data that can take on any value within a specific range. It represents measurements or observations that can be expressed as real numbers. The continuous data are summarized and presented to explore the features of the distribution of the data, including symmetry or asymmetry, outliers, multimodality, and trends, that will guide the researcher to fix further analysis.

3.4.1 Frequency Distribution Table

An illustration of the number of observations inside a certain period in tabular form is called a frequency distribution table. The data being examined and the analyst's objectives determine the period's size. Although they do not have to be of identical size, the intervals must be exhaustive and mutually exclusive. It helps to summarize the data and understand the spread and pattern of the continuous variable, especially for large amounts of data. It is often used as a starting point for further data analysis and visualization.

The first step is to determine the number of classes (k) either based on the objective of interest or using the useful recipe '2 to the k rule' such that $2^k > n$. Then it is necessary to determine the class intervals or width either considering the objective of interest or using the formula: $i \geq \dfrac{(H-L)}{k}$, where i is the interval of a class, H is the highest value, L is the lowest value, and k is the number of the desired classes. Next, establish the boundaries for each class and tally the number of observers in each one. The frequency distribution table can also include relative and cumulative frequencies. Each frequency from a frequency distribution table is added to the total of its predecessors to determine the cumulative frequency. It is used to calculate the proportion of observations in a dataset that fall above (or below) a given value. The frequency is converted to a percentage using a relative frequency distribution. Each class frequency is divided by the total number of observations to transform a frequency distribution into a relative frequency distribution. It displays the probability that each bin will occur.

Example 3.4

We have downloaded a dataset named 'Cardiovascular Diseases Risk Prediction Dataset' from the Kaggle database (https://www.kaggle.com/), which contains 308,854 observations and 19 variables on health status. The description of the variables is presented in Table 3.3.

Let us import the data into R and see the variable type and observation in the dataset using the following R-command:

R-code:

```
data <- read.csv('E:/Book/Chapter 3/Data/CVD_cleaned.csv', header =
TRUE)
attach(data)
str(data)
```

TABLE 3.3

Description of the Variables Included in the Dataset Named 'Cardiovascular Diseases Risk Prediction Dataset'

Variable Name	Description
General_Health	General health status of the respondent [Excellent, Very good, Good, Fair, and Poor] [Categorical]
Checkup	Length of time since last routine checkup [Within past year, Within past 2 years, Within past 5 years, and 5 or more years ago] [Categorical]
Exercise	Exercise in past 30 days [Yes and No] [Categorical]
Heart_Disease	Ever diagnosed with angina or coronary heart disease [Yes and No] [Categorical]
Skin_Cancer	(Ever told) you had skin cancer [Yes and No] [Categorical]
Other_Cancer	(Ever told) you had any other types of cancer [Yes and No] [Categorical]
Depression	(Ever told) you had a depressive disorder [Yes and No] [Categorical]
Diabetes	(Ever told) you had diabetes [Yes and No] [Categorical]
Arthritis	Told had arthritis [Yes and No] [Categorical]
Sex	Sex of the respondent [Male and Female] [Categorical]
Age_Category	Age category in years [18–24, 25–29, 30–34, 35–39, 40–44, 45–49, 50–54, 55–59, 60–64, 65–69, 70–74, 75–79, and 80+] [Categorical]
Height_(cm)	Reported height in cm [Continuous]
Weight_(kg)	Reported weight in kg [Continuous]
BMI	Body mass index [Continuous]
Smoking_History	Smoked at least 100 cigarettes [Yes and No] [Categorical]
Alcohol_Consumption	Total number of alcoholic beverages consumed per week [Continuous]
Fruit_Consumption	Fruit intake in times per day [Continuous]
Green_Vegetables_Consumption	Dark green vegetable intake in times per day [Continuous]
FriedPotato_Consumption	French fry or fried potatoes intake in times per day [Continuous]

Output:

```
Console   Terminal ×   Background Jobs ×
R  R 4.3.2 · ~/
> data <- read.csv("E:/Book/Chapter 3/Data/CVD_cleaned.csv", header = TRUE)
> attach(data)
> str(data)
'data.frame':   308854 obs. of  19 variables:
 $ General_Health              : chr  "Poor" "Very Good" "Very Good" "Poor" ...
 $ Checkup                     : chr  "Within the past 2 years" "Within the past year"
"Within the past year" "Within the past year" ...
 $ Exercise                    : chr  "No" "No" "Yes" "Yes" ...
 $ Heart_Disease               : chr  "No" "Yes" "No" "Yes" ...
 $ Skin_Cancer                 : chr  "No" "No" "No" "No" ...
 $ Other_Cancer                : chr  "No" "No" "No" "No" ...
 $ Depression                  : chr  "No" "No" "No" "No" ...
 $ Diabetes                    : chr  "No" "Yes" "Yes" "Yes" ...
 $ Arthritis                   : chr  "Yes" "No" "No" "No" ...
 $ Sex                         : chr  "Female" "Female" "Female" "Male" ...
 $ Age_Category                : chr  "70-74" "70-74" "60-64" "75-79" ...
 $ Height_.cm.                 : num  150 165 163 180 191 183 175 165 163 163 ...
 $ Weight_.kg.                 : num  32.7 77.1 88.5 93.4 88.5 ...
 $ BMI                         : num  14.5 28.3 33.5 28.7 24.4 ...
 $ Smoking_History             : chr  "Yes" "No" "No" "No" ...
 $ Alcohol_Consumption         : num  0 0 4 0 0 0 0 3 0 0 ...
 $ Fruit_Consumption           : num  30 30 12 30 8 12 16 30 12 12 ...
 $ Green_Vegetables_Consumption: num  16 0 3 30 4 12 8 8 12 12 ...
 $ FriedPotato_Consumption     : num  12 4 16 8 0 12 0 8 4 1 ...
> |
```

There are many ways or packages for making frequency distribution. Herein, we use `descriptr` package. Let us `install` and `library` the `descriptr` package using the following R-code:

R-code:

```
install.packages('descriptr')
library('descriptr')
```

Now, let us do the frequency distribution table for BMI with seven classes using the following R-command:

R-code:

```
fdt<-ds_freq_table(data, BMI, 7)
fdt
```

Output:

```
Console    Terminal    Background Jobs

R  R 4.3.1 · ~/
> fdt <- ds_freq_table(data, BMI, 7)
> fdt
                                    Variable: BMI
|-------------------------------------------------------------------------------------------|
|            Bins            | Frequency | Cum Frequency |   Percent   | Cum Percent |
|-------------------------------------------------------------------------------------------|
|     12      -     24.5     |   86194   |     86194     |    27.91    |    27.91    |
|-------------------------------------------------------------------------------------------|
|     24.5    -     37       |  192189   |    278383     |    62.23    |    90.13    |
|-------------------------------------------------------------------------------------------|
|     37      -     49.4     |   27364   |    305747     |     8.86    |    98.99    |
|-------------------------------------------------------------------------------------------|
|     49.4    -     61.9     |    2677   |    308424     |     0.87    |    99.86    |
|-------------------------------------------------------------------------------------------|
|     61.9    -     74.4     |     327   |    308751     |     0.11    |    99.97    |
|-------------------------------------------------------------------------------------------|
|     74.4    -     86.9     |      82   |    308833     |     0.03    |    99.99    |
|-------------------------------------------------------------------------------------------|
|     86.9    -     99.3     |      21   |    308854     |     0.01    |     100     |
|-------------------------------------------------------------------------------------------|
|            Total           |  308854   |       -       |   100.00    |      -      |
|-------------------------------------------------------------------------------------------|
>
```

The frequency distribution table reveals that the majority of individuals, comprising approximately 62.23%, exhibit a BMI falling within the range of 24.5 to 37. Following this, around 27.91% of individuals fall within the BMI range of 12 to 24.5.

3.4.2 Frequency Polygon or Frequency Curve

Similar to a histogram, a frequency polygon displays the contour of a distribution. It is made up of a line segment joining the points created by the intersections of the class frequencies and midpoints. Extend the left side to one class width prior to the first class midpoint and the right side to one class width subsequent to the last class midpoint with the graph starting and ending on the horizontal axis. It helps visualize the overall shape of the distribution of data and any trends or patterns that may exist. It is worth noting that frequency polygons are most appropriate when dealing with a reasonably large dataset. For smaller datasets, individual data points may be more appropriately displayed using a line plot or dot plot.

Example 3.5

Recall the dataset from **Example 3.4**; we can see the shape of the distribution of BMI data by frequency polygon or frequency curve. Let us make the frequency polygon using the following R-code. We need to `install` and `library` the `ggplot2` package for using the `ggplot` function. In addition, we need to `install` and `library` the `patchwork` package for combining two or more graphs in `ggplot2` using the following R-code:

R-code:

```
install.packages('ggplot2')
library('ggplot2')
install.packages('patchwork')
library('patchwork')
fre_poly1 <- ggplot(data = data, aes(BMI)) +
  geom_freqpoly(bins = 60, color = 'red') +
  labs(x = 'Bins (BMI)', y = 'Frequency') +
  theme_bw()
fre_poly2 <- ggplot(data = data, aes(BMI)) +
  geom_freqpoly(bins = 60, size = 1, color = 'blue', linetype =
'dashed') +
  labs(x = 'Bins (BMI)', y = 'Frequency') +
  theme_bw()
combined = fre_poly1 + fre_poly2
combined
```

The frequency curve of BMI is illustrated in Figure 3.2. Both panels of Figure 3.2 visually present the frequency curve of BMI with different line types and colors. The frequency curve illustrates that the distribution of individuals' BMI does not follow a symmetrical, or bell-shaped i.e., normal distribution. Instead, it is positively skewed, with a peak around a BMI of 26. This skewness indicates that there are more individuals with lower BMI values and a tail widening toward higher BMI values.

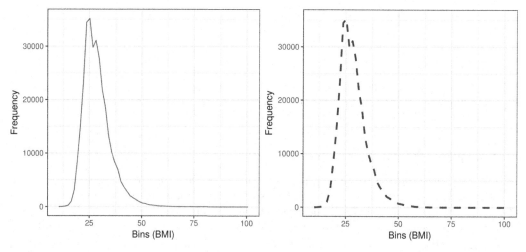

FIGURE 3.2
Frequency curve of BMI.

3.4.3 Cumulative Frequency Curve

A graph that shows the cumulative frequencies for the classes in a frequency distribution is called an ogive or cumulative frequency curve. The cumulative frequency of each class at its upper-class boundary is shown on a line graph. On the horizontal axis, the upper borders are marked. On the vertical axis, the cumulative frequencies are marked. The frequencies of each previous class are added to the frequency of a class in a less than ogive. The frequencies of the classes that come after are added to the frequency of a class in a more than ogive. Starting at the lower edge of the first class (where the cumulative frequency is zero for less than ogive and total frequency for greater than ogive), the graph should end at the upper edge of the last class (where the cumulative frequency is zero for greater than ogive and total frequency for less than ogive). When estimating the numbers that fall below or above a specific variable or value in data, ogives graphs are utilized.

Example 3.6

Recall the dataset from **Example 3.4**. We can draw the cumulative frequency curve for alcohol consumption by using the following R-code:

R-code:

```
fdt<-ds_freq_table(data, Alcohol_Consumption, 15)
cf <- c(0, fdt$cumulative)
breaks <- fdt$breaks
data1 <- cbind(breaks, cf)
data2 <- as.data.frame(data1)
plot1 <- ggplot(data=data2, aes(x=breaks, y=cf)) +
  geom_line() + geom_point() +
  labs(x='Bins (Alcohol Consumption)', y='Cumulative Frequency') +
  theme_bw()
plot2 <- ggplot(data=data2, aes(x=breaks, y=cf)) +
  geom_line(size = 1.5, color = 'darkblue', linetype = 'dotted') +
  geom_point(color = 'darkblue') +
  labs(x='Bins (Alcohol Consumption)', y='Cumulative Frequency') +
  theme_bw()
plot = plot1 + plot2
plot
```

The cumulative frequency curve of alcohol consumption is presented in Figure 3.3. Both panels of Figure 3.3 illustrate the cumulative frequency curve of alcohol consumption with different line types and colors. From the cumulative frequency curve of alcohol consumption, it is evident that nearly three-fourths of individuals consumed ≤10 alcoholic beverages per week.

3.4.4 Histograms

A histogram is a graphical representation of the distribution of continuous or discrete quantitative data. It provides a visual summary of the frequency or relative frequency of values within different intervals or 'bins' along the *x*-axis. The *y*-axis represents the frequency or relative frequency of the data falling into each bin. It provides a quick and intuitive way to assess the distribution of data and identify patterns or anomalies. A histogram is useful for understanding the shape, central tendency, and spread of a dataset. It visually displays whether the data is symmetric, skewed (to the left or right), or has multiple peaks. It can also reveal outliers or gaps in the data. Additionally, histograms can be used to compare the distributions of different datasets or to observe changes in the distribution over time.

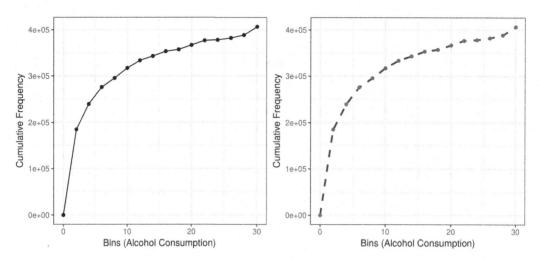

FIGURE 3.3
Cumulative frequency curve of alcohol consumption.

The first step is to divide the range of data values into a set of intervals or bins. Each bin represents a range of values that the data can fall into. The number of bins and their width can vary depending on the data and the desired level of detail. The next step is to count the number of data points (frequency) that fall into each bin. Each bin is represented by a rectangular bar on the histogram. The height of the bar corresponds to the frequency of values in that bin and the width of each bar may vary depending on the range of values covered by the bin. The bars are typically drawn adjacent to each other without any gaps.

Example 3.7

Recall the dataset from **Example 3.4** and suppose we want to look at the distribution of BMI by histogram. Let us make the histogram of BMI using the following R-command:

R-code:

```
his1<-ggplot(data, aes(x=BMI)) +
  geom_histogram(color='white', fill='red') +
  labs(x='BMI', y='Number of Individual') +
  theme_bw()
his2<-ggplot(data, aes(x=BMI)) +
  geom_histogram(aes(y = ..density..), colour = 'white', fill =
'blue') +
  labs(x='BMI', y='Density') +
  theme_bw()
hiscomb1 = his1 + his2
hiscomb1
```

The output of the histogram of BMI is illustrated in Figure 3.4. Similar to the frequency curve of BMI depicted in Figure 3.2, the histogram of BMI shown in both panels of Figure 3.4 with different colors reveals that the distribution of BMI deviates from a bell-shaped, symmetric normal distribution. It appears to peak around a BMI of 26 and is positively skewed. Additionally, the mean of BMI is greater than it's median, which is greater than the mode.

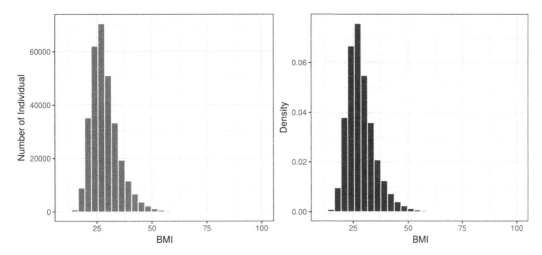

FIGURE 3.4
Histogram of BMI.

Now, suppose we want to see the distribution of BMI of male and female individuals separately by histogram. We can use the following R-code:

R-code:

```
his3<-ggplot(data, aes(x=BMI, fill=Sex, color=Sex)) +
  geom_histogram(fill='white', bins = 30) +
  scale_color_manual(name='Sex', labels=c('Female', 'Male'),
                     values=c('red','blue')) +
  labs(x='BMI', y='Number of Individual') +
  theme_bw()
his4<-ggplot(data, aes(x=BMI, fill=Sex, color = Sex)) +
  geom_histogram(aes(y = ..density..), color = 'white', bins = 30) +
  scale_fill_manual(name='Sex', labels=c('Female', 'Male'),
                    values=c('red','blue')) +
  labs(x='BMI', y='Density') +
  theme_bw()
hiscomb2 = his3 + his4
hiscomb2
```

The histogram of BMI by sex of the individuals is illustrated in Figure 3.5. Both panels of Figure 3.5 depict histograms of BMI by sex, with the left panel featuring the bars without fill color, while the right panel displays the bars with fill color. The histogram depicting BMI by sex reveals deviations from a normal distribution for both sexes, characterized by a peaked and positively skewed shape. Interestingly, the distribution peak for females appears more noticeable compared to males, although both sexes exhibit a peak around the same BMI value of 26.

3.4.5 Probability Density Plot

A density plot is similar to a histogram but represents the distribution as a continuous curve rather than bars. It is handy for visualizing smooth distributions and can be helpful when the data overlaps or we want a smoother representation. The density plot allows you to see the overall pattern of the data distribution, central tendency, spread, identify peaks and clusters, and observe potential skewness or multimodality of the data without the constraints of fixed bins (like in histograms).

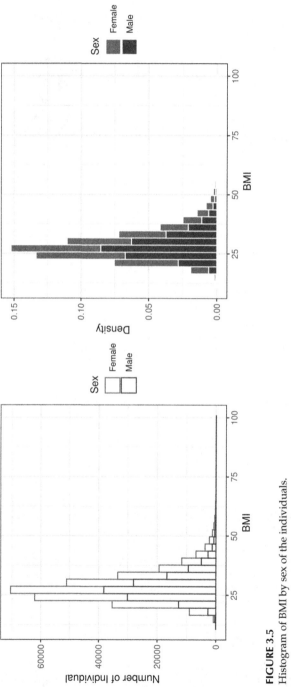

FIGURE 3.5
Histogram of BMI by sex of the individuals.

Example 3.8

Recall the dataset from **Example 3.4**, and suppose we want to see the probability density plot of the alcohol consumption. We need to `install` and `library` the `plyr` package.

R-code:

```
install.packages('plyr')
library('plyr')
```

Let us make the probability density plot of the alcohol consumption using the following R-code:

R-code:

```
plot1 <- ggplot(data = data, aes(x=Alcohol_Consumption)) +
  geom_density(color = 'red') +
  geom_vline(aes(xintercept=mean(Alcohol_Consumption)),
            color='blue', linetype='dashed') +
  labs(x='Alcohol Consumption', y='Density') +
  theme_bw()
mu <- ddply(data, .(Sex), summarize, grp.mean =
mean(Alcohol_Consumption))
plot2 <- ggplot(data = data, aes(x=Alcohol_Consumption, color=Sex)) +
  geom_density(size = 1) +
  geom_vline(data=mu, aes(xintercept=grp.mean, color=Sex),
linetype='dashed') +
  scale_color_manual(name='Sex', labels=c('Female', 'Male'),
                    values=c('red','blue')) +
  labs(x='Alcohol Consumption', y='Density') +
  theme_bw()
plot = plot1 + plot2
plot
```

The probability density plot of the alcohol consumption is shown in Figure 3.6. The probability density plot presented in the left panel of Figure 3.6 demonstrates that the distribution function of alcohol consumption conforms to an exponential distribution, showing a decreasing trend in probability density as alcohol consumption

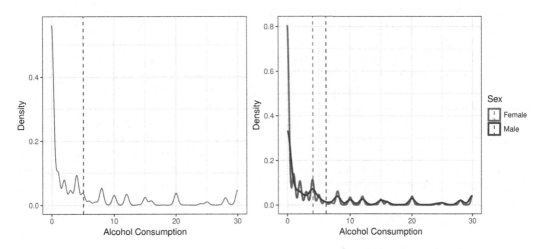

FIGURE 3.6
Probability density plot of the alcohol consumption.

increases. In the right panel of Figure 3.6, the probability density of alcohol consumption for both sexes exponentially decreases. However, there is a notable difference in the density around the range of 0–1 alcohol consumption per week, with females exhibiting approximately double the density compared to males. Furthermore, the average alcohol consumption is observed to be higher among males than females.

3.4.6 Histogram with Density Plot

Combining a histogram with a density plot is a great way to present a single continuous variable, as it provides both the visual representation of the data's distribution through bins (histogram) and the smooth estimate of the underlying probability density (density plot).

Example 3.9

Suppose we want to see the histogram with a density plot of BMI from the dataset from **Example 3.4**. Use the following R-code for making the histogram with a density plot of BMI:

R-code:

```
plot <- ggplot(data, aes(x = BMI)) +
  geom_histogram(aes(y = ..density..), colour = 'blue', fill =
'white') +
  geom_density(color = 'darkorange', size = 1, linetype = 'dashed') +
  geom_vline(aes(xintercept=mean(BMI)),
             color='red', linetype='dashed', size=1) +
  labs(x='BMI', y='Density') +
  theme_bw()
plot
```

The histogram with a density plot of BMI is illustrated in Figure 3.7. The histogram with density plot of BMI depicted in Figure 3.7 also reveals the distribution patterns of BMI similar as observed in Figures 3.2 and 3.4.

3.4.7 Box-and-Whisker Plot

A Box-and-Whisker plot, also known as a box plot, is a five-number summary of the data, which includes the minimum value, the first quartile (Q_1), the median (Q_2 or the second quartile), the third quartile (Q_3), and the maximum value. The Box-and-Whisker plot is particularly useful for visualizing the spread, skewness, and presence of outliers in a dataset. It provides a clear summary of the data distribution without revealing the individual data points. It is particularly useful when dealing with large datasets or when comparing multiple datasets.

The box in the plot represents the interquartile range (IQR), which is the range between the Q_1 and Q_3. The box covers the middle 50% of the data, and its length is a visual indicator of the spread or dispersion of the data. The whiskers are lines that extend from the edges of the box and reach the minimum and maximum values within a certain range. The length of the whiskers can vary, and their exact length is determined by the data and any outliers present. Inside the box, there is a line that represents the median of the data, which is the value that separates the lower 50% from the upper 50% of the data. Individual data points that fall outside the range covered by the whiskers are often displayed as individual points or dots and are considered outliers.

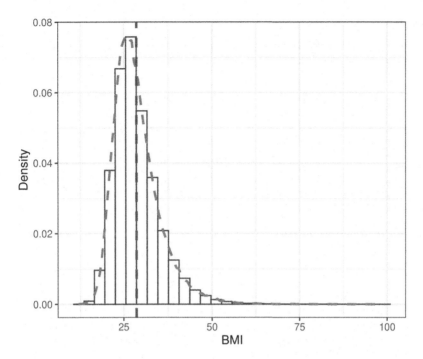

FIGURE 3.7
Histogram with a density plot of BMI.

The first step is to order the data in ascending order. The next step is to calculate the quartiles and IQR. Then determine the minimum and maximum values within a certain range (usually 1.5 times the IQR). Finally, plot the box, which extends from Q_1 to Q_3, with a line at the median. Extend the whiskers from the box to the minimum and maximum values, excluding outliers. Display any outliers as individual points beyond the whiskers. Data values further than 1.5×IQR from a lower/upper quartile are called outliers and data beyond 3×IQR indicate extreme outliers. If the median line deviated from the center of the 50% portion of the distribution or the whiskers are not the same in length, some degree of skew or asymmetry is indicated (Figure 3.8).

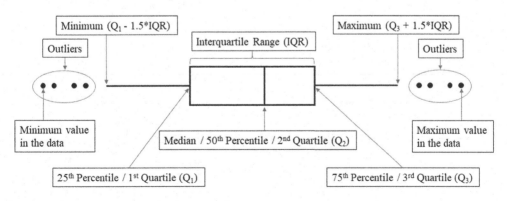

FIGURE 3.8
Concept of box plot.

Example 3.10

Suppose we want to see the variability in fruit consumption in the dataset from **Example 3.4**.
We can make a box plot using the following R-code:

R-code:

```
boxp1 <- ggplot(data = data, aes(y = Fruit_Consumption)) +
  geom_boxplot(color = 'black', fill = '#E8E8E8') +
  labs(y='Fruit Consumption') +
  theme_bw()
boxp2 <- ggplot(data = data, aes(x = General_Health, y = Fruit_
Consumption)) +
  geom_boxplot(color = 'black', fill = '#E8E8E8') +
  stat_summary(fun = "mean", geom = "point", shape = 8,
               size = 2, color = "black") +
  labs(x='General Health', y='Fruit Consumption') +
  theme_bw()
boxp = boxp1 + boxp2
boxp
```

To see the variability in fruit consumption, the box plot of fruit consumption is illustrated in Figure 3.9. The box plot of fruit consumption, as depicted in the left panel
of Figure 3.9, highlights many outliers in the dataset, exceeding the maximum value.
Half of the individuals consume fruits between 12 to 30 times daily, with a median consumption of around 30 times per day. Conversely, the right panel of Figure 3.9 shows no
outliers in fruit consumption among individuals who reported excellent health status.
Moreover, individuals with excellent health status exhibit higher daily fruit consumption compared to others, with half of those individuals consuming fruits between 12 and
60 times daily. Additionally, the median fruit consumption among individuals reporting
poor health status is lower compared to others.

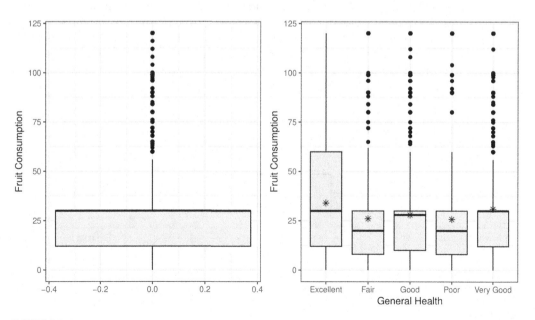

FIGURE 3.9
Box plot of fruit consumption.

3.4.8 Line Graph/Line Plot/Time Series Plot

Line plots are suitable for continuous data collected over time or in a specific sequence. It shows the trend, pattern, and changes in the data over the independent variable (e.g., time) using lines connecting the data points. Line plots can also help make predictions and understand the data's patterns and behavior.

Example 3.11

We have a dataset of BCG, MCV1, Pol3, and DTP3 vaccine coverage in Bangladesh from 1982 to 2021. The dataset was retrieved from the database of Our World in Data (https:// ourworldindata.org/). Let us import the data into R and see the head of the data using the following R-command:

R-code:

```
data <- read.csv('E:/Book/Chapter 3/Data/Vaccine Coverage.csv', header
= TRUE)
attach(data)
head(data)
```

Output:

```
Console  Terminal    Background Jobs

R  R 4.3.2 · E:/Book/Chapter 3/Data/
> data <- read.csv("E:/Book/Chapter 3/Data/Vaccine Coverage.csv", header = TRUE)
> attach(data)
> head(data)
       Entity Code Year BCG MCV1 Pol3 DTP3
1 Bangladesh  BGD 1982   1    1    1    1
2 Bangladesh  BGD 1983   2    1    1    1
3 Bangladesh  BGD 1984   2    1    1    1
4 Bangladesh  BGD 1985   2    1    2    2
5 Bangladesh  BGD 1986   4    3    4    4
6 Bangladesh  BGD 1987  15    6    8    9
>
```

Now, let us make the line plot of vaccine coverage in Bangladesh over the period from 1982 to 2021 using the following R-code:

R-code:

```
plot1 <- ggplot(data=data, aes(x=Year, y=BCG)) +
  geom_line()+
  geom_point()+
  labs(x='Year', y='Coverage of BCG Vaccine') +
  theme_bw()
plot2 <- ggplot(data=data, aes(x=Year)) +
  geom_line(aes(y=BCG, color='BCG')) +
  geom_point(aes(y=BCG, color='BCG')) +
  geom_line(aes(y=MCV1, color='MCV1')) +
  geom_point(aes(y=MCV1, color='MCV1')) +
  geom_line(aes(y=Pol3, color='Pol3')) +
  geom_point(aes(y=Pol3, color='Pol3')) +
  geom_line(aes(y=DTP3, color='DTP3')) +
  geom_point(aes(y=DTP3, color='DTP3')) +
  scale_color_manual(name='Vaccine', labels=c('BCG', 'MCV1', 'Pol3',
'DTP3'), values=c('red', 'blue', 'cyan', 'orange')) +
  labs(x='Year', y='Coverage') +
  theme_bw()
```

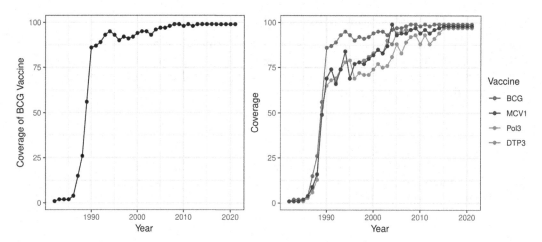

FIGURE 3.10
Line plot of vaccine coverage in Bangladesh over the period from 1982 to 2021.

```
plot = plot1 + plot2
plot
```

The line plot of vaccine coverage in Bangladesh over the period from 1982 to 2021 is shown in Figure 3.10. The left panel of Figure 3.10 displays the percentage coverage trend of the BCG vaccine over the years using both lines and dots, while the right panel illustrates the coverage trends of BCG, MCV1, Pol3, and DTP3 vaccines simultaneously, each represented by a different color. The coverage of all the mentioned vaccines exhibits a sharp increase up to 1990, followed by a gradual rise with slight fluctuations up to 2010. However, the coverage rates are almost similar around 100% after 2010.

3.5 Presentation of a Categorical Variable

A categorical variable is a type of variable that represents data in categories or groups. It has discrete, distinct categories with no inherent numerical meaning or ordering. Categorical, discrete, or qualitative variables can only take one of a limited number of distinct values (categories) or labels, and each observation in the dataset falls into one of these categories. A categorical variable can be nominal, ordinal, or discrete. Different visual methods for presenting categorical variables will be discussed in this section.

3.5.1 Frequency Distribution Table

A frequency distribution table for categorical data presents the frequency or count of each category in the dataset. It helps summarize the data and understand the distribution of the categorical variable. Firstly, determine the unique categories present in the data. Each unique value in the categorical variable will become a category in the frequency distribution table. Next, count the occurrences of each category in the dataset and construct a table with two columns: one for the categories and the other for the frequency or count of each category.

Example 3.12

Consider the general health variable in the dataset from **Example 3.4**, which is a categorical variable. We can summarize this variable by frequency distribution table using the following R-code:

R-code:

```
General.Health = as.factor(General_Health)
df = as.data.frame(General.Health)
fdtc <- ds_freq_table(df, General.Health)
fdtc
```

Output:

```
Console   Terminal ×   Background Jobs ×                                                          ━ ☐

 R 4.3.1 · ~/
> General.Health = as.factor(General_Health)
> df = as.data.frame(General.Health)
> fdtc <- ds_freq_table(df, General.Health)
> fdtc
                     variable: General.Health
-------------------------------------------------------------------------------
 Levels      Frequency    Cum Frequency        Percent        Cum Percent
-------------------------------------------------------------------------------
Excellent      55954           55954            18.12            18.12
-------------------------------------------------------------------------------
  Fair         35810           91764            11.59            29.71
-------------------------------------------------------------------------------
  Good         95364          187128            30.88            60.59
-------------------------------------------------------------------------------
  Poor         11331          198459             3.67            64.26
-------------------------------------------------------------------------------
Very Good     110395          308854            35.74             100
-------------------------------------------------------------------------------
  Total        308854            -             100.00              -
-------------------------------------------------------------------------------

> |
```

The frequency distribution table output indicates that a significant portion, around 36%, of individuals reported their health status as very good. Following this, approximately 31% reported their health status as good, while about 18% indicated excellent health. Conversely, only nearly 4% of individuals reported their health status as poor.

3.5.2 Bar Chart

The bar chart is a graph in which the classes are reported on the horizontal axis and the class frequencies on the vertical axis. The class frequencies are proportional to the heights of the bars. A distinguishing characteristic of a bar chart is the distance or gap between the bars and the bars are not adjacent to each other because of the qualitative nature.

A bar chart provides a visual representation of the frequency distribution of different categories in the dataset and helps to compare the frequencies or counts of different categories (e.g. Rahman et al., 2024). A bar chart over time or ordered categories can identify patterns and trends in the data. It can effectively highlight categories with exceptionally high or low frequencies, which might be outliers or warrant further investigation. Grouped or stacked bar charts can explore relationships between different categories, subcategories, and their combinations.

Example 3.13

Again, consider the general health categorical variable in the dataset from **Example 3.4**, which can be summarized by a bar chart using the following R-code:

R-code:

```
data1 <- as.data.frame(table(data$General_Health))
colnames(data1) <- c('General_Health_Status', 'Frequency')
attach(data1)
bar1 <- ggplot(data = data1, aes(x = General_Health_Status, y =
Frequency)) +
  geom_bar(stat='identity', color = 'red', fill = 'blue') +
  geom_text(aes(label=Frequency), vjust=1.3, color='white', size=3.5) +
  labs(x = 'General Health', y = 'Number of Individuals') +
  theme_bw()
Frequency1 = round(((Frequency/(sum(Frequency)))*100), 1)
data2 = cbind(data1, Frequency1)
data3 = as.data.frame(data2)
attach(data3)
bar2 <- ggplot(data = data3, aes(x = General_Health_Status, y =
Frequency1)) +
  geom_bar(stat='identity', color = 'red', fill = 'blue', width=0.5) +
  geom_text(aes(label=paste0(Frequency1, '%')), hjust=-0.1,
color='red', size=3.5) +
  ylim(0, 40) +
  labs(x = 'General Health', y = '% of Individuals') +
  coord_flip() +
  theme_bw()
data4 <- as.data.frame(table(data$General_Health, data$Sex))
colnames(data4) <- c('General_Health_Status1', 'Sex1', 'Frequency2')
attach(data4)
Frequency3 = round(((Frequency2/(sum(Frequency2)))*100), 1)
data5 = cbind(data4, Frequency3)
data6 = as.data.frame(data5)
attach(data6)
bar3 <- ggplot(data = data6, aes(x = General_Health_Status1, y =
Frequency3,
                                 fill = Sex1)) +
  geom_bar(stat='identity', color = 'white') +
  geom_text(aes(label=paste0(Frequency3, '%')), vjust=1.3,
    color='white',
            size=3.5, position = 'stack') +
  ylim(0, 40) +
  scale_fill_manual(name='Sex', labels=c('Female', 'Male'),
                    values=c('red','blue')) +
  labs(x = 'General Health', y = '% of Individuals') +
  theme_bw()
bar4 <- ggplot(data = data6, aes(x = General_Health_Status1, y =
  Frequency3,
                                 fill = Sex1)) +
  geom_bar(stat='identity', color = 'white', width=0.7,
position=position_dodge(width=0.8)) +
  geom_text(aes(label=paste0(Frequency3, '%')), angle = 90, hjust=1.2,
    color='white',
            size=3.5, position = position_dodge(0.9)) +
  ylim(0, 40) +
  scale_fill_manual(name='Sex', labels=c('Female', 'Male'),
                    values=c('red','blue')) +
```

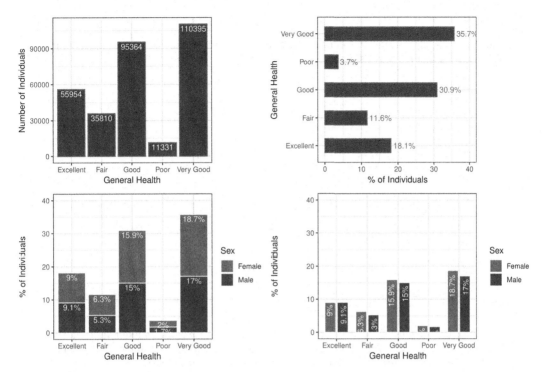

FIGURE 3.11
Bar chart of the general health status.

```
        labs(x = 'General Health', y = '% of Individuals') +
        theme_bw()
   bar = bar1 + bar2 + bar3 + bar4
   bar
```

The bar chart of the general health status is shown in Figure 3.11. Figure 3.11 displays the percentage distribution of general health status through both vertical and horizontal bar charts in panels 1 and 2, respectively. Additionally, it visualizes the disparity between sexes in general health status via stacked and multiple bar charts in panels 3 and 4, respectively. The panel 1 and 2 reveal similar conclusions as described in the frequency distribution table of Section 3.5.1. However, panels 3 and 4 conclude that there is no noticeable difference in the distribution of general health status between males and females.

3.5.3 Pie Chart

A pie chart is a circular chart divided into slices, with each slice representing a category and its proportion (percentage) of the whole. The circle represents 100% of the data or the total, and the individual slices represent the components or categories that make up that total. The angle of each slice is proportional to the percentage it contributes to the whole, making it easy to visualize the relative sizes of different categories. Therefore, pie charts are useful when we want to show the composition of a whole and the relative contribution of each category. It helps the researcher to show the distribution of categorical data or compare the parts of a whole.

Example 3.14

We can also summarize the general health categorical variable in the dataset from **Example 3.4** by pie chart using the following R-code:

R-code:

```
data1 <- as.data.frame(table(data$General_Health))
colnames(data1) <- c('General_Health1', 'Frequency')
attach(data1)
Percent = round(((Frequency/(sum(Frequency)))*100), 1)
data2 = cbind(data1, Percent)
data3 = as.data.frame(data2)
attach(data3)
pie <- ggplot(data = data3, aes(x=", y=Percent, fill=General_Health1)) +
  geom_col(color='white') +
  geom_text(aes(label=paste0(Percent, '%')), color='darkblue',
size=3.5,
            position = position_stack(vjust = 0.5)) +
  guides(fill = guide_legend(title = 'General Health')) +
  coord_polar(theta = 'y') +
  theme_void()
pie
```

The pie chart of general health status is illustrated in Figure 3.12. As like bar chart, the pie chart depicted in Figure 3.12 also visually represents the percentage distribution of health statuses outlined in the frequency distribution table of Section 3.5.1. The observations drawn from this chart align with the conclusions outlined in the frequency distribution table of Section 3.5.1.

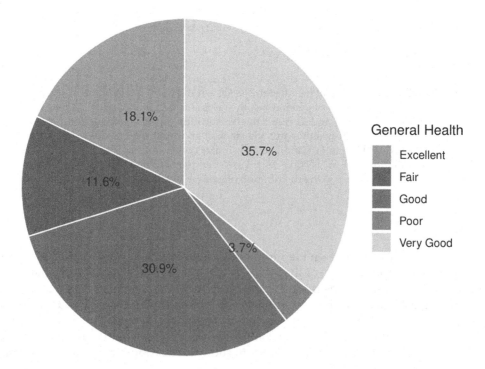

FIGURE 3.12
Pie chart of general health status.

3.6 Presenting Association of Continuous Data

Several graphical representations help in identifying potential correlations or associations between continuous variables, aiding researchers, analysts, and decision-makers in making informed interpretations.

3.6.1 Scatter Plot

Visualizing the relationship between two continuous variables can be achieved with scatter plots, which are a simple and powerful tool. With one variable plotted on the x-axis and the other on the y-axis, each point is represented as a dot in a scatter plot and corresponds to the scores (x, y) for a specific case (a person, for example), where x is the score that person was assigned or obtained on one variable and y is the score they received on the other. We can see the general pattern of the data and notice any possible patterns or correlations by using scatter plots. The closer the scatter plot's set of points resembles a diagonal 'line' across the graph, the greater the linear relationship between x and y.

Example 3.15

We have the 'Sleep Health and Lifestyle Dataset' downloaded from Kaggle, which consists of 13 variables. The description of the variables included in the 'Sleep Health and Lifestyle Dataset' is given in Table 3.4.

TABLE 3.4

Description of the Variables Included in the 'Sleep Health and Lifestyle Dataset'

Variable Name	Description
Person.ID	An identifier for each individual
Gender	The gender of the person (Male/Female)
Age	The age of the person in years
Occupation	The occupation or profession of the person
Sleep.Duration	The number of hours the person sleeps per day
Quality.of.Sleep	A subjective rating of the quality of sleep, ranging from 1 to 10
Physical.Activity.Level	The number of minutes the person engages in physical activity daily
Stress.Level	A subjective rating of the stress level experienced by the person, ranging from 1 to 10
BMI.Category	The BMI category of the person (e.g., Underweight, Normal, Overweight)
Blood.Pressure	The blood pressure measurement of the person, indicated as systolic pressure over diastolic pressure
Heart.Rate	The resting heart rate of the person in beats per minute
Daily.Steps	The number of steps the person takes per day
Sleep.Disorder	The presence or absence of a sleep disorder in the person (None, Insomnia, Sleep Apnea)

Let us import the data file into R and see the first five observations using the following R-code:

R-code:

```
data <- read.csv('E:/Book/Chapter 3/Data/Sleep_health_and_lifestyle_
dataset.csv', header = TRUE)
attach(data)
head(data, 5)
```

Output:

```
Console   Terminal ×   Background Jobs ×                                                    — □
R R 4.3.2 · ~/
> data <- read.csv("E:/Book/Chapter 3/Data/Sleep_health_and_lifestyle_dataset.csv", header = TRUE)
> attach(data)
> head(data, 5)
  Person.ID Gender Age            Occupation Sleep.Duration Quality.of.Sleep Physical.Activity.Level
1         1   Male  27     Software Engineer            6.1                6                      42
2         2   Male  28                Doctor            6.2                6                      60
3         3   Male  28                Doctor            6.2                6                      60
4         4   Male  28 Sales Representative            5.9                4                      30
5         5   Male  28 Sales Representative            5.9                4                      30
  Stress.Level BMI.Category Blood.Pressure Heart.Rate Daily.Steps Sleep.Disorder
1            6   Overweight         126/83         77        4200           None
2            8       Normal         125/80         75       10000           None
3            8       Normal         125/80         75       10000           None
4            8        Obese         140/90         85        3000    Sleep Apnea
5            8        Obese         140/90         85        3000    Sleep Apnea
>
```

Suppose we want to investigate the relationship between variables 'physical activity' and 'duration of sleep' of individuals using a scatter plot. Let us make the scatter plot using the following R-code:

R-code:

```
scatter1 <- ggplot(data = data, aes(x=Physical.Activity.Level,
y=Sleep.Duration)) +
  geom_point(color = 'darkblue') +
  labs(x = 'Physical Activity Level (Minutes/Day)', y = 'Duration of
Sleep (Hours/Day')
scatter1
```

The relationship between 'physical activity' and 'duration of sleep' of individuals is illustrated in Figure 3.13. The scatter plot suggests a positive association between the daily duration of sleep (in hours) and the number of minutes individuals engage in physical activity each day. However, there are some outliers where individuals exhibit either extreme sleep duration with low physical activity or extreme physical activity with low sleep duration.

Now, we can group the scatter plot by gender and sleep disorder using the following R-code.

R-code:

```
scatter2 <- ggplot(data = data, aes(x = Physical.Activity.Level, y =
Sleep.Duration, color = Gender)) +
  geom_point() +
  scale_color_manual(name='Gender', labels=c('Female', 'Male'),
                     values=c('red', 'blue')) +
  labs(x = 'Physical Activity Level (Minutes/Day)', y = 'Duration of
Sleep (Hours/Day)') +
  theme_bw()
scatter3 <- ggplot(data = data, aes(x = Physical.Activity.Level, y =
Sleep.Duration, color = Sleep.Disorder)) +
```

```
    geom_point() +
    scale_color_manual(name='Sleep Disorder', labels=c('Insomnia',
'None', 'Sleep Apnea'),
                    values=c('red', 'blue', 'orange')) +
    labs(x = 'Physical Activity Level (Minutes/Day)', y = 'Duration of
Sleep (Hours/Day)') +
    theme_bw()
scatter = scatter2 + scatter3
scatter
```

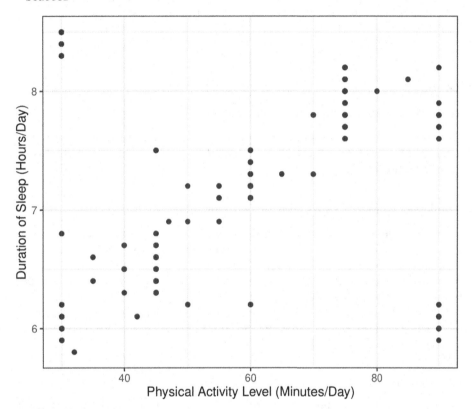

FIGURE 3.13
Scatter plot between 'physical activity' and 'duration of sleep'.

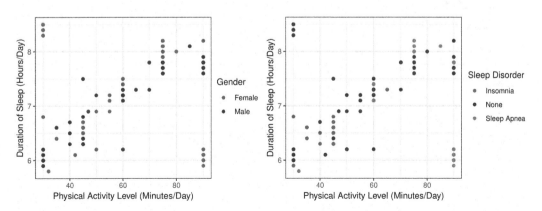

FIGURE 3.14
Scatter plots between physical activity and duration of sleep by gender and sleep disorder.

The scatter plots between 'physical activity' and 'duration of sleep' by gender and sleep disorder are shown in Figure 3.14. The scatter plot in the left panel illustrates the relationship between physical activity and duration of sleep by gender. It is evident that both males and females show a positive association between daily sleep duration (in hours) and the number of minutes engaged in physical activity each day. The outliers mentioned previously in Figure 3.13 are predominantly among females. On the right panel, the scatter plot depicts the relationship between physical activity and sleep duration by sleep disorder. It is observed that there is a positive association between physical activity and sleep duration among those individuals who have no sleep disorder.

3.7 Presenting Association of Categorical Data

Similar to continuous data, visualizations provide a clear and intuitive way to depict the relationships between different categorical variables, making it easier to identify associations, patterns, and trends.

3.7.1 Contingency Table or Cross Table or Two-Way Table

A contingency table, also known as a cross table or two-way table, is used to display the frequency distribution of two categorical variables. It shows the number of occurrences of each combination of categories for the two variables being studied (Allan et al., 2022). The purpose of a contingency table is to explore the relationship between the variables and identify any patterns or associations. The table is organized into rows and columns, with one variable's categories forming the rows and the other variable's categories forming the columns. Each cell in the table represents the count of occurrences for a specific combination of categories.

Example 3.16

Suppose we want to investigate the relationship between BMI category and sleep disorder in the dataset from **Example 3.15**. A contingency table, cross table, or two-way table is one method to summarize the relationship between BMI category and sleep disorder using the following R-code:

R-code:

```
cont = table(data$BMI.Category, data$Sleep.Disorder)
cont
cont1 = addmargins(cont)
cont1
cont2 = prop.table(cont, margin = 1)
cont2
```

Output:

```
Console   Terminal ×   Background Jobs ×

R  R 4.3.1 · ~/
> cont = table(data$BMI.Category, data$Sleep.Disorder)
> cont

                Insomnia None Sleep Apnea
  Normal               7  183           5
  Normal weight        2   17           2
  Obese                4    0           6
  Overweight          64   19          65
> cont1 = addmargins(cont)
> cont1

                Insomnia None Sleep Apnea Sum
  Normal               7  183           5 195
  Normal weight        2   17           2  21
  Obese                4    0           6  10
  Overweight          64   19          65 148
  Sum                 77  219          78 374
> cont2 = prop.table(cont, margin = 1)
> cont2

                  Insomnia       None Sleep Apnea
  Normal        0.03589744 0.93846154  0.02564103
  Normal weight 0.09523810 0.80952381  0.09523810
  Obese         0.40000000 0.00000000  0.60000000
  Overweight    0.43243243 0.12837838  0.43918919
> |
```

The contingency table above reveals that insomnia rates are higher among overweight individuals, followed by obese individuals. Conversely, sleep apnea is more prevalent among obese individuals, followed by overweight individuals.

3.7.2 Stacked or Multiple Bar Chart

A stacked or multiple bar chart is useful to visualize the relationship between two categorical variables because both charts allow us to compare multiple subcategories (categories of one variable) with bars for each category (categories of another variable).

Example 3.17

The relationship between the BMI category and sleep disorder in the dataset from **Example 3.15** can also be visualized by a stacked or bar chart instead of a contingency table using the following R-code:

R-code:

```
data1 <- as.data.frame(table(data$BMI, data$Sleep.Disorder))
colnames(data1) <- c('BMI1', 'Sleep_Disorder', 'Frequency')
attach(data1)
Frequency1 = round(((Frequency/(sum(Frequency)))*100), 1)
data2 = cbind(data1, Frequency1)
data3 = as.data.frame(data2)
```

```
attach(data2)
bar1 <- ggplot(data = data3, aes(x = BMI1, y = Frequency1,
                              fill = Sleep_Disorder)) +
  geom_bar(stat='identity', color = 'white') +
  geom_text(aes(label=paste0(Frequency1, '%')), vjust=1.3,
color='black',
              size=3.5, position = 'stack') +
  scale_fill_manual(name='Sleep Disorder', labels=c('Insomnia',
'None', 'Sleep Apnea'),
                       values=c('#FF6347','green', '#E69F00')) +
  labs(x = 'BMI', y = '% of Individuals') +
  theme_bw()
bar2 <- ggplot(data = data3, aes(x = BMI1, y = Frequency1,
                              fill = Sleep_Disorder)) +
  geom_bar(stat='identity', color = 'white', width=0.7,
position=position_dodge(width=0.8)) +
  geom_text(aes(label=paste0(Frequency1, '%')), angle = 90,
hjust=1.15, color='black',
              size=3.5, position = position_dodge(0.8)) +
  scale_fill_manual(name='Sleep Disorder', labels=c('Insomnia',
'None', 'Sleep Apnea'),
                       values=c('#FF6347','green', '#E69F00')) +
  labs(x = 'BMI', y = '% of Individuals') +
  theme_bw()
bar = bar1 + bar2
bar
```

The stacked and multiple bar charts between BMI category and sleep disorder are illustrated in Figure 3.15. Figure 3.15 provides a visual representation of the relationship between BMI category and sleep disorder, as summarized in the contingency table in Section 3.7.1. Both panels illustrate the same information, but they utilize different types of bar charts. The left panel employs a stacked bar chart, while the right panel utilizes a multiple bar chart.

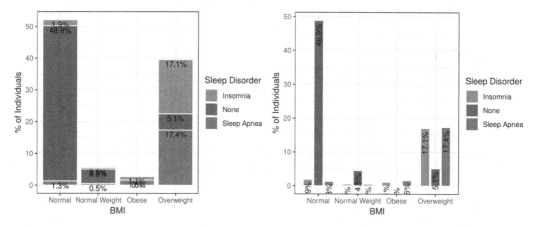

FIGURE 3.15
Stacked and multiple bar charts between BMI category and sleep disorder.

3.8 Conclusion and Exercises

This section concludes the chapter with some practical problem-solving exercises. These concepts and measures are used in the subsequent chapters.

3.8.1 Concluding Remarks

Data visualization and summarization serve as a synthesis of important ideas and a commentary on the importance of the subjects covered. It seeks to give readers a comprehensive grasp of the significance of efficient data visualization and summarization across a range of fields. Our journey through various visualization methods has highlighted their capacity to reveal patterns, trends, and outliers that might otherwise remain obscured in the raw data. The visual representation of data enhances comprehension and facilitates knowledge transfer. Readers will be able to summarize and visualize their dataset with the help of understanding the discussion presented in this chapter. Different illustrations are also used in the subsequent chapters.

3.8.2 Practice Questions

Exercise 3.1 Consider the dataset named 'Heart Disease Dataset' downloaded from the UCI Machine Learning Repository that contains 1,025 observations under 14 variables. The description of the variables included in the dataset is presented below:

Variable Name	Description
age	The age of the individual in years
sex	Gender of the individual, usually represented as binary (0 for female, 1 for male)
cp	A categorical variable indicating the type of chest pain (four types) experienced by the individual (1 for typical angina, 2 for atypical angina, 3 for non-anginal pain, and 4 for asymptomatic)
trestbps	The blood pressure of the individual at rest (in mm Hg) on admission to the hospital
chol	Serum cholesterol in mg/dl
fbs	A binary variable indicating whether fasting blood sugar is greater than 120 mg/dL (1 for true, 0 for false)
restecg	Results of the resting electrocardiogram, categorized as 0 for normal, 1 for having ST-T wave abnormality (T wave inversions and/or ST elevation or depression of >0.05 mV), and 2 for showing probable or definite left ventricular hypertrophy by Estes' criteria
thalach	The maximum heart rate achieved during exercise
exang	A binary variable indicating whether angina (chest pain) was induced by exercise (0 for no and 1 for yes)
oldpeak	ST depression induced by exercise relative to rest
slope	The slope of the peak exercise ST segment (1 for upsloping, 2 for flat, and 3 for downsloping)
ca	The number of major blood vessels (0–3) colored by fluoroscopy
thal	Results of the thallium stress test, categorized as 0 = normal; 1 = fixed defect; 2 = reversible defect
target	The presence or absence of heart disease, typically binary (1 for presence, 0 for absence)

Download this dataset and read it into R under the name 'heart_disease' and then,

a. Summarize and present the serum cholesterol in mg/dl using the frequency distribution table.
b. Make the frequency polygon or frequency curve and cumulative frequency curve for the serum cholesterol in mg/dl.
c. Visualize the resting blood pressure using a histogram, probability histogram, and histogram with density plot.
d. Present the maximum heart rate achieved using a box-and-whisker plot.
e. Show graphically the relationship between age and serum cholesterol in mg/dl using a scatter plot.
f. Demonstrate graphically the relationship between age and resting blood pressure using a scatter plot.

Bibliography

Allan, J., Kleinschafer, J., Saksena, T., Rahman, A., Lawrence, J., & Lock, M. (2022). A comparison of rural Australian First Nations and Non-First Nations survey responses to COVID-19 risks and impacts: Implications for health communications. *BMC Health Services Research*, 22(1), 1276. https://doi.org/10.1186/s12889-022-13643-6

Andrews, M. (2021). *Doing data science in R: an introduction for social scientists*. Sage Publications Ltd.

Bivand, R. S., Pebesma, E. J., Gómez-Rubio, V., & Pebesma, E. J. (2008). *Applied spatial data analysis with R* (Vol. 747248717, pp. 237–268). Springer.

Chambers, J. M. (2018). *Graphical methods for data analysis*. CRC Press.

Chen, C. H., Härdle, W. K., & Unwin, A. (Eds.). (2007). *Handbook of data visualization*. Springer Science & Business Media.

Cook, D., Swayne, D. F., & Buja, A. (2007). *Interactive and dynamic graphics for data analysis: with R and GGobi* (Vol. 1). Springer.

Evergreen, S. D. (2019). *Effective data visualization: The right chart for the right data*. SAGE publications.

Fischetti, A. (2018). *Data analysis with R: A comprehensive guide to manipulating, analyzing, and visualizing data in R*. Packt Publishing Ltd.

Friendly, M., & Meyer, D. (2015). *Discrete data analysis with R: visualization and modeling techniques for categorical and count data* (Vol. 120). CRC Press.

Healy, K. (2018). *Data visualization: A practical introduction*. Princeton University Press.

Heiberger, R. M., & Neuwirth, E. (2009). *R through Excel: A spreadsheet interface for statistics, data analysis, and graphics* (pp. 165–191). Springer.

Horton, N. J., & Kleinman, K. (2015). *Using R and RStudio for data management, statistical analysis, and graphics*. CRC Press.

Iliinsky, N., & Steele, J. (2011). *Designing data visualizations: Representing informational Relationships*. O'Reilly Media, Inc.

Jacoby, W. G. (1997). *Statistical graphics for univariate and bivariate data* (Vol. 117). Sage.

Kabacoff, R. (2022). *R in action: Data analysis and graphics with R and Tidyverse*. Simon and Schuster.

Kassambara, A. (2019). *GGPlot2 essentials: Great data visualization in R*. Datanovia.

Keen, K. J. (2018). *Graphics for statistics and data analysis with R*. CRC Press.

Knell, R. J. (2014). *Introductory R: A Beginner's guide to data visualisation, statistical analysis and programming in R*. Robert Knell.

Kolaczyk, E. D., & Csárdi, G. (2014). *Statistical analysis of network data with R* (Vol. 65). Springer.

Long, J. D., & Teetor, P. (2019). *R cookbook: Proven recipes for data analysis, statistics, and graphics*. O'Reilly Media.

Maindonald, J., & Braun, J. (2006). *Data analysis and graphics using R: An example-based approach* (Vol. 10). Cambridge University Press.

McGrath, M. (2023). *R for data analysis in easy steps: Covers R programming essentials*. Easy Steps Limited.

Megabiaw, B., & Rahman, A. (2013). Prevalence and determinants of chronic malnutrition among under-5 children in Ethiopia. *International Journal of Child Health and Nutrition*, 2(3), 230–236.

Mirkin, B. (2019). *Core data analysis: Summarization, correlation, and visualization*. Springer International Publishing.

Moraga, P. (2019). *Geospatial health data: Modeling and visualization with R-INLA and shiny*. CRC Press.

Murrell, P. (2005). *R graphics*. Chapman and Hall/CRC.

Nordmann, E., McAleer, P., Toivo, W., Paterson, H., & DeBruine, L. M. (2022). Data visualization using R for researchers who do not use R. *Advances in Methods and Practices in Psychological Science*, 5(2), 25152459221074654.

Oyana, T. J. (2020). *Spatial analysis with r: Statistics, visualization, and computational methods*. CRC Press.

Rahlf, T. (2019). *Data visualisation with R: 111 examples*. Springer Nature.

Rahman, A., & Harding, A. (2013). Prevalence of overweight and obesity epidemic in Australia: Some causes and consequences. *JP Journal of Biostatistics*, 10(1), 31–48.

Rahman, A., Othman, N., Kuddus, M. A., & Hasan, M. Z. (2024). Impact of the COVID-19 pandemic on child malnutrition in Selangor, Malaysia: A pilot study. *Journal of Infection and Public Health*, 17(5), 833–842. https://doi.org/10.1016/j.jiph.2024.02.019

Reimann, C., Filzmoser, P., Garrett, R., & Dutter, R. (2011). *Statistical data analysis explained: Applied environmental statistics with R*. John Wiley & Sons.

Schwabish, J. (2021). *Better data visualizations: A guide for scholars, researchers, and wonks*. Columbia University Press.

Sievert, C. (2020). *Interactive web-based data visualization with R, plotly, and shiny*. CRC Press.

Teetor, P. (2011). *R cookbook: Proven recipes for data analysis, statistics, and graphics*. O'Reilly Media, Inc.

Theus, M., & Urbanek, S. (2008). *Interactive graphics for data analysis: principles and examples*. CRC Press.

Unwin, A., Theus, M., Hofmann, H., Unwin, A., & Theus, M. (2006). *Graphics of a large dataset* (pp. 227–249). Springer.

Wickham, H. (2016). *ggplot2: Elegant graphics for data analysis*. Springer International Publishing.

Wickham, H., Çetinkaya-Rundel, M., & Grolemund, G. (2023). *R for data science*. O'Reilly Media, Inc.

Wilke, C. O. (2019). *Fundamentals of data visualization: A primer on making informative and compelling figures*. O'Reilly Media.

Young, F. W., Valero-Mora, P. M., & Friendly, M. (2011). *Visual statistics: Seeing data with dynamic interactive graphics*. John Wiley & Sons.

4

Advanced Graphical Presentation of Data

4.1 Single Continuous Variable

4.1.1 Empirical Cumulative Distribution Function

The Empirical Cumulative Distribution Function (ECDF) is a nonparametric statistical tool used to describe the distribution of a dataset. It provides a visual representation of how the data points are spread across the range of possible values. The ECDF essentially allows us to plot a feature of our data in order from least to greatest against their percentiles and see the whole feature as if it is distributed across the dataset. It is a step function that asymptotically approaches 0 and 1 on the vertical y-axis. The ECDF is a helpful tool for quickly understanding the distribution of data, especially when the underlying probability distribution is not known or when dealing with small sample sizes. It can be used to compare datasets, identify percentiles, and make statistical inferences about the population based on the sample data. To determine the range of the data, look for the first and last steps in the step function, but look for the steeper portions of the step function to determine where the most common values occur. Conversely, flatter portions indicate ranges with fewer observations.

> **Example 4.1**
>
> Recall the data from **Example 3.4**. Suppose we want to plot the empirical cumulative distribution function for fruit consumption, the following R-command can be used.
>
> **R-code**:

```
ecdf1 <- ggplot(data = data, aes(Fruit_Consumption)) +
  stat_ecdf(geom = 'step', color='purple', size = 1) +
  labs(x='Fruit Consumption', y='Percentile') +
  theme_bw()
ecdf2 <- ggplot(data = data, aes(Fruit_Consumption, color = Sex)) +
  stat_ecdf(geom = 'step', size = 1) +
  scale_color_manual(name='Sex', labels=c('Female', 'Male'),
                     values=c('red','blue')) +
  labs(x='Fruit Consumption', y='Percentile') +
  theme_bw()
ecdf = ecdf1 + ecdf2
ecdf
```

The plot of the empirical cumulative distribution function for fruit consumption is shown in Figure 4.1.

DOI: 10.1201/9781003426189-4

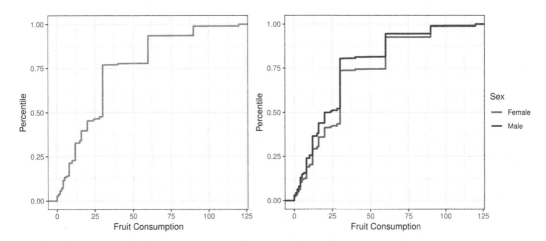

FIGURE 4.1
Plot of empirical cumulative distribution function for fruit consumption.

In the left panel of Figure 4.1, it is evident that approximately 29% of individuals consume fruit 30 times daily, followed by nearly 16% who consume fruit 60 times per day, and almost 10% who consume fruit 12 times each day. Furthermore, there is a sharp increase in fruit consumption in a day among individuals after reaching 30 times. The right panel demonstrates a similar pattern of fruit consumption between males and females.

4.1.2 Violin Plot

A violin plot is another effective way to present a single continuous variable, especially when we want to visualize its distribution along with summary statistics. It combines the features of a box plot and a kernel density plot, providing a richer representation of the data. The width of the violin represents the density of the data at different points, while the box inside shows the quartiles and median. The violin plot allows viewers to see the shape of the distribution, central tendency, variability, any potential multimodality (multiple peaks), and skewness of the data, making it a powerful tool for visualizing continuous data.

Example 4.2

Recall the data from **Example 3.4**. Suppose we want to see the variability and probability density of green vegetable consumption. For this purpose, we can produce a violin plot using the following R-code.

R-code:

```
violin1 <- ggplot(data = data, aes(x = General_Health, y = Green_
Vegetables_Consumption)) +
  geom_violin(color = 'red', fill = 'blue', linewidth = 0.7) +
  labs(x = 'General Health', y='Green Vegetables Consumption') +
  theme_bw()
violin2 <- ggplot(data = data, aes(x = General_Health, y = Green_
Vegetables_Consumption)) +
  geom_violin(color = 'red', fill = 'blue', linewidth = 0.7) +
  geom_jitter(height = 0, width = 0.1, color = 'lightblue') +
  geom_boxplot(width = .2, color = 'purple') +
  stat_summary(fun = 'mean', geom = 'point', shape = 8,
```

```
                    size = 2, color = 'red') +
    labs(x = 'General Health', y='Green Vegetables Consumption') +
    theme_bw()
violin = violin1 + violin2
violin
```

The violin plot of green vegetable consumption is illustrated in Figure 4.2. Both panels of Figure 4.2 depict violin plots of green vegetable consumption, with varying styles. The left panel illustrates only the violins, while the right panel presents the violin plot alongside a box plot, summary statistics by mean, and jitter points. In the left panel, it is evident that individuals with excellent, very good, and good general health status show the highest likelihood of consuming green vegetables 30 times per day. Furthermore, individuals with fair general health status are most likely to consume green vegetables 4 times per day. Conversely, those with poor health status show the highest probability of consuming no (0 times) green vegetables in a day. The right panel illustrates that the median and mean consumption of green vegetables among individuals with excellent, very good, and good health statuses are higher compared to those with fair and poor health statuses. Additionally, outlier points are observed in the consumption data of green vegetables across all health status groups.

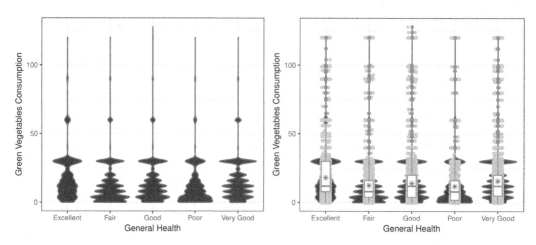

FIGURE 4.2
Violin plot of green vegetable consumption.

4.2 Single Categorical Variable

4.2.1 Donut Chart

A donut chart, in its simplest form, is a pie chart with its center cut out to look like a donut. It typically shows the proportions of categorical data where the size of each piece of the donut communicates the proportion of each category. At first glance, this may not seem to serve a much greater purpose than to create aesthetic variety. However, a donut chart helps avoid confusion around the area parameter that often trips people up in a pie chart. It encourages the reader to focus on the length of the arc instead and not compare it with the total area a circle would represent. Another visual benefit of a donut chart is that the space inside the donut can be used to represent data, labels, and such to make reading the chart easy.

Example 4.3

The general health categorical variable in the dataset from **Example 3.4** can be summarized by a donut chart instead of bar or pie charts using the following R-code.

R-code:

```
data1 <- as.data.frame(table(data$General_Health))
colnames(data1) <- c('General_Health1', 'Frequency')
attach(data1)
Percent = round(((Frequency/(sum(Frequency)))*100), 1)
data2 = cbind(data1, Percent)
data3 = as.data.frame(data2)
attach(data3)
donut <- ggplot(data = data3, aes(x=2, y=Percent, fill=General_
Health1)) +
  geom_col(color='white') +
  xlim(0.5, 2.5) +
  geom_text(aes(label=paste0(Percent, '%')), color='darkblue',
      size=3.5,
            position = position_stack(vjust = 0.5)) +
  scale_fill_manual(values=c('#808000','#C71585', '#00BFFF',
'#9370DB', '#F4A460')) +
  guides(fill = guide_legend(title = 'General Health')) +
  coord_polar(theta = 'y') +
  theme_void()
donut
```

The donut chart of the general health status is presented in Figure 4.3. The donut chart representing general health status in Figure 4.3 conveys the same information as the pie chart presented in Section 3.5.3 of Chapter 3.

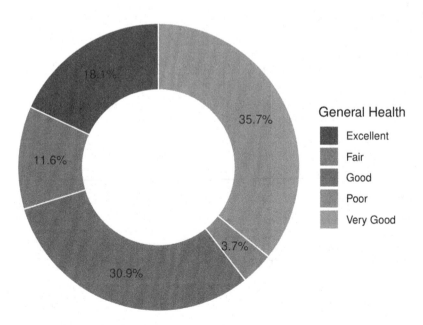

FIGURE 4.3
Donut chart of the general health status.

4.2.2 Waffle Chart

A waffle chart is essentially a square grid where each cell represents a portion or percentage of the whole. It is often used to display the proportion of different categories within a dataset and is an alternative to pie or donut charts when visualizing proportions. It is a great way of visualizing data about a whole, highlighting progress against a given threshold, or dealing with populations too varied for pie charts. But it is useful for displaying small sizes of data.

Example 4.4

Suppose we want to summarize a subset of size 175 observation of the general health variable in the dataset from **Example 3.4**. It can be summarized by a waffle chart instead of a bar, pie, or donut chart. We need to install and library the ggwaffle package for making a waffle chart in ggplot2 using the following R-code.

R-code:

```
install.packages('devtools')
library('devtools')
devtools::install_github('liamgilbey/ggwaffle')
library('ggwaffle')
```

Now, we can make the waffle chart using the following R-code.

R-code:

```
data1 = data[1:175,]
waffle_data <- waffle_iron(data1, rows = 10, aes_d(group =
General_Health))
waffle1 = ggplot(data = waffle_data, aes(x, y, fill = group)) +
  geom_waffle() +
  coord_equal() +
  scale_colour_waffle() +
  scale_fill_manual(name='General Health', labels=c('Excellent',
'Fair', 'Good', 'Poor', 'Very Good'),
                    values=c('red', 'darkblue', 'darkorange', 'cyan',
'darkgreen')) +
  theme_waffle()
waffle2 = ggplot(data = waffle_data, aes(x, y, colour = group)) +
  geom_waffle(tile_shape = 'circle', size = 5) +
  coord_equal() +
  scale_colour_waffle() +
  scale_colour_manual(name='General Health', labels=c('Excellent',
'Fair', 'Good', 'Poor', 'Very Good'),
                    values=c('red', 'darkblue', 'darkorange',
'cyan', 'darkgreen')) +
  theme_waffle()
waffle = waffle1 + waffle2
waffle
```

The waffle chart of the general health status is presented in Figure 4.4.

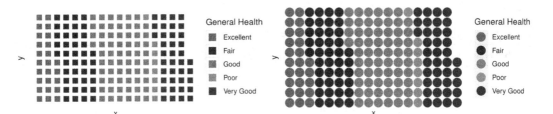

FIGURE 4.4
Waffle chart of the general health status.

Both panels in Figure 4.4 depict the general health status of the initial 175 individuals using waffle charts with different styles. It is evident that the highest proportion of individuals exhibit a good health status, followed by fair, very good, excellent, and poor health status.

4.3 Association of Continuous Variables

4.3.1 Contour Plot or 2D Density Plot

A contour plot, also known as a 2D density plot, is a two-dimensional generalization of a histogram that resembles a contour plot but is computed by grouping a set of points specified by their x and y coordinates into bins and applying an aggregation function such as count or sum to estimate the value to be used to compute contours. This visualization is often used to manage over-plotting or situations where showing large datasets as scatter plots would result in overlapping points and hiding patterns. Contour plots display lines, called contours, which connect points of equal value, providing a way to visualize the variations and patterns in the data. In a contour plot, the x and y axes represent the input variables, and the function's output values are represented by contour lines or filled color regions.

Example 4.5

We have a dataset named 'Diabetes Healthcare: Comprehensive Dataset-AI' downloaded from Kaggle. Description of the variables included in the Diabetes Healthcare: Comprehensive Dataset-AI dataset is presented in Table 4.1.

Let us import the data and see the first five observations using the following R-code

R-code:

```
data <- read.csv('E:/Book/Chapter 4/Data/health care diabetes.csv',
header = TRUE)
attach(data)
head(data, 5)
```

TABLE 4.1

Description of the Variables Included in the Diabetes Healthcare: Comprehensive Dataset-AI Dataset

Variable Name	Description
Pregnancies	Number of times pregnant
Glucose	Plasma glucose concentration in an oral glucose tolerance test
BloodPressure	Diastolic blood pressure (mm Hg)
SkinThickness	Triceps skinfold thickness (mm)
Insulin	Two hour serum insulin
BMI	Body Mass Index
DiabetesPedigreeFunction	Diabetes pedigree function
Age	Age in years
Outcome	Class variable (either 0 or 1). 268 of 768 values are 1, and the others are 0

Output:

```
Console   Terminal    Background Jobs

R 4.3.2 · E:/Book/Chapter 4/Data/
> data <- read.csv("E:/Book/Chapter 4/Data/health care diabetes.csv", header = TRUE)
> attach(data)
> head(data, 5)
  Pregnancies Glucose BloodPressure SkinThickness Insulin  BMI DiabetesPedigreeFunction Age Outcome
1           6     148            72            35       0 33.6                    0.627  50       1
2           1      85            66            29       0 26.6                    0.351  31       0
3           8     183            64             0       0 23.3                    0.672  32       1
4           1      89            66            23      94 28.1                    0.167  21       0
5           0     137            40            35     168 43.1                    2.288  33       1
> |
```

Now, let us make the contour or 2D density plot between blood pressure and BMI from the 'Diabetes Healthcare: Comprehensive Dataset-AI' dataset (both of the variables are continuous) using the following R-code.

R-code:

```
cp1 = ggplot(data, aes(x = BMI, y = BloodPressure)) +
  geom_density_2d(color = 'red') +
  labs(x = 'BMI', y = 'Blood Pressure') +
  theme_bw()
cp1
```

The contour or 2D density plot between blood pressure and BMI is shown in Figure 4.5. In Figure 4.5, the central contour signifies a high density of data points, while the outer contour indicates a lower density. Hence, contours with high density represent individuals with a BMI ranging from 31 to 34 and a blood pressure between 67 and 74. The overall contour pattern suggests a slight positive correlation between BMI and blood pressure.

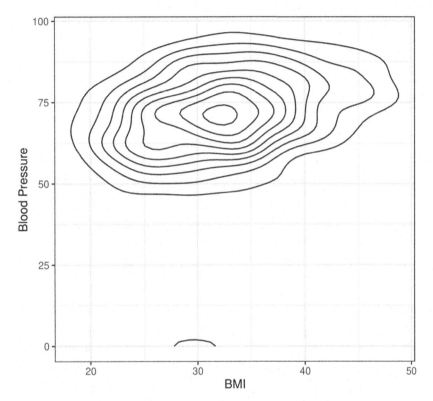

FIGURE 4.5
Contour or 2D density plot between blood pressure and BMI.

We can format the contour plot as below.

R-code:

```
cp2 = ggplot(data, aes(x = BMI, y = BloodPressure)) +
  geom_point(color = 'blue') +
  geom_density_2d(bins = 15, color = 'red') +
  labs(x = 'BMI', y = 'Blood Pressure') +
  theme_bw()
cp3 = ggplot(data, aes(x = BMI, y = BloodPressure)) +
  geom_point() +
  geom_density_2d(bins = 15, aes(color = ..level..)) +
  scale_color_viridis_c() +
  labs(x = 'BMI', y = 'Blood Pressure') +
  theme_bw()
cp4 = ggplot(data, aes(x = BMI, y = BloodPressure)) +
  geom_density_2d_filled() +
```

```
      geom_density_2d(colour = 'red1') +
      guides(fill = guide_legend(title = 'Level')) +
      labs(x = 'BMI', y = 'Blood Pressure') +
      theme_bw()
    cp5 = ggplot(data, aes(x = BMI, y = BloodPressure)) +
      geom_point(color = 'blue') +
      geom_density_2d_filled(alpha = 0.4) +
      geom_density_2d(colour = 'red1') +
      guides(fill = guide_legend(title = 'Level')) +
      labs(x = 'BMI', y = 'Blood Pressure') +
      theme_bw()
    cp = cp2 + cp3 + cp4 + cp5
    cp
```

Figure 4.6 illustrates the different styles of contour or 2D density plots between blood pressure and BMI.

4.3.2 Joint Histogram or Hexbin Plot

A joint histogram, also known as a hexbin plot, is a 2D visual representation of the data distribution between two continuous variables. It is similar to a scatter plot but particularly useful when dealing with a large number of data points, where individual points may overlap, and the underlying structure and patterns are difficult to understand. In a joint histogram or hexbin plot, the data points are aggregated into hexagonal bins, and each bin is colored according to the number of data points it contains. The color intensity represents the density of data points within each bin, allowing us to visualize the data distribution more effectively.

Example 4.6

Let us consider the blood pressure and BMI continuous variables in the dataset from **Example 4.5**. We can present the relationship between these two variables by a joint histogram or hexbin plot. We need to install and library the hexbin package for making hexbin plots using ggplot2. Let us install and library the hexbin package using the following R-code.

R-code:

```
    install.packages('hexbin')
    library('hexbin')
```

Now, let us make the joint histogram or hexbin plot between blood pressure and BMI using the following R-code.

R-code:

```
    hexbinplot = ggplot(data, aes(x = BMI, y = BloodPressure)) +
      geom_hex(bins = 30, color = 'white') +
      scale_fill_viridis_c(option = 'H') +
      guides(fill = guide_colourbar(title = 'Count')) +
      labs(x = 'BMI', y = 'Blood Pressure') +
      theme_bw()
    hexbinplot
```

Figure 4.7 shows the joint histogram or hexbin plot between blood pressure and BMI.

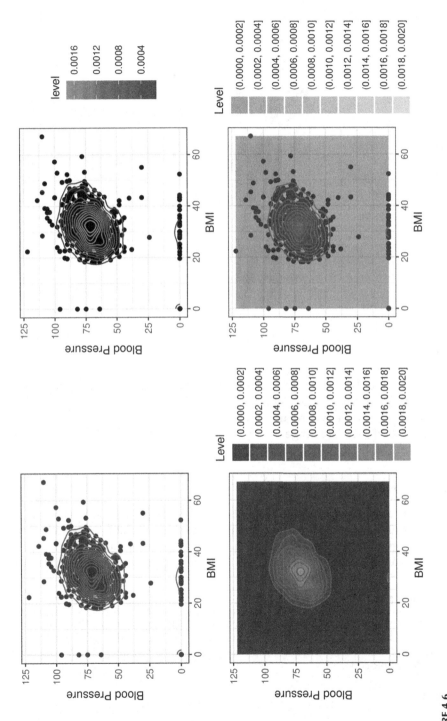

FIGURE 4.6
Different styles of contour or 2D density plot between blood pressure and BMI.

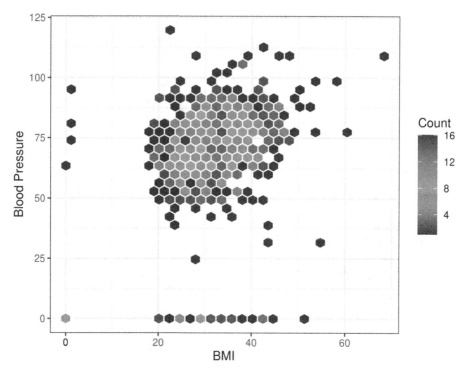

FIGURE 4.7
Joint histogram or hexbin plot between blood pressure and BMI.

Figure 4.7 provides the same information as Figures 4.5 and 4.6. As depicted in Figure 4.7, the majority of data points are concentrated within the BMI range of 25 to 37 and the blood pressure range of 60 to 84. Approximately 90% of the data falls within the BMI range of 20 to 40 and the blood pressure range of 50 to 88. Furthermore, a slight positive correlation is evident between BMI and blood pressure.

4.3.3 Bubble Chart

A bubble chart, also known as a bubble plot or bubble graph, is an extension of the scatter plot used to look at relationships between three numeric variables in a two-dimensional visual presentation. In a bubble chart, the position of each bubble is determined by its x and y coordinates, and the size of each bubble represents a third numerical variable.

Example 4.7

Suppose we want to visualize the relationship between the physical activity level, quality of sleep, and age of the individuals in the dataset from **Example 3.15**. Further, we also want to add another factor called sleep disorder. It can be possible by bubble plot using the following R-code.

R-code:

```
bubble1 = ggplot(data = data, aes(x = Physical.Activity.Level,
                                   y = Quality.of.Sleep, size = Age)) +
    geom_point(alpha = 0.7, color = 'blue') +
    scale_size(range = c(2, 7), name='Age') +
```

```
    labs(x = 'Physical Activity Level (Minutes/Day)', y = 'Quality of
        Sleep') +
    theme_bw()
  bubble2 = ggplot(data = data, aes(x = Physical.Activity.Level,
                                    y = Quality.of.Sleep)) +
    geom_point(aes(color = Sleep.Disorder, size = Age), alpha = 0.7) +
    scale_color_manual(name = 'Sleep Disorder',
                       values = c('red1', 'blue', 'orange')) +
    scale_size(range = c(2, 7)) +
    labs(x = 'Physical Activity Level (Minutes/Day)', y = 'Quality of
        Sleep') +
    theme_bw()
  bubble = bubble1 + bubble2
  bubble
```

The output of the bubble plot is shown in Figure 4.8.

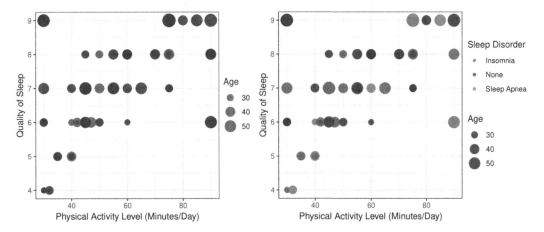

FIGURE 4.8
The left panel depicts the relationship between the physical activity level, quality of sleep, and age; the right panel shows the relationship between the physical activity level, quality of sleep, age, and sleep disorder.

In the left panel of Figure 4.8, we observe a positive association between both physical activity level (in minutes per day) and age (in years) with sleep quality. In the right panel, we introduce an additional factor, sleep disorder, alongside sleep quality, age, and physical activity level. Here, we note that sleep apnea is more prevalent among individuals with higher physical activity levels and better sleep quality. Conversely, insomnia appears to be more common among individuals with lower physical activity levels and average sleep quality.

4.3.4 Heatmap

In the case of continuous data, a heatmap produces a grid with multiple attributes of the data frame, representing the relationship between the two attributes taken at a time. The grid is filled with colors where the colors in a heatmap represent the magnitude of the values of the measurement, making it easy to identify patterns and trends. The measurement can be any statistic, such as correlation, probability, or p-value. The heatmap of correlation measurement is also known as a correlation plot. Heatmaps are especially beneficial for displaying and examining relationships and patterns in tabular data.

Example 4.8

Suppose we want to visualize the correlation matrix among the variables age, sleep duration, quality of sleep, physical activity level, stress level, heart rate, daily steps, systolic blood pressure, and diastolic blood pressure in the dataset from **Example 3.15** by heatmap. In order to do this, first of all, we need to calculate the correlation matrix and process it for heatmap. We need to `install` and `library` the `tidyr` and `reshape` packages for using the `separate` and `melt` functions, respectively.

R-code:

```
install.packages('tidyr')
library('tidyr')
install.packages('reshape')
library('reshape')
```

Now, let us process the data for producing a heatmap using the following R-code.

R-code:

```
data1 = data[, c(3, 5, 6, 7, 8, 11, 12, 10)]
data1 = separate(data1, Blood.Pressure, c('SBP','DBP'),sep = '[()/ ]+',
extra = 'drop')
attach(data1)
data1$SBP = as.numeric(SBP)
data1$DBP = as.numeric(DBP)
cm = round(cor(data1, method = c('pearson')),2)
colnames(cm) <- c('Age', 'Sleep Duration', 'Quality of Sleep',
'Physical Activity Level',
                 'Stress Level', 'Heart Rate', 'Daily Steps',
'Systolic BP', 'Diastolic BP')
rownames(cm) <- c('Age', 'Sleep Duration', 'Quality of Sleep',
'Physical Activity Level',
                 'Stress Level', 'Heart Rate', 'Daily Steps',
'Systolic BP', 'Diastolic BP')
df <- melt(cm)
colnames(df) <- c('x', 'y', 'value')
```

Then, we can produce a heatmap of the correlation matrix using the following R-code.

R-code:

```
heatmap = ggplot(df, aes(x = x, y = y, fill = value)) +
  geom_tile(color = 'white',
            lwd = 0.5,
            linetype = 1) +
  geom_text(aes(label = value), color = 'white', size = 3) +
  theme(axis.text.x = element_text(angle = 90, vjust = 0.5, hjust=1)) +
  theme(axis.title.x = element_blank()) +
  theme(axis.title.y = element_blank()) +
  scale_fill_gradient(low = 'darkblue', high = 'red', limit = c(-1,1),
                      name = 'Correlation') +
  coord_fixed()
heatmap
```

The output of heatmap is demonstrated in Figure 4.9.

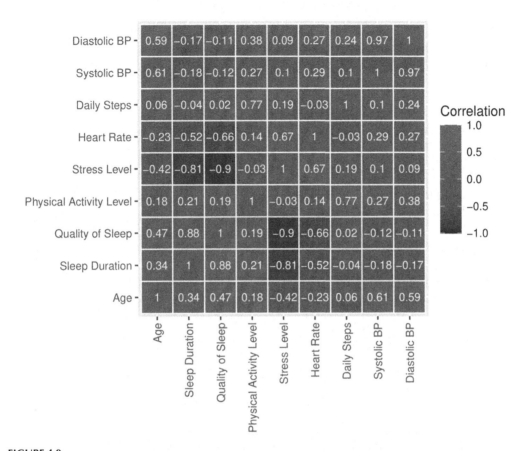

FIGURE 4.9
Heatmap of bivariate correlation matrix among the variables age, sleep duration, quality of sleep, physical activity level, stress level, heart rate, daily steps, systolic blood pressure, and diastolic blood pressure.

As illustrated in Figure 4.9, age exhibits a strong positive correlation with both systolic and diastolic blood pressure, as well as a moderate positive correlation with sleep duration and quality. However, a moderate degree of negative correlation is observed between age and stress level. Sleep duration has a strong positive correlation with sleep quality but a strong negative correlation with stress level and heart rate. Similarly, sleep quality shows a strong negative correlation with stress level and heart rate. Physical activity level is positively associated with sleep duration and quality, albeit to a lower degree of correlation.

4.3.5 Radar Plot

A radar plot, also known as a spider plot or star plot, shows multivariate data of three or more quantitative variables in a two-dimensional format mapped onto an axis starting from the same point. It is particularly useful when comparing several variables across multiple categories and is used to show the overall pattern of data, highlighting the strengths and weaknesses of different categories. In a radar plot, each variable is represented by a spoke, and the data values are plotted along each spoke to create a polygon shape. Comparing the shapes of different polygons allows a quick understanding of how the variables vary across categories.

Example 4.9

Suppose we have generated data of 10 health parameters (H. Para.) of 5 patients from uniform distribution using the following R-code.

R-code:

```
set.seed(1)
df1 <- paste0('Patient', 1:5)
df2 <- data.frame(matrix(runif(50), ncol = 10))
df3 <- cbind(df1, df2)
df4 <- as.data.frame(df3)
colnames(df4) <- c('Group', paste('H. Para.', 1:10))
attach(df4)
df4
```

Output:

```
Console  Terminal ×  Background Jobs ×                                                              ─ □
R R 4.3.1 · ~/
> set.seed(1)
> df1 <- paste0("Patient ", 1:5)
> df2 <- data.frame(matrix(runif(50), ncol = 10))
> df3 <- cbind(df1, df2)
> df4 <- as.data.frame(df3)
> colnames(df4) <- c("Group", paste("H. Para.", 1:10))
> attach(df4)
> df4
      Group H. Para. 1 H. Para. 2 H. Para. 3 H. Para. 4 H. Para. 5 H. Para. 6 H. Para. 7 H. Para. 8 H. Para. 9 H. Para. 10
1 Patient 1  0.2655087 0.89838968  0.2059746  0.4976992  0.9347052 0.38611409  0.4820801  0.6684667  0.8209463   0.7893562
2 Patient 2  0.3721239 0.94467527  0.1765568  0.7176185  0.2121425 0.01339033  0.5995658  0.7942399  0.6470602   0.0233312
3 Patient 3  0.5728534 0.66079779  0.6870228  0.9919061  0.6516738 0.38238796  0.4935413  0.1079436  0.7829328   0.4772301
4 Patient 4  0.9082078 0.62911404  0.3841037  0.3800352  0.1255551 0.86969085  0.1862176  0.7237109  0.5530363   0.7323137
5 Patient 5  0.2016819 0.06178627  0.7698414  0.7774452  0.2672207 0.34034900  0.8273733  0.4112744  0.5297196   0.6927316
> |
```

We want to visualize the data for understanding how the health parameters vary over patients. This can be made possible by using a radar chart. We need to `install` and `library ggradar` package using the following R-code.

R-code:

```
install.packages('devtools')
library('devtools')
devtools::install_github('ricardo-bion/ggradar')
library('ggradar')
```

Now, let us make the radar chart using the following R-code.

R-code:

```
rad = ggradar(
  df4,
  values.radar = c('0%', '50%', '100%'),
  grid.min = 0, grid.mid = 0.5, grid.max = 1,
  grid.label.size = 5,
  axis.label.size = 5,
  group.line.width = 1,
  group.point.size = 3,
  group.colours = c('#00AFBB', '#E7B800', '#FC4E07', 'darkblue',
      'green'),
  background.circle.colour = 'white',
  gridline.min.linetype = 'longdash',
```

```
    gridline.mid.linetype = 'longdash',
    gridline.max.linetype = 'longdash',
    gridline.min.colour = '#007A87',
    gridline.mid.colour = '#007A87',
    gridline.max.colour = '#007A87',
    label.gridline.min = TRUE,
    label.gridline.mid = TRUE,
    label.gridline.max = TRUE,
    legend.position = 'left',
)
rad
```

The output of the radar plot is illustrated in Figure 4.10.

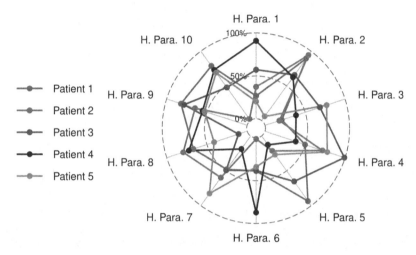

FIGURE 4.10
Radar plot of 10 health parameters (H. Para.) of 5 patients.

The radar plot presented in Figure 4.10 illustrates the variability in the values of health parameters among the patients. Patient-3 notably stands out with a value close to 1 for health parameter-4, significantly differing from the other patients. Patient-2 exhibits the lowest values, close to 0, for health parameters 6 and 10, while showing the highest value for health parameter-2. Conversely, patient-5 demonstrates the lowest value for health parameter-2. Patient-1 showcases the highest values for health parameters 5, 9, and 10. Additionally, patient-4 exhibits the highest values for health parameters 1 and 6.

4.4 Association of Categorical Variables

4.4.1 Heatmap

For categorical data, a heatmap can visualize the relationship between two categorical variables with multiple categories by presenting a contingency table of two attributes or a two-way table of two categorical variables with respect to the interest variable. The grid-filled colors represent the magnitude of count, proportion, and percentage for the respective combination of categories.

Example 4.10

Let us consider again the categorical variables occupation and BMI in the dataset from **Example 3.15**. We can visualize the contingency table between occupation and BMI by heatmap using the following R-code.

R-code:

```
library(reshape)
tab = table(data$Occupation, data$BMI.Category)
tab = round(prop.table(tab, margin = 1), 2)
df <- melt(tab)
colnames(df) <- c('x', 'y', 'value')
heatmap = ggplot(df, aes(x = x, y = y, fill = value)) +
  geom_tile(color = 'white',
            lwd = 0.5,
            linetype = 1) +
  geom_text(aes(label = value), color = 'tomato1', size = 3) +
  theme(axis.text.x = element_text(angle = 90, vjust = 0.5, hjust=1)) +
  theme(axis.title.x = element_blank()) +
  theme(axis.title.y = element_blank()) +
  scale_fill_viridis_c(option = 'D', limit = c(0,1),
                       name = 'Proportion') +
  coord_fixed()
heatmap
```

The output of heatmap is presented in Figure 4.11. The heatmap illustrated in Figure 4.11 indicates that professions such as accountant, doctor, engineer, and lawyer exhibit a higher proportion of individuals with a normal BMI. For software engineers, there is a 50% chance of having a normal weight, a 25% chance of being obese, and a 25% chance of being overweight. Conversely, professions like manager, nurse, salesperson, scientist, and teacher show a higher chance of being overweight. Additionally, sales representatives are more likely to be obese.

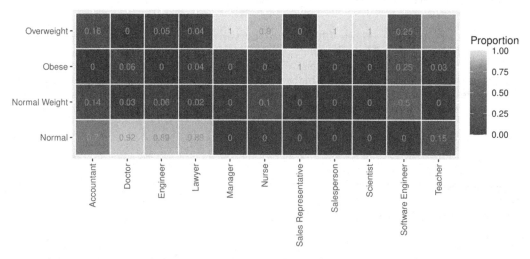

FIGURE 4.11
Heatmap between occupation and BMI.

4.4.2 Lollipop Plot

A lollipop plot is an exciting graph since it is typically a bar plot, where the bar is transformed into a line and a dot at the end of the line. It is used to display and compare categorical data, typically showing the distribution of a numerical value across different categories. Lollipop charts are particularly useful when we want to emphasize individual data points or outliers within a dataset. They allow us to quickly compare values across categories while still retaining the ability to observe specific data points.

Example 4.11

We have country-wise cancer-specific death rates of Asian countries downloaded from the database of Our World in Data. Let us import the data into R and see the first five cases of the data using the following R-code.

R-code:

```
data <- read.csv('E:/Book/Chapter 4/Data/Asian Countries Cancer Death
Rates by Type.csv', header = TRUE)
attach(data)
head(data, 5)
```

Output:

```
Console   Terminal    Background Jobs

R 4.3.2 · E:/Book/Chapter 4/Data/
> data <- read.csv("E:/Book/Chapter 4/Data/Asian Countries Cancer Death Rates by Type.csv", header = TRUE)
> attach(data)
> head(data, 5)
  Asian.Countries Code    Region Year Prostate.cancer Breast.cancer Cervical.cancer Stomach.cancer
1     Afghanistan  AFG South Asia 2019            6.37          8.67            3.90          29.30
2         Armenia  ARM  West Asia 2019            7.00         13.75            3.98          13.01
3      Azerbaijan  AZE  West Asia 2019            6.03          9.84            3.30          22.48
4         Bahrain  BHR  West Asia 2019            8.08         10.61            1.24           6.29
5      Bangladesh  BGD South Asia 2019            4.12          7.20            2.78           6.58
  Pancreatic.cancer Tracheal.bronchus.and.lung.cancer Colon.and.rectum.cancer Liver.cancer
1              2.72                             12.53                    8.43        10.27
2             10.61                             31.62                   14.30         7.07
3              7.45                             23.58                   12.20         4.37
4              7.59                             22.08                   11.42         5.23
5              2.21                              7.81                    4.94         2.75
>
```

Suppose we want to visualize the breast cancer death rate in 2019 of Asian countries and also want to group the countries by region. It can be done easily using a lollipop plot instead of a bar plot. We need to `install` and `library` the ggpubr package using the following R-code.

R-code:

```
install.packages('ggpubr')
library(ggpubr)
```

Now, let us make the lollipop plot for the breast cancer death rate in 2019 of Asian countries grouped by region using the following R-code.

R-code:

```
data1 <- subset(data[, c(1, 3, 6)], Year == '2019')
lollipop1 = ggdotchart(
  data1, x = 'Asian.Countries', y = 'Breast.cancer',
  add = 'segments',
  add.params = list(color = 'Region', size = 0.5),
  sorting = 'descending',
  group = 'Region',
  color = 'Region',
  palette = c('#C71585','#0000FF', 'red', '#9ACD32', '#00BFFF',
'#DAA520'),
  rotate = FALSE,
  x.text.col = TRUE,
  ylab = 'Death Rate of Breast Cancer',
  xlab = 'Asian Countries'
)
lollipop1
```

The output of the lollipop plot is shown in Figure 4.12. As depicted in Figure 4.12, while Pakistan exhibits the highest death rate of breast cancer, the death rate is comparatively higher in West Asian and Southeast Asian countries. Among West Asian countries, Lebanon reports the highest death rate, followed by Georgia, Armenia, Palestine, Cyprus, Israel, and Iraq. In Southeast Asia, Malaysia records the highest death rate, followed by Vietnam, Brunei, and the Philippines. Among East Asian countries, North Korea has the highest death rate of breast cancer, while Uzbekistan and Kazakhstan in Central Asia exhibit the highest death rate.

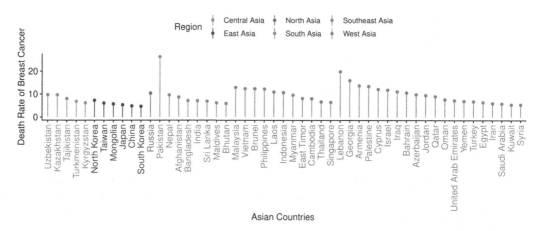

FIGURE 4.12
Lollipop plot for the breast cancer death rate in 2019 of Asian countries grouped by region.

Now, suppose we want to visualize the death rate of prostate cancer, breast cancer, cervical cancer, stomach cancer, pancreatic cancer, and liver cancer in 2019 for only South Asian countries. The R-code is given below.

R-code:

```
install.packages('tidyverse')
library('tidyverse')
data2 <- subset(data[, c(1, 5, 6, 7, 8, 9, 12)], Year == '2019' &
Region %in% c('South Asia'))
colnames(data2) = c('South Asian Countries', 'Prostate cancer',
'Breast cancer', 'Cervical cancer',
                    'Stomach cancer', 'Pancreatic cancer', 'Liver
                        cancer')
data3 <- data2 %>%
  select(colnames(data2)) %>%
  pivot_longer(
    cols = c(colnames(data2[,-1])),
    names_to = 'Cancer',
    values_to = 'Death Rate'
  )
lollipop2 = ggdotchart(
  data3, x = 'South Asian Countries', y = 'Death Rate',
  group = 'Cancer',
  color = 'Cancer',
  add.params = list(color = 'Cancer'),
  palette = c('#C71585','#0000FF', 'red', '#9ACD32', '#00BFFF',
'#DAA520'),
  add = 'segment',
  position = position_dodge(0.3),
  sorting = 'descending',
  ylab = 'Death Rate',
  xlab = 'South Asian Countries',
)
lollipop2
```

The output is demonstrated in Figure 4.13. The death rates from breast cancer in Pakistan and stomach cancer in Afghanistan are significantly higher, exceeding 25%, compared

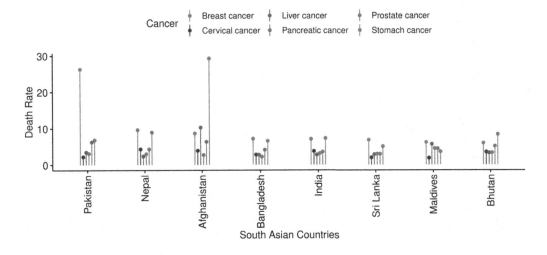

FIGURE 4.13
Lollipop plot of the death rate of prostate cancer, breast cancer, cervical cancer, stomach cancer, pancreatic cancer, and liver cancer in 2019 for South Asian countries.

to other considered cancers and countries. Nepal shows approximately a 10% death rate for both breast and stomach cancer. Cervical and pancreatic cancers exhibit death rates of less than or around 5% across all South Asian countries. The death rate from prostate cancer is the same as that of cervical and pancreatic cancer in South Asian countries, except for Pakistan and Afghanistan, where it is slightly higher. Afghanistan notably records a death rate of over 10% for liver cancer, while other countries report rates lower or around 5%.

We can also format the lollipop plot illustrated in Figure 4.13 using the following R-code.

R-code:

```
lollipop3 = ggdotchart(
  data3, x = 'South Asian Countries', y = 'Death Rate',
  group = 'Cancer',
  color = 'Cancer',
  add.params = list(color = 'Cancer'),
  palette = c('#C71585','#0000FF', 'red', '#9ACD32', '#00BFFF',
'#DAA520'),
  add = 'segment',
  position = position_dodge(0.3),
  sorting = 'descending',
  facet.by = 'Cancer',
  rotate = TRUE,
  legend = 'None',
  ylab = 'Death Rate',
  xlab = 'South Asian Countries'
)
lollipop3
```

The output is shown in Figure 4.14.

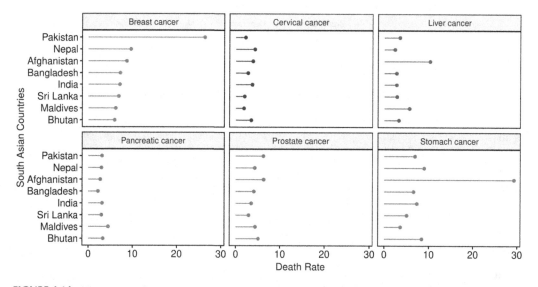

FIGURE 4.14
Lollipop plot of the death rate of prostate cancer, breast cancer, cervical cancer, stomach cancer, pancreatic cancer, and liver cancer in 2019 for South Asian countries.

4.4.3 Sankey Plot

A Sankey plot is a graphical method that displays the flow of information between different elements. It is particularly useful for illustrating the distribution of quantities or the transitions between different states or categories of two or more categorical variables.

The Sankey plot consists of nodes representing the different categories or states and flows representing the connections or transitions between these categories. The width of the flows is typically proportional to the quantity being represented. The plot helps to visualize not only the magnitude of the flows but also the relationships and proportions between different elements in a system.

Example 4.12

Recall the dataset from **Example 3.15**. We want to visualize the relationship among gender, BMI category, stress level, and sleep disorder. It can be easily possible by the Sankey plot. For creating a Sankey plot, we need to `install` and `library` ggsankey, ggplot2, and dplyr packages using the following R-code.

R-code:

```
install.packages('ggsankey')
library(ggsankey)
library(ggplot2)
library(dplyr)
```

Now, let us make the Sankey plot using the following R-code.

R-code:

```
data1 <- data.frame(cbind(Gender, BMI.Category, Stress.Level, Sleep.
Disorder))
TotalCount = nrow(data1)
df <- data1 %>%
  make_long(Gender, BMI.Category, Stress.Level, Sleep.Disorder)
dagg <- df%>%
  dplyr::group_by(node)%>%
  tally()
dagg <- dagg%>%
  dplyr::group_by(node)%>%
  dplyr::mutate(pct = n/TotalCount)
df2 <- merge(df, dagg, by.x = 'node', by.y = 'node', all.x = TRUE)
sankey <- ggplot(df2, aes(x = x,
                    next_x = next_x,
                    node = node,
                    next_node = next_node,
                    fill = factor(node),
                    label = paste0(node,' n=', n, '(', round(pct*
100,1), '%)' )))
sankey <- sankey +geom_sankey(flow.alpha = 0.5, color = 'gray40',
show.legend = TRUE)
sankey <- sankey +geom_sankey_label(size = 2, color = 'black', fill=
'white', hjust = -0.1)
sankey <- sankey + theme_bw()
sankey <- sankey + theme(legend.position = 'none')
sankey <- sankey + theme(axis.title = element_blank(),
                  axis.text.y = element_blank(),
                  axis.ticks = element_blank(),
                  panel.grid = element_blank())
```

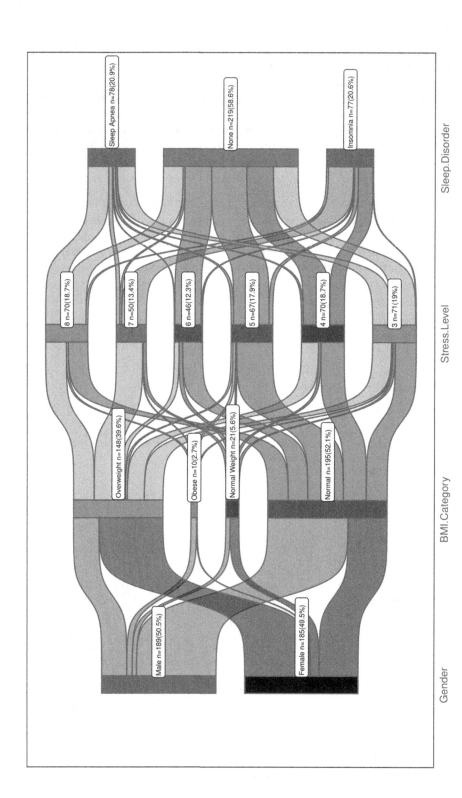

FIGURE 4.15
Sankey plot among gender, BMI category, stress level, and sleep disorder.

```
sankey <- sankey + scale_fill_manual(values = c('Female'    = 'black',
                                    'Male'  = 'red',
                                    'Normal'  = 'green',
                                    'Normal Weight'  = 'blue',
                                    'Obese'  = 'orange',
                                    'Overweight'  = 'yellow',
                                    '3'= 'purple',
                                    '4' = '#800000',
                                    '5' = '#808000',
                                    '6' = '#008080',
                                    '7' = '#DC143C',
                                    '8' = '#FF6347',
                                    'Insomnia' = '#FF0000',
                                    'None' = '#00FA9A',
                                    'Sleep Apnea' = '#FF0000'
  ) )
  sankey
```

The output of the Sankey plot is illustrated in Figure 4.15. As shown in Figure 4.15, females exhibit a higher rate of overweight, whereas males have a higher rate of normal BMI. Overweight individuals are more prone to experiencing high levels of stress; interestingly, sleep apnea and insomnia are more prevalent among individuals with high stress levels. However, a noteworthy number of individuals with lower stress levels also suffer from sleep apnea and insomnia, which could be attributed to some individuals with lower stress levels being overweight.

4.5 Conclusion and Exercises

This section concludes the chapter summary with some practical problem-solving exercises.

4.5.1 Concluding Remarks

By exploring sophisticated visualizations, this chapter has underscored the importance of moving beyond conventional graphical representations to address the challenges faced while dealing with complex datasets in the evolving landscapes of data science and biostatistics. This chapter concludes with the recognition that advanced graphical presentations not only enhance interpretability but also serve as powerful tools for effective communication, fostering innovation and informed decision-making. This chapter serves as a stepping stone for researchers, analysts, and decision-makers to harness the full potential of visual approaches in the dynamic realms of data-driven discovery and biostatistical analysis.

4.5.2 Practice Questions

Exercise 4.1 Consider the dataset from **Exercise 3.1**. Solve the following problems using R-commands:

  a. Check the distribution of the resting blood pressure using an empirical cumulative distribution function.

  b. Check the variability in the maximum heart rate achieved using violin plots.

c. Show the relationship between age and serum cholesterol in mg/dl graphically using a contour plot or 2D density plot and joint histogram or hexbin plot.

d. Demonstrate the relationship between age and resting blood pressure graphically using a contour plot or 2D density plot and joint histogram or hexbin plot.

e. Illustrate graphically the relationship between age, sex, and resting blood pressure using a bubble chart.

Exercise 4.2 Consider the country-wise cancer-specific death rates of Asian countries from **Example 4.11**. Visualize the stomach cancer death rates of Asian countries for 2019 grouped by region using a lollipop plot.

Exercise 4.3 Consider the dataset named 'Diabetes Health Indicators Dataset' from the Kaggle database. After downloading, you will get three datasets. Use the dataset named 'diabetes_012_health_indicators_BRFSS2021' which is a clean dataset of 2,36,378 survey responses under 22 variables to the CDC's BRFSS2021. The description of the variables are given below.

Variable Name	Description
Diabetes_012	0 = no diabetes, 1 = prediabetes, 2 = diabetes
HighBP	0 = no high BP, 1 = high BP
HighChol	0 = no high cholesterol, 1 = high cholesterol
CholCheck	0 = no cholesterol check in 5 years, 1 = yes cholesterol check in 5 years
BMI	Body Mass Index
Smoker	Have you smoked at least 100 cigarettes in your entire life? [Note: 5 packs = 100 cigarettes] 0 = no, 1 = yes
Stroke	(Ever told) you had a stroke. 0 = no, 1 = yes
HeartDiseaseorAttack	Coronary Heart Disease (CHD) or Myocardial Infarction (MI) 0 = no, 1 = yes
PhysActivity	Physical activity in past 30 days - not including job 0 = no, 1 = yes
Fruits	Consume Fruit 1 or more per day 0 = no, 1 = yes
Veggies	Consume Vegetables 1 or more per day 0 = no, 1 = yes
HvyAlcoholConsump	Heavy drinkers (adult men having more than 14 drinks per week and adult women having more than 7 drinks per week) 0 = no, 1 = yes
AnyHealthcare	Have any kind of health care coverage, including health insurance, prepaid plans such as HMO, etc. 0 = no, 1 = yes
NoDocbcCost	Was there a time in the past 12 months when you needed to see a doctor but could not because of cost? 0 = no, 1 = yes
GenHlth	Would you say that in general your health is: scale 1–5: 1 = excellent, 2 = very good, 3 = good, 4 = fair, 5 = poor
MentHlth	Now thinking about your mental health, which includes stress, depression, and problems with emotions, for how many days during the past 30 days was your mental health not good? It is in days, scale will be between 0–30
PhysHlth	Now thinking about your physical health, which includes physical illness and injury, for how many days during the past 30 days was your physical health not good? It is in days, scale will be between 0–30

Variable Name	Description
DiffWalk	Do you have serious difficulty walking or climbing stairs? 0 = no, 1 = yes
Sex	0 = female, 1 = male
Age	13-level age category (_AGEG5YR see codebook21 linked above): scale 1–13: 1 = 18–24, 8 = 55–59, 13 = 80 or older
Education	Education level (EDUCA see codebook21 linked above): scale 1–6: 1 = Never attended school or only kindergarten 2 = Grades 1 through 8
Income	Income scale (INCOME3 see codebook21 linked above): scale 1–8: 1 = less than $10,000 5 = less than $35,000 11 = $200,000 or more

Import this dataset into R under the name 'diabetes_hi' and solve the following problems:

a. Visualize the relationship among Sex, PhysActivity, and GenHlth using a Sankey plot.

b. Make a heatmap of the relationship between GenHlth and PhysActivity.

c. Make a heatmap between PhysActivity and Diabetes_012.

Bibliography

Abdulla, F., Nain, Z., Karimuzzaman, M., Hossain, M. M., & Rahman, A. (2021). A non-linear biostatistical graphical modeling of preventive actions and healthcare factors in controlling COVID-19 pandemic. *International Journal of Environmental Research and Public Health*, 18(9), Article 4491. https://doi.org/10.3390/ijerph18094491

Adler, D., & Kelly, S. T. (2005). vioplot: Violin plot. R package version 0.2. *Zugriff am*, 2, 2017.

Barter, R. L., & Yu, B. (2018). Superheat: An R package for creating beautiful and extendable heatmaps for visualizing complex data. *Journal of Computational and Graphical Statistics*, 27(4), 910–922.

Bessler, L. (2023a). Pie charts and donut charts. In *Visual data insights using SAS ODS graphics: A guide to communication-effective data visualization* (pp. 187–220). Apress.

Bessler, L. (2023b). Bubble plots. In *Visual data insights using SAS ODS graphics: A guide to communication-effective data visualization* (pp. 263–281). Apress.

Bion, R. (2021). ggradar: Create radar charts using ggplot2. *R package version 0.2.* https://github.com/ricardo-bion/ggradar

Carr, D., Lewin-Koh, N., & Maechler, M. (2011). hexbin: Hexagonal binning routines. *R package version, 1260.* https://rdrr.io/cran/hexbin/

Cook, D., Swayne, D. F., & Buja, A. (2007). *Interactive and dynamic graphics for data analysis: with R and GGobi* (Vol. 1). Springer.

Evergreen, S. D. (2019). *Effective data visualization: The right chart for the right data.* SAGE Publications.

Fischetti, A. (2018). *Data analysis with R: A comprehensive guide to manipulating, analyzing, and visualizing data in R.* Packt Publishing Ltd.

Friendly, M., & Meyer, D. (2015). *Discrete data analysis with R: Visualization and modeling techniques for categorical and count data* (Vol. 120). CRC Press.

Grant, R. (2018). *Data visualization: Charts, maps, and interactive graphics.* CRC Press.

Horton, N. J., & Kleinman, K. (2015). *Using R and RStudio for data management, statistical analysis, and graphics*. CRC Press.

Kabacoff, R. (2022). *R in action: Data analysis and graphics with R and Tidyverse*. Simon and Schuster.

Kassambara, A. (2019). *GGPlot2 essentials: Great data visualization in R*. Datanovia.

Kassambara, A. (2020). ggpubr: 'ggplot2' based publication ready plots. *R package version 0.4. 0, 438*. https://github.com/kassambara/ggpubr

Keen, K. J. (2018). *Graphics for statistics and data analysis with R*. CRC Press.

Maindonald, J., & Braun, J. (2006). *Data analysis and graphics using R: An example-based approach* (Vol. 10). Cambridge University Press.

Mangiola, S., & Papenfuss, A. T. (2020). tidyHeatmap: An R package for modular heatmap production based on tidy principles. *Journal of Open Source Software, 5*(52), 2472.

McGrath, M. (2023). *R for Data Analysis in easy steps: Covers R Programming essentials*. Easy Steps Limited.

Mirkin, B. (2019). *Core data analysis: Summarization, correlation, and visualization*. Springer International Publishing.

Moon, K. W. (2016). Violin plot. *Learn ggplot2 using shiny app* (pp. 191–200). Springer.

Moraga, P. (2019). *Geospatial health data: Modeling and visualization with R-INLA and shiny*. CRC Press.

Oyana, T. J. (2020). *Spatial analysis with r: statistics, visualization, and computational methods*. CRC press.

Rahlf, T. (2019). *Data visualisation with R: 111 examples*. Springer Nature.

Rudis, B., & Gandy, D. (2019). Waffle: Create waffle chart visualizations. *R package version, 1(1)*. https://cran.r-project.org/web/packages/waffle/waffle.pdf

Saary, M. J. (2008). Radar plots: A useful way for presenting multivariate health care data. *Journal of Clinical Epidemiology, 61*(4), 311–317.

Schwabish, J. (2021). *Better data visualizations: A guide for scholars, researchers, and wonks*. Columbia University Press.

Sievert, C. (2020). *Interactive web-based data visualization with R, plotly, and shiny*. CRC Press.

Sjoberg, D. (2021). Ggsankey: Sankey, alluvial and Sankey bump plots. *R package version 0.0, 99999*. https://github.com/davidsjoberg/ggsankey

Starbuck, C. (2023). Data visualization. In *The fundamentals of people analytics: With applications in R* (pp. 283–323). Springer International Publishing.

Taylor, R. S., & Mitchell, E. (2016). Minding the gap. *Data visualization: A guide to visual storytelling for libraries* (1st ed., p. 77). Rowman & Littlefield Publishers.

Teetor, P. (2011). *R cookbook: Proven recipes for data analysis, statistics, and graphics*. O'Reilly Media, Inc.

Theus, M., & Urbanek, S. (2008). *Interactive graphics for data analysis: Principles and examples*. CRC Press.

Wexler, S., Shaffer, J., & Cotgreave, A. (2017). *The big book of dashboards: Visualizing your data using real-world business scenarios*. John Wiley & Sons.

Wickham, H. (2016). *ggplot2: Elegant graphics for data analysis*. Springer International Publishing.

Wickham, H. (2022). Flexibly Reshape Data [R package reshape version 0.8. 9]. https://rdrr.io/cran/reshape/

5

Measures of Center and Dispersion

5.1 Central Tendency

Central tendency is a measure that describes the central location of a data distribution. In other words, it refers to the typical point measure around which the data is most likely to cluster. It is used to summarize the data by identifying a single value that best represents the whole dataset. Central tendency is one of the most important concepts in statistics, and it is used in a wide range of fields. In the subsequent sections of this chapter, an overview of central tendency measures, such as mean, median, and mode, along with their applications, is presented.

5.1.1 Arithmetic Mean

The arithmetic mean is a frequently used measure of the center of data that represents the arithmetic average of a set of values. While talking, most people say 'average' instead of 'mean'. It is calculated by adding all the values (positive, negative, or zero) for a variable in a dataset and dividing by the number of observations (values). Suppose the n-data points are represented by $x_1, x_2, ..., x_n$ and arithmetic mean is denoted by $\bar{x}$, and defined as:

$$\bar{x} = \frac{\sum_{i=1}^{n} x_i}{n} = \frac{\text{sum}(x)}{n}.$$

The arithmetic mean is simply understood and is straightforward to compute. There are some nice properties of the arithmetic mean that exist for a given set of data.

In addition to the arithmetic mean, there are two other types of means: the geometric mean and the harmonic mean. The geometric mean is calculated by taking the nth root of the product of n observations, while the harmonic mean is calculated by dividing the number of observations by the sum of the reciprocals of the values. The arithmetic mean is best for data with consistent values, the geometric mean is ideal for comparing rates or growth, and the harmonic mean is suitable for averaging ratios or rates when the quantities are inversely related. However, since the mean is computed using all values in a set of data, each value has an impact on it. Hence, extreme values influence the mean estimation. In some situations, they inflate the mean that cannot be accepted as a valid measure. It is discussed in the later section of this chapter with an example.

Example 5.1

Suppose we are concerned to find the mean weight of 10 people having individual weights given in Table 5.1.

DOI: 10.1201/9781003426189-5

TABLE 5.1

Weight of 10 Individuals

People	1	2	3	4	5	6	7	8	9	10
Weight (kg)	35	42	45	50	48	62	55	47	38	68

In R, the mean can be determined using the mean() function. First of all, the *weight* of 10 people is entered into a single object (or variable) called *weight*. To do this, type the following in an R-script file:

R-code:

```
# create a vector of weights
weight <- c(35, 42, 45, 50, 48, 62, 55, 47, 38, 68)
weight
```

Output:

```
Console   Terminal ×   Background Jobs ×                                    ⚏ ☐
R  R 4.3.2 · ~/ 
> # create a vector of weights
> weight <- c(35, 42, 45, 50, 48, 62, 55, 47, 38, 68)
> weight
 [1] 35 42 45 50 48 62 55 47 38 68
> |
```

The 'arrow', i.e., '< -', indicates that we are assigning the values 35, 42, ..., and 68 to the R object (or variable) *weight*. One can use '=' sign instead of '< -' sign. In order to print values of the variable *weight*, write *weight* in the R-script file.

The [1] indicating the output is a vector (a sequence of numbers or other objects) and that the first printed values represent the first actual values in *weight*.

The following R-code can be used for calculating the average using the mean() function.

R-code:

```
# calculate the mean
mean(weight)
```

Output:

```
Console   Terminal ×   Background Jobs ×                                    ⚏ ☐
R  R 4.3.2 · ~/ 
> # calculate the mean
> mean(weight)
[1] 49
> |
```

Alternatively, one can use the following function to compute the mean of this weight data.

R-code:

```
meanWt <- sum(weight)/length(weight)
meanWt
```

Output:

```
Console   Terminal ×   Background Jobs ×
R R 4.3.2 · ~/
> meanwt <- sum(weight)/length(weight)
> meanwt
[1] 49
>
```

Here, `sum()` function is used to add all entries in a vector, and `length()` function is used to find the number of entries in a vector.

Now, we intend to use a large dataset. Here, we use the 'Pima Indians diabetes' dataset that is internally available to R (Smith et al., 1988); i.e., the dataset can be accessed within R in the following way.

R-code:

```
# install package
install.packages('mlbench')
#make the library of the loaded package
library(mlbench)
#data
data(PimaIndiansDiabetes2)
```

This dataset consists of 768 observations of 9 variables (Table 5.2).

TABLE 5.2

Description of the Variables Included in PimaIndiansDiabetes2

Variable Name	Description
pregnant	Number of times pregnant
glucose	Plasma glucose concentration (glucose tolerance test)
pressure	Diastolic blood pressure (mm Hg)
triceps	Triceps skin fold thickness (mm)
insulin	2-Hour serum insulin (mu U/ml)
mass	Body mass index (i.e. weight in kg/(height in m)^2)
pedigree	Diabetes pedigree function
age	Age (years)
diabetes	Class variable (test for diabetes)

Suppose we want to find out the mean plasma glucose concentration. In R, the following function is employed to obtain the mean.

R-code:

```
# mean of plasma glucose concentration
mean(PimaIndiansDiabetes2$glucose)
```

Output:

```
Console   Terminal   Background Jobs
R  R 4.3.2 · ~/
> # mean of plasma glucose concentration
> mean(PimaIndiansDiabetes2$glucose)
[1] NA
>
```

No worries, you know the mean function does not work if the variable contains missing observations. In the presence of the missing value, we can apply the following function to compute the mean:

R-code:

```
# Check NA available
nacheck=is.na(PimaIndiansDiabetes2$glucose)
which(nacheck==TRUE)
# mean of plasma glucose concentration after omitting missing values
mean(PimaIndiansDiabetes2$glucose, na.rm=TRUE)
```

Output:

```
Console   Terminal   Background Jobs
R  R 4.3.2 · ~/
> # Check NA available
> nacheck=is.na(PimaIndiansDiabetes2$glucose)
> which(nacheck==TRUE)
[1]   76 183 343 350 503
> # mean of plasma glucose concentration after omitting missing values
> mean(PimaIndiansDiabetes2$glucose, na.rm=TRUE)
[1] 121.6868
>
```

Here, is.na() function is used to identify the missing value; if the corresponding value is missing, it is set at TRUE; otherwise FALSE. Hence, a which(nacheck==TRUE) function is employed to discover the position of the missing values in the vector of the variables. From the output, it is observed that there are 5 missing observations in the 76th, 183rd, 343rd, 350th, and 503rd positions of the *glucose* variable. Moreover, na.rm=TRUE is used because in the dataset there are some missing observations that are replaced by NA. One can be interested to find the mean of other variables in a similar fashion.

Example 5.2

Suppose Table 5.3 represents the income from the child department during a month (30 days) in a community hospital. What is the average income per day of that department in a community hospital?

In this circumstance, we have applied the following formula to determine the mean:

TABLE 5.3

Income from the Child Department during a Month (30 Days) in a Community Hospital

Income ($000)	3	4	3	6	5	2	7
Number of Days	2	3	4	5	6	7	8

$$\bar{x} = \frac{f_1x_1 + f_2x_2 + \ldots + f_kx_k}{f_1 + f_2 + \ldots + f_k} = \frac{\sum_{i=1}^{k} f_i x_i}{\sum_{i=1}^{k} f_i} = \frac{\text{sum}(f * x)}{\text{sum}(f)}$$

where $f_1, f_2, \ldots, f_k$ represent the frequencies of the corresponding group/category, and k indicates the number of groups/categories.

In R, the following functions help to compute the mean income.

R-code:

```
n.days<-c(3,      4,      3,      6,      5,      2,      7)
income<-c(2,      3,      4,      5,      6,      7,      8)
prod<-n.days*income
mean.income<-sum(prod)/sum(n.days)
mean.income*1000
```

Output:

```
Console   Terminal ×   Background Jobs ×

R  R 4.3.2 · ~/
> n.days<-c(3,    4,      3,      6,      5,      2,      7)
> income<-c(2,    3,      4,      5,      6,      7,      8)
> prod<-n.days*income
> mean.income<-sum(prod)/sum(n.days)
> mean.income*1000
[1] 5333.333
>
```

Therefore, the average income of the child department data is $5,333.33.

Properties of the arithmetic mean

a. The sum of the deviations of each observation from the arithmetic mean is zero. Mathematically, $\sum_{i=1}^{n}(x_i - \bar{x}) = 0$. We can prove it mathematically in the following way. $\sum_{i=1}^{n}(x_i - \bar{x}) = \sum_{i=1}^{n} x_i - n\bar{x} = n\bar{x} - n\bar{x} = 0$, since $\bar{x} = \frac{\sum_{i=1}^{n} x_i}{n} \Rightarrow \sum_{i=1}^{n} x_i = n\bar{x}$.

b. If a variable x is linearly transformed as $y_i = \alpha + \beta x_i$; $i = 1, 2, \ldots, n$, where α and β are known constants, then we can write $\bar{y} = \alpha + \beta\bar{x}$. We prove it easily as

$$\bar{y} = \frac{\sum_{i=1}^{n} y_i}{n} = \frac{1}{n}\sum_{i=1}^{n}(\alpha + \beta x_i) = \frac{1}{n}\sum_{i=1}^{n}\alpha + \frac{1}{n}\sum_{i=1}^{n}\beta x_i = \frac{n\alpha}{n} + \frac{\beta}{n}\sum_{i=1}^{n} x_i = \alpha + \beta\bar{x}.$$

5.1.2 Median

The median of a finite set of values is the middle value in a dataset, i.e., an equal number of sample points be placed on either side of the sample median. It is computed by ordering the values in the dataset and selecting the middle value. If there is an even number of values in a dataset, there is no single middle value. In such situations, the median is calculated by taking the average of the two middle values.

Consider a sample that contains n observations. First, we need to order these observations from the smallest to the largest. Then, the median is obtained in the following ways:

a. The $\left(\dfrac{n+1}{2}\right)$th largest observation if n is odd.

b. The average of the $\left(\dfrac{n}{2}\right)$th and $\left(\dfrac{n}{2}+1\right)$th largest observations if n is even.

Example 5.3

We recall, the weights of the people presented in **Example 5.1**. Suppose we want to compute the sample median weight of the first seven people. First, we order the weight as:

Weight (kg)	35	42	45	48	50	55	62

Here, the number of observations (n=7) is an odd number. The middle value is the $\left(\dfrac{7+1}{2}\right) = 4$th one. Count up from the smallest up to the 4th value, it is observed that it is 48. Thus, the median weight of the 7 people is 48 kg.

Again, we wish to find the median weight of ten people. Arranging the ten weights in the order of magnitude from the smallest to the largest provides as:

35	38	42	45	47	48	50	55	62	68

Since we have an even number of weights, there is no single middle value. However, we find the two middle values: 47 and 48. So, the median is (47+48)/2 = 47.5 kg.

In R, we can calculate the median using the median() function. Here, interestingly, there is no need to put data in an ordered form. To calculate the median using R, we can use the following code.

R-code:

```
# Create a vector of numbers
x <- c(35, 42, 45, 50, 48, 62, 55)
# Median
median(x)
```

Output:

```
Console   Terminal    Background Jobs
R R 4.3.2 · ~/
> # Create a vector of numbers
> x <- c(35, 42, 45, 50, 48, 62, 55)
> # Median
> median(x)
[1] 48
>
```

If we consider weights of ten people, i.e., in the dataset the number of observations is even, the median is calculated as the average of the two middle values. The R-code and output are given below.

R-code:

```
# Create a vector of numbers
x <- c(35, 42, 45, 50, 48, 62, 55, 47, 38, 68)
# Median
median(x)
```

Output:

```
Console   Terminal ×   Background Jobs ×                                    ▭ ☐
R  R 4.3.2 · ~/
> # Create a vector of numbers
> x <- c(35, 42, 45, 50, 48, 62, 55, 47, 38, 68)
> # Median
> median(x)
[1] 47.5
>
```

Moreover, we can compute the median from an uploaded data file in R. Suppose we wish to compute the median of the variable *age* of the dataset 'Pima Indians Diabetes Database'. We then assign the variable *age* of the dataset the name *age*. Finally, we use the `median` commands to compute the median, and the result is given below.

R-code:

```
age<-PimaIndiansDiabetes2$age
median(age)
```

Output:

```
Console   Terminal ×   Background Jobs ×                                    ▭ ☐
R  R 4.3.2 · ~/
> age<-PimaIndiansDiabetes2$age
> median(age)
[1] 29
>
```

The key advantage of the median is that, unlike the mean, it is not influenced by extreme values. For example, let the 10th value of the dataset of **Example 5.1** is 120. Now, we need to run the following R-code.

R-code:

```
# Create a vector of numbers
x <- c(35, 42, 45, 50, 48, 62, 55, 47, 38, 120)
# Median
median(x)
```

Output:

```
Console   Terminal ×   Background Jobs ×                                    ▭ ☐
R  R 4.3.2 · ~/
> # Create a vector of numbers
> x <- c(35, 42, 45, 50, 48, 62, 55, 47, 38, 120)
> # Median
> median(x)
[1] 47.5
>
```

Hence, the sample median remains unchanged because the fifth and sixth ordered values are still 47 and 48, respectively.

However, the main drawback of the sample median is that it is computed mainly considering only the middle points of a sample.

5.1.3 Mode

The mode represents the value that appears most frequently in a dataset, or mode is the value of a variable that occurs with the highest frequency. Also, the mode is the highest point of the frequency distribution curve. If all the values occur exactly once, then no mode exists. However, multiple modes exist in some distributions. The distribution with one mode is called unimodal, two modes is bimodal, three modes is trimodal, and so on.

Example 5.4

Suppose we want to know the number of siblings of ten students selected from a class, and the responses were: 3, 2, 2, 1, 3, 6, 3, 3, 4, 2. Find the mode for the number of siblings.

Here, the mode for the number of siblings is 3 because 4 students (maximum) have 3 siblings.

Example 5.5

Let us consider a research laboratory having 10 employees whose ages are 22, 28, 30, 25, 34, 25, 32, 27, 32, and 29 years. Hence, we observed that these data have two modes, 25 and 32. Again, the sample consisting of 5 employees ages 22, 28, 30, 25, and 34 has no Mode because all the values are different.

To calculate the mode of a dataset in R, we can define a Mode() function as in the following R-code. Please note that if there are multiple modes in the dataset, the Mode() function will only return the first mode.

R-code:

```
# Define a function to calculate mode
Mode <- function(x) {
  ux <- unique(x)
  ux[which.max(tabulate(match(x, ux)))]
}
# Calculate the mode of the vector
y <- c(3, 2, 2, 1, 3, 6, 3, 3, 4, 2)
Mode(y)
z <- c(22, 28, 30, 25, 34, 25, 32, 27, 32, 29)
Mode(z)
```

Output:

```
Console   Terminal ×   Background Jobs ×

R  R 4.3.2 . ~/
> # Define a function to calculate mode
> Mode <- function(x) {
+     ux <- unique(x)
+     ux[which.max(tabulate(match(x, ux)))]
+ }
> # Calculate the mode of the vector
> y <- c(3, 2, 2, 1, 3, 6, 3, 3, 4, 2)
> Mode(y)
[1] 3
> z <- c(22, 28, 30, 25, 34, 25, 32, 27, 32, 29)
> Mode(z)
[1] 25
>
```

TABLE 5.4

Blood Grouping of 70 Individuals

Blood Type	A	B	O	AB
Number of People	12	15	27	16

In the second example of the above output, we observed that among two modes R provide only the first mode 25.

The mode may also apply for describing qualitative data. For example, in a blood grouping event, we got the following (Table 5.4):

Here, blood type 'O' most frequently occurred (27 times which is maximum) in the group of people.

5.1.4 Weighted Mean

The weighted mean is a type of mean that considers the weights of each observation in a dataset. In calculating the weighted mean, we use the same weight for each observation in the dataset. However, if the weights vary from observation to observation, then the weighted mean is applicable.

The weighted mean of n observation $x_1, x_2, \ldots, x_n$ with their corresponding weights $w_1, w_2, \ldots, w_n$ respectively is obtained by the following formula:

$$\overline{w} = \frac{\sum_{i=1}^{n} w_i x_i}{\sum_{i=1}^{n} w_i} = \frac{\text{sum}(w * x)}{\text{sum}(w)}.$$

To calculate the weighted mean of a dataset in R, we can use the function `weighted. mean()`.

Example 5.6

The grades of a student's semester courses are given in Table 5.5. Calculate the Grade Point Average (GPA).

Here, the credits of the courses are different which is considered as weights. The GPA is actually the weighted mean.

TABLE 5.5

Distribution of Grade Point Average

Course	Grade	Points (x)	Credits (w)
Math	A	4	3
Economics	B	3	2
Statistics	A	4	3
Computing	A	4	2

In R, we need to write the following code to calculate the weighted mean.

R-code:

```
# Create a vector of numbers
x <- c(4, 3, 4, 4)
# Create a vector of weights
w <- c(3, 2, 3, 2)
# Calculate the weighted mean of the vector
weighted.mean(x, w)
```

Output:

```
Console   Terminal ×   Background Jobs ×

R  R 4.3.2 · ~/
> # Create a vector of numbers
> x <- c(4, 3, 4, 4)
> # Create a vector of weights
> w <- c(3, 2, 3, 2)
> # Calculate the weighted mean of the vector
> weighted.mean(x, w)
[1] 3.8
>
```

How Extreme Value Influence the Mean?

We say previously that the mean is influenced by the extreme values (outliers). The following scenario serves as an illustration of how extreme values might impact the mean. Suppose the five patients visited a doctor in a hospital for their physical checkup and their charges for certain procedures are: $50, $70, $65, $75, and $270. It is discovered that the five patients' mean charges came to $106, a figure that is not particularly representative of the date set. In this case, the single number ($270) caused the mean to be inflated.

Typically, the outlier is eliminated from the dataset for subsequent analysis; however, it reduces the degrees of freedom or someone suggests to use another measure of central tendency such as the median, which may be a better description of central tendency in such circumstances. However, the main disadvantage of median is that it is not based on all observation in dataset.

In this circumstance, Hossain (2016) proposed a formula to complete the weighted mean in the presence of outliers in the dataset. Suppose we have a dataset containing the values $x_1, x_2, \ldots, x_n$. It is assumed that the dataset contains at least one outlier, namely, x_O. In such situation the following formula is used to compute the mean,

$$\bar{x}_R = \frac{\sum_{i=1}^{n} w_i x_i}{\sum_{i=1}^{n} w_i} = \frac{\text{sum}(w * x)}{\text{sum}(w)}$$

where $w_i = \dfrac{1}{|x_i - \bar{x}|} = \dfrac{1}{abs(x - \text{mean}(x))}$ and $\bar{x}$ denote the arithmetic mean is defined by the

formula $\bar{x} = \dfrac{\sum_{i=1}^{n} x_i}{n} = \dfrac{\text{sum}(x)}{n}$.

In R, first we need to define a function `weightedMean()` to compute the weighted mean in the presence of outlier in the dataset. We can then use the `weightedMean()` function in this purpose as follows.

R-code:

```
#Define a function to calculate weighted mean
weightedMean<- function(x){
  wt<-1/(abs(x-mean(x)))
  xm<-sum(x*wt)/sum(wt)
  return(xm)
}
# Create a vector of numbers
c<-c(50, 70, 65, 75, 270)
# Calculate the weighted mean of the vector
weightedMean(c)
```

Output:

```
Console  Terminal ×  Background Jobs ×

R  R 4.3.2 · ~/
> #Define a function to calculate weighted mean
> weightedMean<- function(x){
+    wt<-1/(abs(x-mean(x)))
+    xm<-sum(x*wt)/sum(wt)
+    return(xm)
+ }
> # Create a vector of numbers
> c<-c(50, 70, 65, 75, 270)
> # Calculate the weighted mean of the vector
> weightedMean(c)
[1] 78.31982
>
```

Compared to the arithmetic mean, the proposed weighted mean is less impacted by the outlier. Interestingly, here no need to supply the weights of the observation since in the given R function `weightedMean`, it calculates automatically.

5.1.5 Trimmed Mean

The trimmed mean is a type of mean that excludes a certain percentage of the highest and lowest values in a dataset. To determine the trimmed mean of a dataset in R, we can apply the `mean()` function along with the `trim` parameter as argument.

Example 5.7

Suppose the waiting time in minutes of 10 patients for visiting a doctor are 25, 32, 20, 27, 35, 18, 29, 37, 45, and 42. The R-code and output are given below:

R-code:

```
# Create a vector of values
time <- c(25, 32, 20, 27, 35, 18, 29, 37, 45, 42)
# Calculate the trimmed mean (trimming 10% from both ends)
trimmed.mean <- mean(time, trim = 0.1)
# Print the result
print(trimmed.mean)
```

Output:

```
Console   Terminal ×   Background Jobs ×                                    ▬☐
R  R 4.3.2 · ~/ 
> # Create a vector of values
> time <- c(25, 32, 20, 27, 35, 18, 29, 37, 45, 42)
> # Calculate the trimmed mean (trimming 10% from both ends)
> trimmed.mean <- mean(time, trim = 0.1)
> # Print the result
> print(trimmed.mean)
[1] 30.875
> 
```

In the above example, we have a vector of waiting time containing 10 numeric values. We then use the mean() function with the trim parameter set to 0.1, indicating that we want to trim 10% from both ends of the distribution. The function computes the trimmed mean and stores it in the *trimmed.mean* object. Finally, we print the result, which gives us the trimmed mean of the values.

Note that the trim parameter accepts values between 0 and 0.5. A value of 0.2 means that the function will remove 20% of the data from both the lower and upper tails of the distribution before calculating the mean. Adjust the trim value as you need for your purpose.

5.1.6 Other Measures of Location

Percentiles and Quartiles

In statistics, a percentile is a measure that indicates the value below a certain percentage of the data falls. Percentiles are used to understand the relative position of a specific value within a dataset. A percentile is defined as follows:

Suppose we have a dataset consisting of n observations $x_1, x_2, ..., x_n$, then p^{th} percentile (P) indicates the value of X such that $p\%$ of the observations are less than P, and $(100 - p)\%$ of the observations are greater than P. These sorts of measures are frequently used in contemporary data analysis, including quantile regression (Rahman & Hossain, 2022; Hossain et al., 2022; Abdulla et al., 2023).

In R, you can calculate percentiles using the quantile() function.

The following example demonstrates how to calculate percentiles using R:

Example 5.8

Recall **Example 5.1**, now we want to calculate percentiles of Weight (kg). The R-code and output are given below.

R-code:

```
# create a vector of weights
weight <- c(35, 42, 45, 50, 48, 62, 55, 47, 38, 68)
# Calculate the 25th and 50th percentiles
percentiles <- quantile(weight, probs = c(0.25, 0.5))
print(percentiles)
percentiles75 <- quantile(weight, probs = 0.75)
percentiles75
```

Output:

```
Console    Terminal ×    Background Jobs ×
R  R 4.3.2 · ~/
> # create a vector of weights
> weight <- c(35, 42, 45, 50, 48, 62, 55, 47, 38, 68)
> # Calculate the 25th and 50th percentiles
> percentiles <- quantile(weight, probs = c(0.25, 0.5))
> print(percentiles)
  25%   50%
42.75 47.50
> percentiles75 <- quantile(weight, probs = 0.75)
> percentiles75
  75%
53.75
>
```

In the above example, we have vector values containing weights of 10 people. We use the quantile() function to calculate the percentiles specified in the probs argument. The probs argument takes a vector of probabilities to calculate the percentiles. In this case, we calculate the 25th and 50th percentiles. The function returns a numeric vector containing the computed percentiles, which we store in the *percentiles* object. Finally, we print the result.

The output shows the 25th and 50th percentiles of weights. This means that 25% of the weight falls below the value of 42.75 kg, and 75% of the weight falls below the value of 47.50 kg.

Quartiles

Moreover, sometimes, we are interested in calculating quartiles. In statistics, quartiles are values that divide a dataset into four (4) equal parts and each part representing 25% of the data. Quartiles are often used to understand the distribution and spread of a dataset. In R, you can calculate quartiles using the quantile() function.

The 25th percentile is indicated as the first quartile and denoted by Q_1. The 50th percentile (the median) and 75th percentile are referred as the second or middle quartile (Q_2), and third quartile (Q_3), respectively. The quartiles for a set of data are calculated using the following formulas:

$$Q_1 = \frac{n+1}{4} \text{ th ordered observation}$$

$$Q_2 = \frac{n+1}{2} \text{ th ordered observation, and}$$

$$Q_3 = \frac{3(n+1)}{4} \text{ th ordered observation.}$$

Example 5.9

Recall 'PimaIndiansDiabetes2' dataset that is internally available to R. Suppose we want to find the 1st, 2nd and 3rd quartiles of a variable *insulin*. Then we use the following commands:

R-code:

```
#create variable from the loaded data
insulin <- PimaIndiansDiabetes2$insulin
#calculate quartiles
quartiles <- quantile(insulin, probs = c(0.25, 0.5, 0.75), na.rm=TRUE)
#print result
quartiles
```

Output:

```
Console   Terminal ×   Background Jobs ×                                    ▭ ☐
R  R 4.3.2 · ~/
> #create variable from the loaded data
> insulin <- PimaIndiansDiabetes2$insulin
> #calculate quartiles
> quartiles <- quantile(insulin, probs = c(0.25, 0.5, 0.75), na.rm=TRUE)
> #print result
> quartiles
    25%    50%    75%
 76.25 125.00 190.00
> |
```

The output shows the 25th, 50th (median), and 75th percentiles (first, second, and third quartiles) of the variable *insulin*.

5.2 Measures of Dispersion

As mentioned before, measures of central tendency offer us an overall understanding of where the majority of the data is concentrated. However, while having identical arithmetic means for the measure of central tendency in two separate datasets, they could have different variability around the mean. In this instance, the distribution of the data may not be sufficiently described by the location measures. Hence, variability or dispersion of data may help to make reasonable decisions.

The variation that a group of observations exhibits is referred to as its dispersion. The degree of variability in a single dataset can be determined using a measure of dispersion. There is no dispersion if all the values are the same; if they are not all the same, there is dispersion in the data. When the values, though dissimilar, are close together, the degree of dispersion may be little. The dispersion is bigger if the values are widely dispersed. Variation, spread, and scatter are additional words that are used synonymously with dispersion.

Example 5.10

Suppose two individuals Mr X and Mr Y performed physical exercise regularly. We now want to know the individual who practice physical exercise in more consistent manner. Their weekly duration of physical exercise (in minutes) data are presented in Table 5.6.

TABLE 5.6

Weekly Duration of Physical Exercise in Minutes by Mr. X and Mr. Y

Mr X	185	135	200	185	155	250
Mr Y	182	185	188	185	180	190

We now compute the different measures of central tendency to know who do more physical exercise. The results of Mr. X and Mr. Y are displayed below (Table 5.7).

TABLE 5.7

Measures of Central Tendency

Individuals	Mean	Median	Mode
Mr X	185	185	185
Mr Y	185	185	185

Here, all three measures of central tendency are the same for two individuals, which does not help us to find the individual who practice physical exercise in more consistent manner. However, if we look at the weekly durations of physical exercise, we notice that Mr. Y's weekly duration of physical exercise are very consistent, and they do not vary much around his average. On the other hand, Mr. X has wildly varying weekly durations of physical exercise.

Example 5.11

Consider another example in which a man who opened a new clinic and purchased 10 wooden doors with an exact width of 30 inches. The manufacturer provides 5 doors that are 30.5 inches wide and the remaining 5 doors that are 29.5 inches wide. The arithmetic mean of 10 doors is, therefore, 30 inches. The arithmetic mean can lead one to believe that all the doors are good, but in reality, none of them can be used because they won't fit in the space.

The aforementioned examples demonstrate that a measure of dispersion is required in addition to a measure of central tendency in order to characterize a variable's distribution. Therefore, we are going to discuss about several measures of dispersion.

5.2.1 Range and Interquartile Range

Calculating the range is a relatively easy method of determining how much a group of numbers varies. The difference between the greatest and smallest value in a set of observations is called the range. The range is denoted by R, and suppose x_L and x_S are the largest and the smallest values respectively then the range is determined in the following way:

$$R = x_L - x_S = \max(x) - \min(x).$$

The difference between the 75th and 25th quartiles is known as the interquartile range (IQR) and can be defined as:

$$\text{IQR} = x_{0.75} - x_{0.25} = Q_3 - Q_1 = \text{quantile}(x, 0.75) - \text{quantile}(x, 0.25).$$

It includes 50% of the observations and covers the middle of the distribution.

Example 5.12

Let the levels of hemoglobin (g/dl) in elderly women in a geriatric hospital were as follows:

12.2,11.1,14.0,11.3,10.8,12.5,12.2,11.9,13.6,12.7,13.4,13.7,11.9,10.7,
12.3,13.9,11.1,11.2,13.3,11.4,12.0,11.1.

Now, we compute the range and interquartile range of the hemoglobin levels. If we arrange the dataset from smallest to largest, then we have

10.7,10.8,11.1,11.1,11.2,11.3,11.4,11.9,11.9,12.2,
12.2,12.3,12.5,12.7,13.3,13.4,13.6,13.7,13.9,14.0

Hence, the range is $R = 14.0 - 10.7 = 3.3$ g/dl which means that the levels of hemoglobin is varying at most by 3.3 g/dl and IQR $= 13.325 - 11.275 = 2.05$ g/dl that can be interpreted as 50% of the data is centered between 11.275 and 13.325 g/dl. The following R-code can be used.

R-code:

```
# Create vector
h <- c(12.2, 11.1, 14.0, 11.3, 10.8, 12.5, 12.2, 11.9, 13.6, 12.7,
13.4, 13.7, 11.9, 10.7, 12.3, 13.9, 11.1, 11.2, 13.3, 11.4)
#Range
r <- max(h)-min(h)
r
x_0.75 <- quantile(h, 0.75)
x_0.25 <- quantile(h, 0.25)
IQR <- x_0.75-x_0.25
IQR
```

Output:

```
Console   Terminal    Background Jobs
R  R 4.3.2 · ~/
> # Create vector
> h <- c(12.2, 11.1, 14.0, 11.3, 10.8, 12.5, 12.2, 11.9, 13.6, 12.7, 13.4, 13.7, 11.9, 10.7, 12.
3, 13.9, 11.1, 11.2, 13.3, 11.4)
> #Range
> r <- max(h)-min(h)
> r
[1] 3.3
> x_0.75 <- quantile(h, 0.75)
> x_0.25 <- quantile(h, 0.25)
> IQR <- x_0.75-x_0.25
> IQR
 75%
2.05
>
```

The range has some drawbacks including (Figure 5.1),

i. It ignores the way in which data are distributed. For example, suppose we have two dataset presented in the following way:
   Here, the ranges of two dataset $R = 12 - 7 = 5$ are same; however, the distributions are not alike.

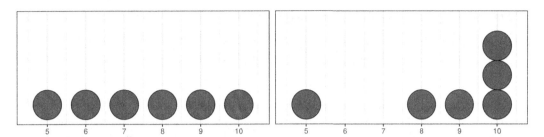

FIGURE 5.1
Demonstration of the distribution of two datasets.

    ii. It is sensitive to outliers. For instance, the range of the dataset
        1,1,2,2,2,3,3,3,3,4,5 is 5. However, for the dataset 1,1,2,2,2,2,3,3,3,3,4,50, the
        range is 49. Hence, all values of two datasets are same except the last value of
        the second dataset which influences the range a lot.

5.2.2 Mean Deviation, Variance, and Standard Deviation

A set of observations that are near to their mean have less dispersion than those that are dispersed widely. This being the case, it would make sense to assess dispersion with the values scattered around their mean.

Suppose we have a set of observations $x_1, x_2, ..., x_n$, now the deviations of n observations around the mean $\bar{x}$ and find the average of them in the following way:

$$MD = \frac{1}{n}\sum_{i=1}^{n}(x_i - \bar{x}) = \frac{1}{n} * sum(x - mean(x)).$$

However, the disadvantage of this measure is that since the deviations $(x_i - \bar{x}); i = 1, 2, ..., n$, might be positive or negative and if we take their sum then it will be very small or even zero. Therefore, it is not wise to use MD as a measure of variability because MD could be small even for the presence of a substantial variability in the data.

Interestingly, if we use absolute values of these deviations then the problem is solved. The absolute mean deviation (AMD) is therefore defined as:

$$AMD(\bar{x}) = \frac{1}{n}\sum_{i=1}^{n}|x_i - \bar{x}| = \frac{1}{n} * sum(abs(x - mean(x))).$$

Consideration of the squares of deviations $(x_i - \bar{x})$, as an alternative to using the absolute value, offers a further means of avoiding the positive and negative signs of deviations. This introduces another measure of dispersion, i.e., variance.

For instance, we subtract the mean from each value, square the resulting differences, and then add the squared differences to determine the variance of a sample of values. The total of the squared deviations of the values from their mean divided by the sample size minus one yields the sample variance. The value of variance ranges from 0 to $+\infty$. If all values are the same, then the variance would be zero.

Mathematically, the sample variance is defined as:

$$s^2 = \frac{\sum_{i=1}^{n} (x_i - \bar{x})^2}{n-1} = \frac{\text{sum}(x - \text{mean}(x))^\wedge 2}{n-1}.$$

However, the steps mentioned above are followed when computing the variance from a finite population of N values, deduct μ from each x with the exception that we divide by N instead of $N - 1$. The population variance is denoted by σ^2 and the formula is as follows:

$$\sigma^2 = \frac{\sum_{i=1}^{N} (x_i - \mu)^2}{N} = \frac{\text{sum}(x - \text{mean}(X))^\wedge 2}{N}.$$

The square of the units of the observations is the unit of variance, which may be more challenging to interpret. For example, we are interested in estimating the variance of a dataset consisting of the age of a particular group of people in years. Hence, the unit of variance is year^2 which is difficult to interpret. In such circumstances, we simply take the square root of the variance to get a measure of dispersion in the original unit.

The sample standard deviation is the positive square root of the variance, defined as:

$$s = \sqrt{s^2} = \sqrt{\frac{\sum_{i=1}^{n} (x_i - \bar{x})^2}{n-1}} = \text{sqrt}(s^2).$$

The standard deviation measures how widely distributed the data are from the arithmetic mean or how much they deviate from it. When the standard deviation is low, the values are tightly clustered around the mean. Moreover, when the standard deviation is high, the observations are less concentrated around the mean, and may be some observed values could be substantially far away from the mean.

Example 5.13

Assume that the waiting time of 20 patients to get health-care from a consulting doctor are given in Table 5.8 and two consultant doctors X and Y are available.

Calculate absolute mean deviation, variance, and standard deviation of waiting as well as interpret the results.

We know that, absolute mean deviation, variance, and standard deviation can be obtained by

$$\text{AMD}(\bar{x}) = \frac{1}{n} \sum_{i=1}^{n} |x_i - \bar{x}|, s^2 = \frac{\sum_{i=1}^{n} (x_i - \bar{x})^2}{n-1}, \text{and } s = \sqrt{s^2}$$

TABLE 5.8

Waiting Time of 20 Patients for Consulting Doctor

Consultant Doctor	Waiting Time (minutes)
X	05, 12, 15, 03, 04, 18, 37, 14, 07, 35
Y	12, 15, 17, 18, 14, 15, 16, 14, 13, 16

In R, we can use the var and sd commands to compute the variance and standard deviation, respectively. In this case, we use the following command and get the output below.

R-code:

```
# Doctor X
D_X <- c(05, 12, 15, 03, 04, 18, 37, 14, 07, 35)
sum(D_X)
mean(D_X)
#absolute mean deviation
sum(abs(D_X-mean(D_X)))
#variance
var(D_X)
#Standard deviation
sd(D_X)
# Doctor Y
D_Y <- c(12, 15, 17, 18, 14, 15, 16, 14, 13, 16)
sum(D_Y)
mean(D_Y)
#absolute mean deviation
sum(abs(D_Y-mean(D_Y)))
#variance
var(D_Y)
#Standard deviation
sd(D_Y)
```

Output:

```
Console   Terminal ×   Background Jobs ×
R  R 4.3.2 · ~/
> # Doctor X
> D_X <- c(05, 12, 15, 03, 04, 18, 37, 14, 07, 35)
> sum(D_X)
[1] 150
> mean(D_X)
[1] 15
> #absolute mean deviation
> sum(abs(D_X-mean(D_X)))
[1] 90
> #variance
> var(D_X)
[1] 148
> #Standard deviation
> sd(D_X)
[1] 12.16553
> # Doctor Y
> D_Y <- c(12, 15, 17, 18, 14, 15, 16, 14, 13, 16)
> sum(D_Y)
[1] 150
> mean(D_Y)
[1] 15
> #absolute mean deviation
> sum(abs(D_Y-mean(D_Y)))
[1] 14
> #variance
> var(D_Y)
[1] 3.333333
> #Standard deviation
> sd(D_Y)
[1] 1.825742
> |
```

Interpretation: Here, it is observed that the average waiting time is 15 minutes for both consultant doctors. However, absolute mean deviation, variance, and standard deviation of waiting times are quite different. We see that the variability in the waiting time is smaller for Doctor Y than Doctor X. For Doctor Y, $s = 1.826$ minutes, which means that the average difference of the observations from the arithmetic mean is 1.826 minutes.

Variance of Transformed Data

Suppose, we consider a linear transformation $y_i = bx_i (b \neq 0)$ of the original data x_i, $i = 1, 2,$ ..., n. Now, the variance of the transformed data can be written as $s_y^2 = b^2 s_x^2$.

5.2.3 Coefficient of Variation (CV)

The standard deviation is used to compare the existing variability of two or more datasets. However, the measurement unit of the two datasets may be different. In that case, the use of standard deviation does not make sense. Moreover, the two means could differ even though they utilize the same unit of measurement. For instance, if we compare the primary school students' standard deviation of weights to the high school students' standard deviation of weights, we might notice that the standard deviation of the high school students is numerically larger than that of the primary school students. This is because the high school students' weights are typically heavier and not because of greater dispersion.

Again, let us want to compare the variability of expenditure on health measured in US dollars and British pounds. How can we make a fair comparison? It is not very useful to directly compare the standard deviations because the prices are expressed in different units and, consequently, most likely have arithmetic means that differ significantly.

The coefficient of variation (CV) is a comparison-friendly measure of dispersion that considers both the standard deviation and the mean. It is a unit-free measure of dispersion. Interestingly, CV can be used to compare the datasets having the same unit as well as different units of measurement. The CV is expressed as the standard deviation as a percentage of the mean. The formula of CV is given by

$$CV = \frac{s}{\bar{x}} \times 100\% = \frac{sd(x)}{mean(x)} * 100\%.$$

Example 5.14

Recall **Example 5.12**, suppose we wish to compare the variability of the waiting times for two doctors X and Y. We have,

$$CV(X) = \frac{s_X}{\bar{x}_X} \times 100\% = \frac{12.165}{15} \times 100\% = 81.10\% \text{ and}$$

$$CV(Y) = \frac{s_Y}{\bar{x}_Y} \times 100\% = \frac{1.825}{15} \times 100\% = 12.17\%$$

From these results, it is clear that variation is considerably higher in the sample of waiting time of doctor X than in the sample of waiting time of doctor Y. Less CV indicate good choice, so people prefer to visit doctor Y instead of doctor X.

Example 5.15

We consider a dataset described by Hand and Crowder (1996), which is available in the R library named 'BodyWeight: Rat weight over time for different diets'. This dataset consists of 176 rows and 4 columns for the following 4 variables:

Weight: body weight of the rat (grams),
Time: time at which the measurement is made (days),
Rat: rat whose weight is measured,
Diet: diet that the rat receives.

Suppose we want to calculate different measures of dispersion described above using the variable *Weight* using R. For this purpose, we perform the following commands.

R-code:

```
#Data can be loaded in R in the following way
# install package
install.packages('nlme')
#make the library of the loaded package
library(nlme)
#data
data(BodyWeight)
#select weight from data
W=BodyWeight$weight
#mean
mean(w)
# absolute mean deviation
sum(abs(w-mean(w)))/length(w)
# variance
var(w)
# standard deviation
sd(w)
# Coefficient of variation
(sd(w)/mean(w))*100
```

Output:

```
Console   Terminal ×   Background Jobs ×
R  R 4.3.2 · ~/
> #make the library of the loaded package
> library(nlme)
> #data
> data(Bodyweight)
> #select weight from data
> W=Bodyweight$weight
> #mean
> mean(w)
[1] 2.5
> # absolute mean deviation
> sum(abs(w-mean(w)))/length(w)
[1] 0.5
> # variance
> var(w)
[1] 0.3333333
> # standard deviation
> sd(w)
[1] 0.5773503
> # Coefficient of variation
> (sd(w)/mean(w))*100
[1] 23.09401
> |
```

One can upload his/her own dataset in R and easily calculate various measures of dispersion.

5.2.4 Skewness

We learn nothing about the shape of the distribution from both the center and dispersion measures. Skewness is, therefore, used to know the shape of the distribution of a variable in the dataset. In statistical data science, it helps us understand how much a dataset deviates from a symmetrical, bell-shaped distribution. A symmetric distribution has a skewness value of zero. Positive skewness indicates that the tail of the distribution is stretched to the right. In contrast, negative skewness indicates that the tail is stretched to the left (Figure 5.2).

If the left tail is longer (left panel) than the right tail, then it is said to be left skewed, and the majority of data points are concentrated on the right side. In that case, the mean of the dataset is smaller than the median and mode. Conversely, suppose the plot is skewed to the right. In that case, it means that the right tail of the distribution is longer than the left tail (right panel). Most of the data points are concentrated on the left side; i.e., the mean is larger than the median and mode. In the case of symmetric distribution, the mean, median, and mode are equal.

There are different ways of measuring the skewness. The most commonly used measure of skewness is Pearson's moment coefficient of skewness, also known as simply the skewness. The formula to calculate skewness is as follows:

$$S_k = \frac{\sqrt{n}\sum_{i=1}^{n}(x_i - \bar{x})^3}{\left(\sum_{i=1}^{n}(x_i - \bar{x})^2\right)^{3/2}} = \frac{\mathrm{sqrt}\big(\mathrm{length}(x)\big)*\mathrm{sum}\big(x - \mathrm{mean}(x)\big)^3}{\big(\mathrm{sum}(x - \mathrm{mean}(x))\wedge 2\big)^{(3/2)}}.$$

The value of $S_k = 0$ indicate the absence of skewness, i.e., symmetric about mean, if $S_k > 0$ then we say the distribution of the data is positively skewed or skewed to the right and $S_k < 0$ indicates the distribution of the data is skewed to the left, i.e., negatively skewed.

Other measures of skewness suggested by Karl Pearson are also used. The Pearson mode skewness or first skewness coefficient is obtained using $S_k = \dfrac{\mathrm{Mean} - \mathrm{Mode}}{\mathrm{Standard\,deviation}}$ and the Pearson median skewness or second skewness coefficient is defined as

$$S_k = \frac{3(\mathrm{Mean} - \mathrm{Mode})}{\mathrm{Standard\,deviation}}.$$

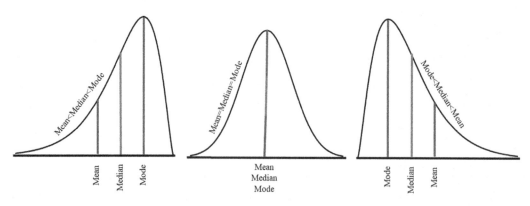

FIGURE 5.2
Left or negatively skewed (left panel), symmetric (middle panel), and right or positively skewed (right panel).

Example 5.16

The body mass index (BMI), a popular health indicator, evaluates a person's weight in relation to their height. Consider the following dataset with BMI measurements for a population of 15 individuals.

18.5, 22.0, 24.7, 25.5, 30.2, 27.8, 23.6, 29.1, 21.3, 33.6, 26.4, 20.1, 35.0, 19.8, 22.9

In R, to calculate the skewness simply we use a library called e1071 where the function skewness is defined. Now, we use the following commands.

R-code:

```
install.packages('e1071')
library(e1071)
bmi_data <- c(18.5, 22.0, 24.7, 25.5, 30.2, 27.8, 23.6, 29.1, 21.3,
33.6, 26.4, 20.1, 35.0, 19.8, 22.9)
skewness(bmi_data)
```

Output:

```
Console  Terminal ×   Background Jobs ×
R  R 4.3.2 · ~/
> library(e1071)
> bmi_data <- c(18.5, 22.0, 24.7, 25.5, 30.2, 27.8, 23.6, 29.1, 21.3, 33.6, 26.4, 20.1, 35.0, 1
9.8, 22.9)
> skewness(bmi_data)
[1] 0.4641787
>
```

Visualization of Skewness: We can also visually represent the skewness of the datasets using density plots. The R-code is given below.

R-code:

```
# Load the required library for plotting
install.packages('ggplot2')
library(ggplot2)
# Density plot for hospital stay data
ggplot(data.frame(BMI = bmi_data), aes(x = BMI)) +
  geom_density(fill = 'red', alpha = 0.1) +
  theme_bw()+
  theme(axis.text = element_text(size = 12))   +
  theme(axis.title = element_text(size = 12))    +
  theme(plot.margin = margin(t = 0.75,  # Top margin
                             r = 0.75,  # Right margin
                             b = 0.75,  # Bottom margin
                             l = 0.75, unit='cm'))+ #Left margin
  xlab('BMI')+
  ylab('Density')+
  ggtitle('Skewness')
```

In the density plots, if the right tail of the distribution is longer than the left tail, it indicates positive skewness, and if the left tail is longer than the right tail, it indicates negative skewness.

Figure 5.3 illustrates that the BMI data is positively skewed. It implies that there are relatively few people with extremely high BMI, which indicate obesity or overweight. The majority of the population might have BMIs concentrated in the normal or lower ranges.

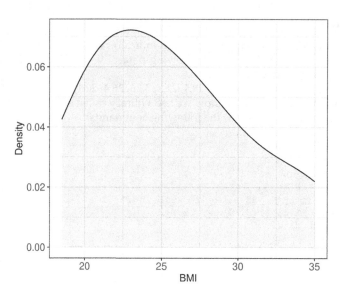

FIGURE 5.3
Density plot of BMI data given in **Example 5.16**.

5.2.5 Kurtosis

Kurtosis is a measure of the 'peakedness' or 'flatness' of a probability distribution compared to the normal distribution (also known as the Gaussian distribution). It is often used in conjunction with skewness to fully characterize the distribution of data (e.g. Rahman et al., 2024). There are three types of kurtosis (Figure 5.4):

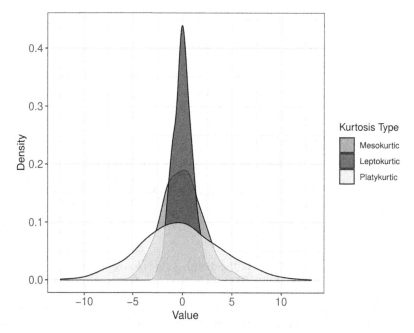

FIGURE 5.4
Different types of kurtosis.

a. **Mesokurtic**: If the kurtosis value is close to 0, the distribution has similar peaked-ness as the normal distribution (bell-shaped).

b. **Leptokurtic**: If the kurtosis value is positive, the distribution has heavier tails and a more peaked center compared to the normal distribution.

c. **Platykurtic**: If the kurtosis value is negative, the distribution has lighter tails and a flatter peak compared to the normal distribution.

The most commonly used measure of kurtosis is the Pearson's moment coefficient of kurtosis, also known as simply kurtosis and is defined as

$$\text{kurtosis} = \frac{n \sum_{i=1}^{n} (x_i - \bar{x})^4}{\left(\sum_{i=1}^{n} (x_i - \bar{x})^2 \right)^2} - 3.$$

The positive kurtosis value (approximately 0.312) suggests that the reaction time distribution is also leptokurtic.

Example 5.17

Suppose blood pressure of twenty people are given below.

$$120, 130, 140, 150, 160, 140, 130, 125, 135, 142, 145,$$
$$155, 150, 138, 128, 142, 146, 135, 152, 130$$

In R, to calculate the kurtosis, we use the `kurtosis()` function from the `moments` package as follows.

R-code:

```
install.packages('moments')
library(moments)
# Sample blood pressure data
blood_pressure <- c(120, 130, 140, 150, 160, 140, 130, 125, 135, 142,
145, 155, 150, 138, 128, 142, 146, 135, 152, 130)
kurtosis(blood_pressure)
```

Output:

```
Console   Terminal ×   Background Jobs ×
R  R 4.3.2 · ~/
> library(moments)
> # Sample blood pressure data
> blood_pressure <- c(120, 130, 140, 150, 160, 140, 130, 125, 135, 142, 145, 155, 150, 138, 128,
142, 146, 135, 152, 130)
> kurtosis(blood_pressure)
[1] 2.216053
> |
```

5.3 Conclusion and Exercises

This section concludes the chapter with some practical problem-solving exercises. These concepts and measures are used in the subsequent chapters.

5.3.1 Concluding Remarks

In this chapter, we've covered fundamental concepts related to measures of central tendency, which are critical in statistical analysis. We explored primary measures such as the mean, median, and mode, emphasizing their relevance in decision-making across various fields. The selection of a specific measure depends on the data's nature and the research goals. Additionally, we delved into measures of dispersion, shedding light on data variability and spread, crucial for understanding dataset distribution and reliability. Key measures discussed include the range, interquartile range, variance, standard deviation, and coefficient of variation. Skewness and kurtosis, along with examples, were also examined. Mastering these measures equips you to uncover insights within your data, forming a robust foundation for advanced statistical techniques.

5.3.2 Practice Questions

Exercise 5.1 Number of maternal deaths in 2020 collected on 2 August 2023 from 'The World Bank', available online at http://www.worldbank.org, of 30 selected countries is given below.

15, 41, 42, 44, 18, 26, 29, 30, 61, 67, 33, 34, 56, 56, 57, 34, 36, 18, 19, 20, 24, 38, 74, 44, 52, 52, 54, 54, 67, 69

a. Calculate the arithmetic mean, median, mode, and minimum, maximum for number of maternal deaths.

b. Calculate the first quartile, third quartile, and interquartile range. Interpret your results.

c. Calculate the absolute mean deviation and standard deviation for number of maternal deaths. Make a conclusion about the variability of the data.

d. Compute the coefficient of skewness of the data. Comment on your result.

e. Use R to compute the results of (a), (b), (c), and (d).

Exercise 5.2 Suppose a student got the following percentage grade corresponding to the assessment items in the course.

Item	Weights (%)	Grade (%)
Assignment #1	5	75
Assignment #2	5	70
Midterm exam	25	65
Short project	15	75
Final exam	50	80

Calculate his/her percentage grade in a course.

Exercise 5.3 The following data represent the total deaths of COVID-19 in ten selected countries from Asia and Europe on 2 August 2023 from 'Worldometer', available online at https://www.worldometers.info/coronavirus/.

Asia	531917, 74694, 35313, 102174, 43206, 19005, 146311, 161894, 37164, 12585
Europe	167642, 174352, 191012, 228144, 399814, 121760, 22992, 119635, 37089, 22542

a. Calculate the appropriate mean, median, minimum, and maximum of total deaths for Asian and European selected countries.

b. Find range and interquartile range for both datasets.

c. Calculate the variance, standard deviation, and coefficient of variation of total deaths for Asian and European selected countries. What is your conclusion about the variability of the two datasets?

d. Use R to reproduce the results of (a), (b), and (c).

e. Compute the coefficient of skewness for both datasets. Make density plots of two datasets and comment on the shape of the data.

f. Calculate the coefficient of kurtosis and comment on your result.

Bibliography

Abdulla, F., El-Raouf, M. M. A., Rahman, A., Aldallal, R., Mohamed, M. S., & Hossain, M. M. (2023). Prevalence and determinants of wasting among under-5 Egyptian children: Application of quantile regression. *Food Science and Nutrition*, 11(2), 1073–1083. https://doi.org/10.1002/fsn3.3144

Agresti, A., & Kateri, M. (2021). *Foundations of statistics for data scientists: With R and Python*. CRC Press.

Braun, W. J., & Murdoch, D. J. (2021). *A first course in statistical programming with R*. Cambridge University Press.

Campbell, M. J., Machin, D., & Walters, S. J. (2010). *Medical statistics: A textbook for the health sciences*. John Wiley & Sons.

Dalgaard, P. (2008). *Introductory statistics with R*. Springer Publication.

Daniel, W. W., & Cross, C. L. (2018). *Biostatistics: A foundation for analysis in the health sciences*. Wiley.

Demidenko, E. (2019). *Advanced statistics with applications in R* (Vol. 392). John Wiley & Sons.

Glover, T., & Mitchell, K. (2008). *An introduction to biostatistics*. Waveland Press.

Gordon, R. A. (2012). *Applied statistics for the social and health sciences*. Routledge.

Groeneveld, R. A., & Meeden, G. (1984). Measuring skewness and kurtosis. *Journal of the Royal Statistical Society Series D: The Statistician*, 33(4), 391–399.

Hand, D. and Crowder, M. (1996). *Practical longitudinal data analysis*. Chapman and Hall.

Heumann, C., & Shalabh, M. S. (2016). *Introduction to statistics and data analysis*. Springer International Publishing.

Hoffmann, J. P. (2021). *Linear regression models: Applications in R*. CRC Press.

Hossain, M. M. (2016). Proposed mean (robust) in the presence of Outlier. *Journal of Statistics Applications & Probability Letters*, 3(3), 103–107.

Hossain, M. M. (2022). Linear regression models: Applications in R. *Journal of the Royal Statistical Society Series A: Statistics in Society*, 185(Supplement_2), S771.

Hossain, M. M., Abdulla, F., & Rahman, A. (2022). Prevalence and determinants of wasting of under-5 children in Bangladesh: Quantile regression approach. *PLoS One*, 17(11 November), 1–16. Article e0278097. https://doi.org/10.1371/journal.pone.0278097

Irizarry, R. A. (2019). *Introduction to data science: Data analysis and prediction algorithms with R*. CRC Press.

Joanes, D. N., & Gill, C. A. (1998). Comparing measures of sample skewness and kurtosis. *Journal of the Royal Statistical Society: Series D (The Statistician)*, 47(1), 183–189.

Le, C. T., & Eberly, L. E. (2016). *Introductory biostatistics*. John Wiley & Sons.

Mann, P. S. (2007). *Introductory statistics*. John Wiley & Sons.

Rahman, A., & Hossain, M. (2022). Quantile regression approach to estimating prevalence and determinants of child malnutrition. *Journal of Public Health: From Theory to Practice*, 30(2), 323–339. https://doi.org/10.1007/s10389-020-01277-0

Rahman, A., Othman, N., Kuddus, M. A., & Hasan, M. Z. (2024). Impact of the COVID-19 pandemic on child malnutrition in Selangor, Malaysia: A pilot study. *Journal of Infection and Public Health*, 17(5), 833–842. https://doi.org/10.1016/j.jiph.2024.02.019

Rosner, B. (2015). *Fundamentals of biostatistics*. Cengage Learning.

Rossi, R. J. (2022). *Applied biostatistics for the health sciences*. John Wiley & Sons.

Saiz, Z., & Rosteck. (2020). *An introduction to data analysis in R*. Springer International Publishing.

Smith, J. W., Everhart, J. E., Dickson, W. C., Knowler, W. C., & Johannes, R. S. (1988). Using the ADAP learning algorithm to forecast the onset of diabetes mellitus. *Symposium on Computer Applications and Medical Care* (pp. 261–265).

Vexler, A., & Hutson, A. (2018). *Statistics in the health sciences: Theory, applications, and computing*. CRC Press.

Villanueva, R. A. M., & Chen, Z. J. (2019). ggplot2: Elegant graphics for data analysis. *Measurement: Interdisciplinary Research and Perspectives*, 17(3), 160–167. https://doi.org/10.1080/15366367.2019.1565254

Wells, A. (2019). *Statistics: An introduction using R*. Scientific e-Resources.

Wilcox, R. R. (2016). *Understanding and applying basic statistical methods using R*. John Wiley & Sons.

Witten, D., & James, G. (2013). *An introduction to statistical learning with applications in R*. Springer Publication.

6

Probability, Random Variables, and Distributions

6.1 Probability

The concept of probability (chance) has been around for a long time, particularly in gambling arena. The mathematical theory of probability was started around the 17th century by Galilei, Pascal, and Fermat to solve gambling problems. Probability is a fundamental statistical concept that quantifies the likelihood or chance of an event occurring. It is a numerical measure that ranges from 0 to 1, where a probability of '0' indicates that an event is impossible and will not occur, and '1' means that an event is certain and will definitely happen. Moreover, a probability between 0 and 1 represents the likelihood of an event occurring, with higher values indicating a greater likelihood.

Example 6.1: The probability of a false-positive or false-negative result is significant in medical testing. For instance, a test for a certain disease can have a 5% false-positive rate, which means that there is a 5% probability of the test indicating the illness when it's not present.

6.1.1 Key Concepts of Probability

Set: A set is a collection of distinct and specific elements or data points that share a common characteristic or property relevant to a particular study or analysis. Sets are typically denoted by capital letters and using braces {}. The elements are listed within these braces, separated by commas. For example, suppose you are studying the occurrence of hypertension in a particular region. In this context, the age (in years) of 5 individuals with hypertension may be considered as a set and denoted by $B = \{42, 37, 51, 48, 64\}$.

Experiment: In probability theory, an experiment refers to a controlled process or procedure that generates a set of possible outcomes. These outcomes are often subject to chance or randomness. The set of outcomes forms the sample space. For instance, we want to predict the weather for a particular day. Then the possible outcomes could be 'Sunny,' 'Partly Cloudy,' 'Rainy,' or 'Snowy' and the sample space can be $S = \{$Sunny, Partly Cloudy, Rainy, Snowy$\}$.

Event: An event in probability is a subset of the sample space, representing one or more possible outcomes of a probability experiment. Events can be a single outcome, a combination of outcomes, or any other subset of the sample space. Consider a bioinformatics scenario where researchers are analyzing DNA sequences to study the presence of a specific genetic mutation in a population. In this context, we may define an event as the occurrence of a specific genetic mutation in an individual's DNA sequence, i.e., event A can be denoted as $A = \{$mutation present$\}$. The probability of event A is denoted by $P(A)$.

DOI: 10.1201/9781003426189-6

Mutually exclusive events: Mutually exclusive events refer to a pair or a set of events that cannot occur simultaneously. They are also known as disjoint events. The probability of the intersection (both events occurring) of mutually exclusive events is zero. For example, consider a researcher analyzing genetic variants within a specific region of a genome. Hence, two mutually exclusive events can be defined: (i) Event X: presence of variant X and (ii) Event Y: presence of variant Y. In this context, variant X and variant Y are considered mutually exclusive because they refer to different genetic variants at the same genetic locus. If variant X is present in a given individual's DNA sequence in that region, it means that variant Y cannot be present in the same location, and vice versa. These two events cannot co-occur within the same genomic region.

6.1.2 Types of Probability

Classical probability or priori probability: This is a type of probability used for simple and well-defined experiments where all possible outcomes are equally likely. It's calculated by dividing the number of favorable outcomes by the total number of equally likely outcomes. Mathematically, if an experiment has N mutually exclusive and equally likely outcomes, and if k of these outcomes were favorable to an event (A), then the probability of the occurrence of event A is equal to $\frac{k}{N}$, i.e., $P(A) = \frac{k}{N}$. Here we read $P(A)$ as 'the probability of A'.

Empirical probability or relative frequency probability: This type of probability is based on observed data or historical frequencies. Suppose a process is repeated a large number of times, say, n, and if a predefined event A occurs k times, the relative frequency of occurrence of A, i.e., $\frac{k}{n}$ will be approximately equal to the probability of A. Simply, we write as $P(A) = \frac{k}{n}, n \rightarrow \infty$. However, we must remember that $\frac{k}{n}$ is only an estimate of $P(A)$.

Subjective probability: Subjective probability is based on an individual's personal judgment or belief regarding the likelihood of an event. It's not necessarily grounded in empirical data or mathematical principles. For example, a weather forecaster might assign a subjective probability to the likelihood of rain.

6.1.3 Properties of Probability

The main properties of probability include the following.

a. Probabilities are always non-negative; they cannot be negative. For any event A, the probability $P(A)$ is greater than or equal to 0, i.e., $0 \leq P(A) \leq 1$.

b. For any two mutually exclusive events A_i and A_j (i.e., events that cannot occur simultaneously), the probability of the occurrence of either A_i or A_j is equal to the sum of their individual probabilities, i.e., $P(A_i \cup A_j) = P(A_i) + P(A_j)$. However, if two events are not mutually exclusive then the probability of occurrence of either A_i or A_j is calculated as the sum of their individual probabilities minus the probability of their intersection.

c. The sum of the probabilities of all possible outcomes in a sample space is always equal to 1. Mathematically, $\Sigma P(A_i) = 1$, where A_i represents all the possible outcomes in the sample space.

TABLE 6.1

Distribution of Blood Types of 50 Patients

Blood Type	Number of Patients
A	15
B	11
AB	8
O	16
Total	50

Example 6.1

Suppose we have blood types of 50 patients and Table 6.1 displays the distribution of blood types among them.

Now, we calculate the following probabilities:

i. If we pick a patient at random, what is the probability that the patient will be a blood type 'A'.
ii. If we pick a patient at random, what is the probability that the patient will be a blood type 'AB'.
iii. If we pick a patient at random, what is the probability that the patient will be a blood type 'O'.

To calculate the probability in R, we use the following command:

R-code:

```
## probability calculation
#B
P_B=11/50
print(P_B)
#AB
P_AB=8/50
print(P_AB)
#O
P_O=16/50
print(P_O)
```

Output:

```
Console   Terminal ×   Background Jobs ×
R  R 4.3.2 · ~/
> ## probability calculation
> #B
> P_B=11/50
> print(P_B)
[1] 0.22
> #AB
> P_AB=8/50
> print(P_AB)
[1] 0.16
> #O
> P_O=16/50
> print(P_O)
[1] 0.32
> |
```

Therefore, the estimated probabilities are 0.22, 0.16, and 0.32.

6.1.4 Joint and Conditional Probability

There are situations where we wish to determine the probability that a subject selected randomly from a group of subjects would have two characteristics simultaneously. This kind of probability a referred to as joint probability and is denoted by $P(A \cap B)$ for two events A and B.

Conditional probability is the probability of an event occurring given that another event has occurred and it is denoted as $P(A \mid B)$, here the vertical line is read as 'given'.

Let $P(A) > 0$, then the conditional probability of occurring event B, given that event A has already occurred, is

$$P(B \mid A) = \frac{P(A \cap B)}{P(A)}.$$

Similarly, let $P(B) > 0$ then the conditional probability of A given B is

$$P(A \mid B) = \frac{P(A \cap B)}{P(B)}.$$

Example 6.2

Let us consider that we have the blood types of 100 patients (50 men and 50 women), and the distribution of blood types is shown in Table 6.2:

TABLE 6.2

Distribution of Blood Types of 100 Patients (50 Men and 50 Women)

Blood Type	Men	Women	Total
A	15	9	24
B	11	14	25
AB	8	7	15
O	16	20	36
Total	50	50	100

Assume that we randomly select a single participant from the 100 participants and find that the patient we have chosen is a man. What is the probability that this subject will have blood type 'A'?

Here, we have to choose only men, so we are no longer interested in the total patient count. Therefore, the desired probability may be defined as follows: What is the probability that the selected patient has blood type 'A', given that the selected patient is a man? This is a conditional probability estimation problem.

Here, the probability of men is $P(M) = \frac{50}{100} = 0.5$ and the joint probability of men and blood type 'A' is $P(A \cap M) = \frac{15}{100} = 0.15$ since we have 50 male patients out of 100 patients and 15 patients have blood type 'A' and male patients.

Therefore, our estimated conditional probability is

$$P(A \mid M) = \frac{P(A \cap M)}{P(M)} = \frac{15/100}{50/100} = \frac{15}{50} = 0.3.$$

In R, simply we use the following command:

R-code:

```
# Conditional probability
P_AM=(15/100)/(50/100)
print(P_AM)
```

Output:

Similarly, we can find different conditional probabilities from Table 6.2.

6.1.5 The Multiplication Rule

Joint probability is also calculated by using conditional probability. We can use the following formula for any two events A and B:

$$P(A \cap B) = P(B) \times P(A \mid B) \text{ if } P(B) > 0.$$

For the same two events A and B, the multiplication rule may also be written as:

$$P(A \cap B) = P(A) \times P(B \mid A) \text{ if } P(A) > 0.$$

Example 6.3

Recall **Example 6.2**, suppose we want to determine the probability of blood type 'O' of women, i.e., we want to compute

$$P(\text{blood type "O" and women}) = P(O \cap W).$$

Here, the marginal probability of the selected patient will be a woman is $P(W) = \dfrac{50}{100} = 0.5$

and the conditional probability blood type 'O' given women is $P(O \mid W) = \dfrac{20}{50} = 0.4.$
Now, we have

$$P(\text{blood type "O" and women}) = P(O \cap W) = P(W) \times P(O \mid W) = 0.5 \times 0.4 = 0.2.$$

6.1.6 The Addition Rule

For two events '*A*' and '*B*', the probability the occurrence of event '*A*', or event '*B*', or both occur is obtained by the probability of occurrence event '*A*', plus the probability of occurrence event '*B*', minus the probability the occurrence of both events simultaneously. Mathematically, the addition rule of probability can be written as

$$P(A \text{ or } B) = P(A \cup B) = P(A) + P(B) - P(A \cap B).$$

If both events are mutually exclusive then $P(A \cap B) = 0$.

Example 6.4

Recall **Example 6.2**, now let us select a patient at random from the 100 patients represented in table, what is the probability that this patient will be women (*W*) or will have blood type '*O*' (*O*) or both?

Here, we have to find the probability $P(W \cup O)$. By using the addition rule, we can write as

$$P(W \cup O) = P(W) + P(O) - P(W \cap O).$$

Here, $P(W) = \dfrac{50}{100} = 0.5$, $P(O) = \dfrac{36}{100} = 0.36$, and $P(W \cap O) = P(O \cap W) = 0.2$.

$$\text{Therefore}, P(W \cup O) = 0.5 + 0.36 - 0.2 = 0.66.$$

Note: The 20 patients are included in women and blood type '*O*', i.e., 20 have been added into the numerator twice. To calculate probability, we need to subtract out once to ignore the effect of overlapping or duplication.

6.1.7 Independent Events

In probability, independent events refer to events where one event's occurrence or non-occurrence does not affect another event's occurrence or non-occurrence. In other words, the probability of one event happening is not influenced by the presence or absence of the other event. Mathematically, two events, '*A*' and '*B*', are considered independent if and only if:

$$P(A \text{ and } B) = P(A \cap B) = P(A) \times P(B).$$

Hence, it can be observed that in the case of two independent events, the probability of their joint occurrence is determined by multiplying the probabilities of each event separately.

It should be noted that two events have non-zero probabilities are said to be independent when following assertions are true.

$$P(A \mid B) = P(A), P(B \mid A) = P(B), \text{and}, P(A \cap B) = P(A) \times P(B).$$

Two events are not independent unless all these statements are true.

Example 6.5

Suppose we are conducting a health study to assess the probability of two individuals, Mr. X and Mr. Y, testing positive for two different diseases, Disease A and Disease B. Suppose the probability of Mr. X testing positive for Disease A is

$$P\left(\text{Mr X tests positive for Disease } A\right) = 0.1.$$

Also, let the probability of Mr. Y testing positive for Disease B is

$$P\left(\text{Mr Y tests positive for Disease } B\right) = 0.2.$$

To find the probability that both Mr. X and Mr. Y test positive for their respective diseases, we can use the formula for the probability of independent events:

$$P\left(\text{Mr X tests positive for Disease } A \text{ and Mr.Y tests positive for Disease } B\right)$$
$$= P\left(\text{Mr X tests positive for Disease } A\right) \times P\left(\text{Mr Y tests positive for Disease } B\right).$$

By substituting the given probabilities, we have

$$P\left(\text{Mr X tests positive for Disease } A \text{ and Mr Y tests positive for Disease } B\right)$$
$$= 0.1 \times 0.2 = 0.02.$$

So, the probability that both Mr. X and Mr. Y test positive for their respective diseases is 0.02 or 2%.

In this example, the independence of the events is assumed because the probability of one individual testing positive for a specific disease does not influence the probability of the other individual testing positive for a different disease.

6.1.8 Complementary Events

Complementary events, also known as complementary outcomes or mutually exclusive events, are two events in probability theory that are related in such a way that if one event occurs, the other cannot occur. In other words, they are events that cover all possible outcomes in a given sample space, and if one event happens, the other event is guaranteed not to happen. Suppose the complement of an event 'A' is denoted by 'A^C'. The sum of probabilities of an event 'A' and its complement 'A^C' is equal to 1, i.e., $P(A) + P(A^C) = 1$. Therefore, we can write as $P(A^C) = 1 - P(A)$.

Example 6.6

Recall **Example 6.1**, now we want to compute the probability of the complement of patients have blood type 'O'. Here, the sample space can be written as $S = \{A, B, AB, O\}$.

The probability of a randomly selected patients having bold type 'O' is $P(O) = \dfrac{16}{50} = 0.32$.

So, the probability of a randomly selected patient having bold type other than 'O' is $P(O^C) = 1 - P(O) = 1 - 0.32 = 0.68$.

6.1.9 Bayes' Theorem

Bayes' theorem, named after the Reverend Thomas Bayes, is a fundamental concept in probability theory and statistics. Bayes' theorem is particularly useful in situations where you have prior beliefs or knowledge and want to incorporate new data to make more informed decisions. A comprehensive account of Bayesian concepts is provided in the literature (e.g., see Rahman and Upadhyay, 2015).

Let A_1, A_2, ..., A_k are mutually exclusive and exhaustive events with prior probabilities $P(A_1)$, $P(A_2)$, ..., $P(A_k)$ respectively. If an event 'E' occurs, the posterior probability of S_i given that 'E' occurred is

$$P(S_i \mid E) = \frac{P(S_i)P(E \mid S_i)}{P(S_1)P(E \mid S_1) + P(S_2)P(E \mid S_2) + \ldots + P(S_k)P(E \mid S_k)}, \text{for } i = 1, 2, \ldots k.$$

Example 6.7

Suppose in a certain population few people are identified to be at high risk of having a heart attack. It is also assumed that 49% of the population is female. Among the population, 9% of patients who are female and 12% of patients who are male have a high risk of having a heart attack. A single person is selected at random and found to be high risk. (i) What is the probability that it is a male? (ii) What is the probability that it is a female?

Suppose 'M' and 'F' represent the male and female respectively. Also, 'H' indicates a person is selected who have a high risk of having a heart attack. Now, we have

$$P(M) = 51\% = 0.51,$$

$$P(F) = 49\% = 0.49,$$

$$P(H \mid F) = 9\% = 0.09, \text{and}$$

$$P(H \mid M) = 12\% = 0.12.$$

Now, we have (i) the probability of the selected person who have a high risk of heart attack that it is a male can be obtained as:

$$P(M \mid H) = \frac{P(M)P(H \mid M)}{P(M)P(H \mid M) + P(F)P(H \mid F)}$$
$$= \frac{0.51 \times 0.12}{0.51 \times 0.12 + 0.49 \times 0.09} = 0.581.$$

Hence, the estimated probability is 0.581 i.e., about 58.1% chance of a selected person who has a high risk of heart attack is a male person.

Again, (ii) the probability of the selected person who have a high risk of heart attack that it is a female can be found using the following way,

$$P(F \mid H) = \frac{P(F)P(H \mid F)}{P(M)P(H \mid M) + P(F)P(H \mid F)}$$
$$= \frac{0.49 \times 0.09}{0.51 \times 0.12 + 0.49 \times 0.09} = 0.419.$$

Hence, the estimated probability is 0.581, i.e., about 41.9% chance of a selected person who has a high risk of heart attack is a female person.

6.2 Random Variable

A random variable can take on various values as a result of a random experiment, process, or event. Random variables are used to model and analyze uncertainty and variability in data and are a key component of probability distributions. There are two main types of random variables that follow.

6.2.1 Discrete Random Variable

A discrete random variable is one that can take on a countable set of distinct values. Each value has an associated probability or likelihood. Examples of discrete random variables include the number of patients arriving at the emergency room of a hospital within one hour, the number of babies born in a hospital in a day, and the number of customers entering a pharmacy in an hour.

To compute the mean (expected value) and variance can be calculated as follows:

Mean: $\mu = \sum_{i=1}^{n} x_i P(X = x_i)$ and variance: $\sigma^2 = \sum_{i=1}^{n} (x_i - \mu)^2 P(X = x_i)$, where x_i represents each possible value of the random variable X, $P(X = x_i)$ is the probability of X taking the value x_i.

Example 6.8

Suppose we are working with a dataset of DNA sequence lengths in bioinformatics. We want to calculate the mean and variance of the sequence lengths based on the data given in Table 6.3:

TABLE 6.3

Probability Distribution of DNA Sequence of Different Lengths

Sequence Lengths	100,	200,	150,	300,	250,	180,	220
Probabilities	0.1,	0.25,	0.2,	0.05,	0.1,	0.15	0.15

Now in R, we use the following command to get the mean and variance:

R-code:

```
# Define sequence lengths and their probabilities
seq_lengths <- c(100, 200, 150, 300, 250)
prob <- c(0.2, 0.3, 0.2, 0.1, 0.2)
sum(prob)
# Calculate the mean (expected value)
mean <- sum(seq_lengths * prob)
cat('Mean (Expected Value):', mean, '\n')
# Calculate the variance
variance <- sum((seq_lengths - mean)^2 * prob)
cat('Variance:', variance, '\n')
```

Output:

```
Console   Terminal ×   Background Jobs ×
R  R 4.3.2 · ~/
> # Define sequence lengths and their probabilities
> seq_lengths <- c(100, 200, 150, 300, 250)
> prob <- c(0.2, 0.3, 0.2, 0.1, 0.2)
> sum(prob)
[1] 1
> # Calculate the mean (expected value)
> mean <- sum(seq_lengths * prob)
> cat("Mean (Expected Value):", mean, "\n")
Mean (Expected Value): 190
> # Calculate the variance
> variance <- sum((seq_lengths - mean)^2 * prob)
> cat("Variance:", variance, "\n")
Variance: 3900
>
```

Here, the mean sequence length is 190 base pairs and the variance of the sequence lengths is 5,100 square base pairs.

6.2.2 Continuous Random Variable

A continuous random variable is one that can take on an uncountable number of possible values within a given range. Instead of probabilities associated with specific values, it defines probabilities for intervals or ranges of values. Common examples of continuous random variables include the height of individuals in a population, the time it takes for a computer to execute a task, and the temperature in a specific location at a particular time.

The mean or expected value and variance of a continuous random variable X with probability density function (PDF) $f(x)$, $-\infty \le x \le \infty$ are obtained using the following formula:

$$\text{Mean: } \mu = \int_{-\infty}^{\infty} x f(x) dx \text{ and variance: } \sigma^2 = \int_{-\infty}^{\infty} (x-\mu)^2 f(x) dx.$$

Example 6.9

Let, a test instrument needs to be calibrated periodically to prevent measurement errors. After some time of use without calibration, it is known that the probability density function of the measurement error (millimeters) is:

$$f(x) = 1 - 0.5x, 0 \le x \le 2.$$

Now, for finding the mean and variance we use the following commands in R. We need to define a function of the probability density function first as follows:

R-code:

```
# Define the probability density function (PDF)
pdf_function <- function(x){
  1-0.5*x
}
```

```
# Calculate the mean (expected value) by integrating the PDF
mean_value <- integrate(function(x) x*pdf_function(x), lower = 0,
    upper = 2)$value
cat('Mean (Expected Value):', mean_value, '\n')
# Calculate the variance by integrating the square of the difference
    from the mean
variance_value_calculated <- integrate(function(x) (x - mean_
    value)^2 * pdf_function(x), lower = 0, upper = 2)$value
cat('Variance:', variance_value_calculated, '\n')
```

Output:

```
Console   Terminal ×   Background Jobs ×

R  R 4.3.2 · ~/
> # Define the probability density function (PDF)
> pdf_function <- function(x){
+    1-0.5*x
+ }
> # Calculate the mean (expected value) by integrating the PDF
> mean_value <- integrate(function(x) x*pdf_function(x), lower = 0, upper = 2)$value
> cat("Mean (Expected Value):", mean_value, "\n")
Mean (Expected Value): 0.6666667
> # Calculate the variance by integrating the square of the difference from the mean
> variance_value_calculated <- integrate(function(x) (x - mean_value)^2 * pdf_functi
on(x), lower = 0, upper = 2)$value
> cat("Variance:", variance_value_calculated, "\n")
Variance: 0.2222222
>
```

Here, the mean is 0.667 millimeters and variance is 0.222 square millimeters.

6.3 Probability Distributions

A probability distribution is a mathematical function or a description that specifies the probability of various possible outcomes in a random experiment or event. It assigns probabilities to each possible outcome, providing a complete picture of how likely each outcome is. There are two main types of probability distributions:

Discrete probability distribution:

a. In a discrete probability distribution, the random variable can take on a countable number of distinct values.

b. Each value has a probability associated with it, and the sum of all probabilities equals 1.

Continuous probability distribution:

a. In a continuous probability distribution, the random variable can take on an uncountable number of possible values within a given range.

b. Instead of probabilities for specific values, it defines probabilities for interval or range of values.

6.3.1 Binomial Distribution

The binomial distribution is widely used when you have a finite number of trials; each trial has two possible outcomes (success or failure) with a fixed probability of success (p), and the trials are independent. It provides a way to model and understand the distribution of the number of successes in such scenarios.

The probability mass function (PMF) of the binomial distribution, denoted as $P(X = x)$, provides the probability of observing exactly 'x' successful outcomes in 'n' trials. It is calculated using the following formula:

$$P(X = x) = \binom{n}{x} p^x q^{n-x}; x = 0, 1, \ldots, n.$$

The cumulative distribution function (CDF) of a binomial distribution gives you the probability that a binomial random variable takes on a value less than or equal to a specific value. The CDF for a binomial distribution is often denoted as $F(k; n, p)$ and is defined as:

$$F(k; n, p) = \sum_{x=0}^{k} \binom{n}{x} p^x q^{n-x}$$

where $\binom{n}{x}$ is the binomial coefficient, which represents the number of ways to choose 'x' successes out of 'n' trials and it is calculated as $\dfrac{n!}{x!(n-x)!}$ where '!' denotes factorial; p is the probability of success in a single trial; $q = (1 - p)$ is the probability of failure in a single trial, x is the number of successful outcomes.

The mean (expected value) of the binomial distribution is given by:

$$\mu = np.$$

The variance of the binomial distribution is given by:

$$\sigma^2 = npq = np(1-p).$$

In binomial distribution, mean is larger than variance.

> **Example 6.10**
>
> Let us consider a situation where we are studying the success rate of a new drug treatment. We want to model the number of patients who respond positively to the treatment (success) out of 10 patients in a clinical trial. Consider the probability of a patient responding positively to the treatment is 0.75. Now,
>
> a. Plot the probability mass function and cumulative distribution function.
> b. Compute mean and variance.
> c. Compute $P(X = 5)$, $P(4 < X < 7)$, $P(X < 8)$, and $P(X > 2)$.
>
> To solve these problems, we use the following R code:
>
> i: The following R-code can be used to plot PMF and CDF:

R-code:

```
# Number of trials (patients in the clinical trial)
n <- 10
# Probability of a patient responding positively to the treatment
p <- 0.75
# Number of successful patients (responding to the treatment)
x <- 0:10
# Calculate the probability of observing exactly k successful
patients
prob <- dbinom(x, size = n, prob = p)
par(mfrow=c(1,2))
# Plot the probability mass function (PMF)
plot(x, prob, type = 'h', lwd = 2, col='blue',
     xlab = 'Number of Successful Patients',
     ylab = 'Probability')
points(x, prob, pch=4, col='red')
# Plot cdf
plot(x,cumsum(prob),type='s', col='blue',
     ylim=c(0,1), ylab='CDF',
     xlab= 'Number of Successful Patients')
```

The plots of PMF and CDF are given in Figure 6.1.

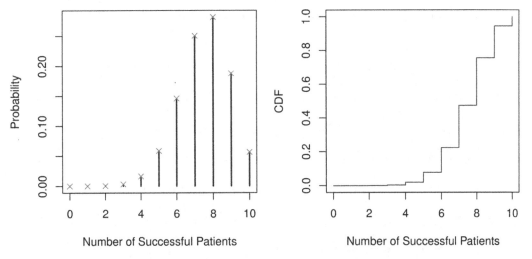

FIGURE 6.1
Plots of PMF (left panel) and CDF (right panel) of the binomial distribution.

ii: To find the mean and variance, we use the following commands in R and get the desired output:

R-code:

```
# Calculate and print the mean (expected number of successes)
mean_successes <- n * p
cat('Mean (Expected Number of Successes):', mean_successes, '\n')
# Calculate and print the variance
variance <- n * p * (1 - p)
cat('Variance:', variance, '\n')
```

Output:

```
Console   Terminal ×   Background Jobs ×                              ─ □
R  R 4.3.2 · ~/
> # Calculate and print the mean (expected number of successes)
> mean_successes <- n * p
> cat("Mean (Expected Number of Successes):", mean_successes, "\n")
Mean (Expected Number of Successes): 7.5
> # Calculate and print the variance
> variance <- n * p * (1 - p)
> cat("Variance:", variance, "\n")
Variance: 1.875
>
```

iii: To compute the following probabilities,

$$P(X=5) = \binom{10}{5} 0.75^5 (1-0.75)^{10-5},$$

$$P(4 < X < 7) = \binom{10}{5} 0.75^5 (1-0.75)^{10-5} + \binom{10}{6} 0.75^6 (1-0.75)^{10-6},$$

$$P(X < 8) = P(X=0) + P(X=1) + \ldots + P(X=7) = F(7),$$

$$P(X > 2) = 1 - \left[P(X=0) + P(X=1) + P(X=2) \right].$$

We use the following commands in R to compute the desired probabilities:

R-code:

```
# P(X = 5) is computed like this:
dbinom(5, 10, 0.75) #x, n, p
# P(4<X<7) is computed like this:
dbinom(5, 10, 0.75)+ dbinom(6, 10, 0.75)
# P(X < 8) Since P(X < 8) = FX(7)
pbinom(7, 10, 0.75)
# P(X >2) Since P(X > 2) = 1-P(X < 3)
1-pbinom(2, 10, 0.75)
#or
1-(dbinom(0, 10, 0.75)+ dbinom(1, 10, 0.75)+dbinom(2, 10, 0.75))
```

Output:

```
Console   Terminal ×   Background Jobs ×                              ─ □
R  R 4.3.2 · ~/
> # P(X = 5) is computed like this:
> dbinom(5, 10, 0.75) #x, n, p
[1] 0.0583992
> # P(4<X<7) is computed like this:
> dbinom(5, 10, 0.75)+ dbinom(6, 10, 0.75)
[1] 0.2043972
> # P(X < 8) Since P(X < 8) = FX(7)
> pbinom(7, 10, 0.75)
[1] 0.4744072
> # P(X >2) Since P(X > 2) = 1-P(X < 3)
> 1-pbinom(2, 10, 0.75)
[1] 0.9995842
> #or
> 1-(dbinom(0, 10, 0.75)+ dbinom(1, 10, 0.75)+dbinom(2, 10, 0.75))
[1] 0.9995842
>
```

6.3.2 Poisson Distribution

The Poisson distribution is a probability distribution that models the number of events occurring within a fixed interval of time or space. It is named after the French mathematician Siméon Denis Poisson. The Poisson distribution is often used when dealing with rare events or situations where events happen independently at a constant average rate.

The PMF and CDF of the Poisson distribution can be written as:

$$P(X = x) = \frac{e^{-\lambda}\lambda^x}{x!}, x = 0,1,\ldots,\infty \text{ and}$$

$$F(k;\lambda) = P(X \le x) = \sum_{x=0}^{k} \frac{e^{-\lambda}\lambda^x}{x!},$$

where $P(X = x)$ is the probability of observing x events, $F(k; \lambda)$ is the cumulative probability of observing a Poisson random variable less than or equal to k with an average rate of occurrence λ; e is the base of the natural logarithm (approximately 2.71828); λ is the average rate of event occurrence within the given interval.

The mean (expected value) of a Poisson distribution is equal to its parameter λ, and the variance is also λ. In other words, both the mean and variance are λ.

Example 6.11

Suppose a psychiatrist works in a hospital and wants to estimate the probabilities of different patient admissions per day to the psychiatric ward. S/he observed that the daily psychiatric patient admissions were an average of 5 admissions per day. Now, s/he want to

a. Draw the PMF and CDF of number of daily admissions.
b. Compute the $P(X = 3)$, $P(1 < X < 4)$, $P(X < 6)$, and $P(X > 3)$.

i: We can write the PMF of Poisson distribution as $P(X = x) = \frac{e^{-5}5^x}{x!}$. Now, to draw the PMF and CDF we use the following command in R:

R-code:

```
# Set the average daily admission rate (lambda)
lambda <- 5
# Define a range of daily admission counts (k)
x_values <- 0:15
# Create a Poisson PMF plot
library(ggplot2)
# Create a data frame with probabilities for various values of x
df <- data.frame(x = 0:10, probability = dpois(0:10, lambda))
# Create the PMF plot
pmf_Poisson=ggplot(df, aes(x = x, y = probability)) +
  geom_bar(stat = 'identity', fill = 'blue', color = 'black') +
  labs(title = 'Poisson PMF',
        x = 'Number of Daily Admissions (x)',
        y = 'Probability') +
  theme_bw()
```

```
# Plot CDF
# Calculate the CDF for each value of x
cdf_values <- ppois(x_values, lambda)
# Create a data frame to store the results
cdf_data <- data.frame(x = x_values, CDF = cdf_values)
cdf_Poisson=ggplot(cdf_data, aes(x = x, y = CDF), col='black') +
  geom_step(size = 1) +
  labs(title = 'Poisson CDF',
       x = 'Number of Daily Admissions (x)',
       y = 'CDF') +
  theme_bw()
# Combined plot
install.packages('patchwork')
library(patchwork)
# Arrange and display the plots using the + operator
(pmf_Poisson + cdf_Poisson)
```

The plots of PMF and CDF are presented in Figure 6.2.

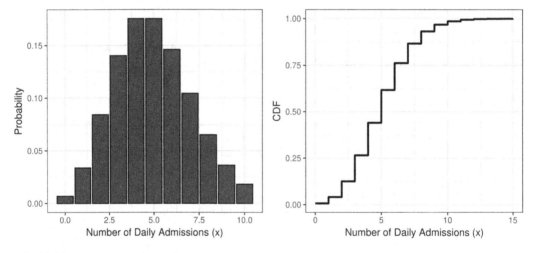

FIGURE 6.2
Plots of PMF (left panel) and CDF (right panel) of the Poisson distribution.

ii: For computing the required probabilities, we use the following R commands:

R-code:

```
# Calculate the probability of observing X=3
dpois(3, lambda)
# Calculate the probability of observing 1<X<4
x <- 2:3
sum(dpois(x, lambda))
#Calculate the probability of observing X<6
ppois(5, lambda)
#Calculate the probability of observing X>3
1-ppois(2, lambda)
```

Output:

```
Console   Terminal ×   Background Jobs ×
R  R 4.3.2 · ~/
> # Calculate the probability of observing X=3
> dpois(3, lambda)
[1] 0.1403739
> # Calculate the probability of observing 1<X<4
> x <- 2:3
> sum(dpois(x, lambda))
[1] 0.2245982
> #Calculate the probability of observing X<6
> ppois(5, lambda)
[1] 0.6159607
> #Calculate the probability of observing X>3
> 1-ppois(2, lambda)
[1] 0.875348
> |
```

6.3.3 Normal Distribution

The normal distribution, often referred to as the Gaussian distribution, is one of the most important probability distributions in statistics. Abraham De Moivre (1667–1754) initially published the formula for this distribution in 1733. It is a continuous probability distribution that is symmetric, bell-shaped, and characterized by two parameters: the mean (μ) and the standard deviation (σ). The PDF of the normal distribution is given by:

$$f(x;\mu,\sigma) = \frac{1}{\sigma\sqrt{2\pi}} e^{-\frac{1}{2}\left(\frac{x-\mu}{\sigma}\right)^2} ; -\infty < x < \infty$$

where x is the random variable, μ is the mean of the distribution, σ is the standard deviation, which measures the spread or dispersion of the distribution, π is the mathematical constant pi (approximately 3.14159), and e is the base of the natural logarithm (approximately 2.71828).

It is a symmetric distribution and its mean, median, and mode are equal.

The standard normal distribution is a special and important case where the mean (μ) is 0, and the standard deviation (σ) is 1. Mathematically, the PDF of the standard normal distribution can be written as:

$$f(z) = \frac{1}{\sqrt{2\pi}} e^{-\frac{z^2}{2}} ; -\infty < z < \infty.$$

In a normal distribution, about 68% of the data falls within one standard deviation of the mean, about 95% within two standard deviations, and about 99.7% within three standard deviations. We can visualize it in Figure 6.3.

The shape of the normal distribution depends on the value of mean and standard deviation. For the same mean with different standard deviations and different means with the same standard deviation are illustrated in Figure 6.4.

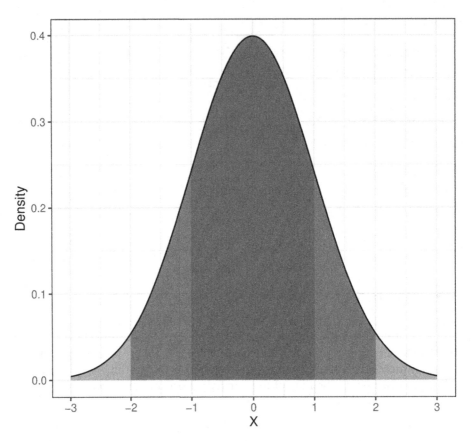

FIGURE 6.3
Normal probability curve illustrating the percent of data falls within one, two, and three standard deviation of the mean (Empirical Rule of Normal Distribution).

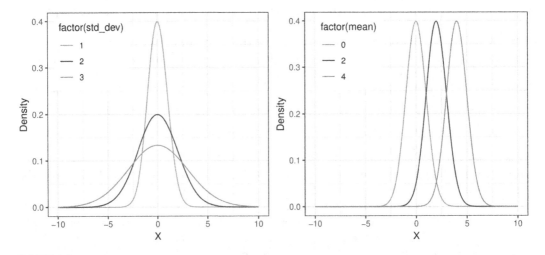

FIGURE 6.4
Normal probability curve with different standard deviation but same mean (left panel) and different mean but same standard deviation (right panel).

Suppose $X \sim N(\mu, \sigma^2)$, now the integrals of the form

$$P(a < X < b) = \int_a^b \frac{1}{\sigma\sqrt{2\pi}} e^{-\frac{1}{2}\left(\frac{x-\mu}{\sigma}\right)^2} dx$$

cannot be evaluated by hand; however, fortunately, R does all these calculations for us.

The cumulative distribution function $F_X(x) = P(X \le x)$ is evaluated using the 'pnorm' function. It has three arguments: x, mean=mu and sd=sigma, so that 'pnorm(x, mean=mu, sd=sigma)' is the probability $F_X(x) = P(X \le x)$ when $X \sim N(\mu, \sigma^2)$.

Example 6.12

Suppose the paracetamol tablets contain a quantity of active ingredients which has a normal distribution $N(500, 100)$, so that the nominal dose is 500 mg. Suppose X is the quantity found in a randomly selected tablet. Now, we want to compute the following probabilities using R,

     i. $P(X \le 490)$
     ii. $P(490 \le X \le 495)$
     iii. $P(X > 485)$.

Here, it is given that $\mu = 500$ and $\sigma^2 = 100$ ($\sigma = 10$). Now we use the following command to compute the required probabilities.

     i. R-code for calculating $P(X \le 490)$ is given below:

R-code:

```
pnorm(490, mean = 500, sd = 10)
```

Output:

```
Console   Terminal ×   Background Jobs ×
R  R 4.3.2 · ~/
> pnorm(490, mean = 500, sd = 10)
[1] 0.1586553
>
```

     ii. We can write as, $P(490 \le X \le 495) = P(X \le 495) - P(X \le 490)$. R-code for calculating this probability is given below:

R-code:

```
pnorm(495, mean = 500, sd = 10)-pnorm(490, mean = 500, sd = 10)
```

Output:

```
Console   Terminal ×   Background Jobs ×
R  R 4.3.2 · ~/
> pnorm(495, mean = 500, sd = 10)-pnorm(490, mean = 500, sd = 10)
[1] 0.1498823
>
```

iii. We have $P(X > 485) = 1 - P(X \leq 485)$. R-code for calculating this probability is given below:

R-code:

```
1-pnorm(485, mean = 500, sd = 10)
```

Output:

```
Console  Terminal ×  Background Jobs ×
R  R 4.3.2 · ~/
> 1-pnorm(485, mean = 500, sd = 10)
[1] 0.9331928
>
```

Moreover, we can illustrate this probability in the normal distribution plot. Hence, we define a function for the plot first and supply values of the required arguments. The code and plots are given below:

R-code:

```
normal_area <- function(mean = 0, sd = 1, lb, ub, acolor =
'lightgray', ylab = ", lwd=1) {
  x <- seq(mean-3*sd, mean+3*sd, length = 1000)
  ## mean: mean of the Normal variable, lb: lower bound of the area,
  # ub: upper bound of the area, # acolor: color of the area
  if (missing(lb)) {
    lb <- min(x)
  }
  if (missing(ub)) {
    ub <- max(x)
  }
  x2 <- seq(lb, ub, length = 1000)
  plot(x, dnorm(x, mean, sd), type = 'n', ylab = ", col=acolor)
  y <- dnorm(x2, mean, sd)
  polygon(c(lb, x2, ub), c(0, y, 0), ylab = ", col = acolor)
  lines(x, dnorm(x, mean, sd), ylab = ", type = 'l', lwd=2)
}
# plots
par(mfrow=c(1,3))
#(i)
normal_area(mean = 500, sd = 10, ub = 490, lwd = 1,
ylab='Probability', acolor = rgb(0, 0, 1, alpha = 0.5))
  text(485, 0.005, '15.87%', srt = 0)
#(ii)
normal_area(mean = 500, sd = 10, lb=490, ub = 495, lwd = 1,
ylab='Probability', acolor = rgb(0, 0, 1, alpha = 0.5))
  text(493, 0.008, '14.99%', srt = 90)
#(iii)
normal_area(mean = 500, sd = 10, lb = 485, lwd = 1,
ylab='Probability', acolor = rgb(0, 0, 1, alpha = 0.5))
  text(500, 0.01, '93.32%', srt = 0)
```

Plots of above-calculated different measures of probability using normal probability curve is shown in Figure 6.5.

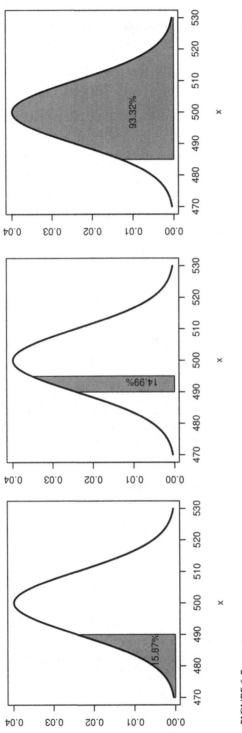

FIGURE 6.5
Plots of different measures of probability using normal probability curve.

6.3.4 Exponential Distribution

The exponential distribution is a probability distribution that describes the time between events in a Poisson process, where events occur continuously and independently at a constant average rate. It is often used to model the time it takes for a specific event to occur after a Poisson process has started. The PDF of the exponential distribution is given by:

$$f(x;\lambda) = \lambda e^{-\lambda x}; x > 0$$

where x is the random variable representing the time between events, λ (lambda) is the rate parameter, which is the average number of events that occur in one unit of time. It is also the reciprocal of the mean $\left(\lambda = \dfrac{1}{\mu}\right)$, where μ is the mean time between events.

Mean and variance of exponential distribution are $\mu = \dfrac{1}{\lambda}$ and $\sigma^2 = \dfrac{1}{\lambda^2}$ respectively.

The exponential distribution has applications in different fields, including modeling the time between arrivals at a service center, the time between the decay of radioactive particles, and the duration of phone calls in a call center, among others. It is commonly used in queuing theory, reliability analysis, and survival analysis.

Example 6.13

Suppose we consider a study on the reliability of a medical device used in a hospital setting. We want to model the time until a specific component of the medical device fails. In this scenario, we can use the exponential distribution to model the time to component failure. Let us assume that historical data suggest that the component has an average failure rate of 0.05 failures per hour i.e., $\lambda = 0.05$ per hour. Now we solve the followings:

   i. Draw the density function of this distribution.
   ii. What is the probability that the component will fail within the next 5 hours?
   iii. What is the expected time until the component fails?
   iv. What is the probability that the component will not fail within the first 15 hours?

In R, we use the following commands to get the desired solutions:

i. **R-code:**

```
library(stats)
library(ggplot2)
# Rate parameter
lambda <- 0.05  # Example rate parameter (average failures per hour)
# Create a data frame with x-values for the plot
x <- seq(0, 50, by = 0.1)
data_exp <- data.frame(x = x)
# Create the ggplot2 plot
ggplot(data_exp, aes(x = x)) +
   stat_function(fun = dexp, args = list(rate = lambda), color =
'blue', size = 1) +
     labs(x = 'Time',
          y = 'Probability Density') +
     theme_bw()
```

The plot of PDF of exponential distribution is illustrated in Figure 6.6.

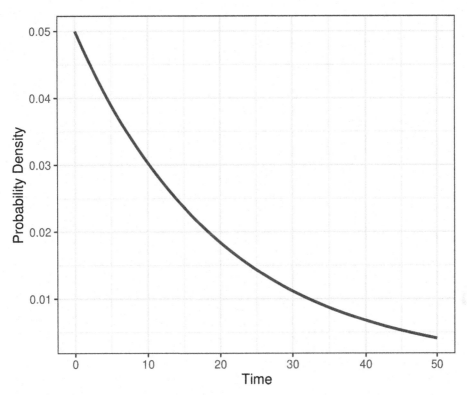

FIGURE 6.6
Plot of PDF of exponential distribution.

ii. **R-code**:

```
# Probability of failure within the next 5 hours
lambda=0.05
1-exp(-lambda*5)
#or
pexp(5, rate = lambda)   # P(X <= 5)
```

Output:

```
Console   Terminal    Background Jobs

R  R 4.3.2 · ~/
> # Probability of failure within the next 5 hours
> lambda=0.05
> 1-exp(-lambda*5)
[1] 0.2211992
> #or
> pexp(5, rate = lambda)   # P(X <= 5)
[1] 0.2211992
>
```

iii. **R-code**:

```
# Expected time until failure
1/lambda
```

Output:

```
Console   Terminal ×   Background Jobs ×                                        
 R  R 4.3.2 · ~/ 
> # Expected time until failure
> 1/lambda
[1] 20
> 
```

iv. **R-code**:

```
# Probability of no failure within the first 20 hours
exp(-lambda*15)
#or
1-pexp(15, rate = lambda)
```

Output:

```
Console   Terminal ×   Background Jobs ×                                        
 R  R 4.3.2 · ~/ 
> # Probability of no failure within the first 20 hours
> exp(-lambda*15)
[1] 0.4723666
> #or
> 1-pexp(15, rate = lambda)
[1] 0.4723666
> 
```

6.4 Sampling Distribution

A sampling distribution refers to the distribution of a statistic, such as the mean or standard deviation, calculated from multiple samples of the same size drawn from a population. Understanding sampling distributions is essential for hypothesis testing. The known properties of sampling distributions, especially when sample sizes are large, allow researchers to make inferences about population parameters based on sample statistics. In addition, knowledge of sampling distributions is applied in determining the required sample size for a study. Larger sample sizes tend to result in smaller standard errors, providing more precise estimates.

The central limit theorem is a fundamental concept related to sampling distributions. It states that, regardless of the shape of the population distribution, the sampling distribution of the sample mean will be approximately normally distributed for a sufficiently large sample size. This is a powerful property that enables the use of normal distribution-based statistical methods. Moreover, the standard error is a measure of the variability of a sample statistic (e.g., mean or proportion) across different samples. It quantifies how much the sample statistic is expected to vary from sample to sample. The standard deviation of the sampling distribution of mean is the standard error of the mean. The standard error is crucial in constructing confidence intervals and conducting hypothesis tests. The standard error is also used to calculate the margin of error in confidence intervals. A wider margin of error indicates greater uncertainty in estimating the population parameter based on the sample statistic.

6.4.1 Student's *t* Distribution

The *t*-distribution is also known as Student's *t*-distribution, is a probability distribution that arises in statistical inference and is used in various statistical tests, particularly for small sample sizes when population standard deviation is unknown. The probability density function of the *t*-distribution is written as follows:

$$f(t) = \frac{1}{\sqrt{n}B\left(\frac{1}{2},\frac{n}{2}\right)}\left(1+\frac{t^2}{n}\right)^{-\frac{(n+1)}{2}} \quad ; -\infty \leq t \leq \infty$$

where B is the Beta function, n is the number of observations.

The mean of the *t*-distribution is 0 when $n > 1$ and the variance is $\frac{n}{n-2}$ for $n > 2$.

Figure 6.7 represents the shape of the PDF of *t*-distributions for different degrees of freedom.

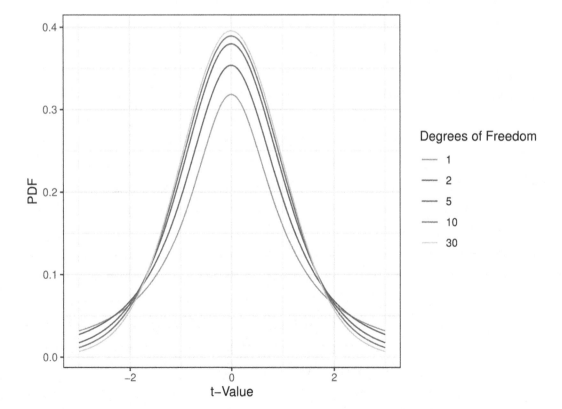

FIGURE 6.7
Plot of PDF of t distribution for different degrees of freedom.

6.4.2 Chi-square Distribution

The chi-square distribution, denoted by χ^2 (chi-squared), is often used in various statistical tests, particularly in hypothesis testing and estimating population variance.

Suppose Z_1, Z_2, ..., Z_k are independent standard normal random variables then the sum of their squares, i.e., $Q = \sum_{i=1}^{k} Z_i^2$ is distributed as chi-squared with k degrees of freedom and denoted as $Q \sim \chi_k^2$. The probability density function of the chi-square distribution is written as:

$$f\left(\chi^2;k\right) = \frac{1}{2^{k/2}\,\Gamma\left(\dfrac{k}{2}\right)}\left(\chi^2\right)^{\frac{k}{2}-1} e^{-\frac{\chi^2}{2}} ; \chi^2 > 0$$

where χ^2 is a non-negative real number, k is the degrees of freedom, and $\Gamma\left(\dfrac{k}{2}\right)$ is the gamma function evaluated at $\dfrac{k}{2}$.

The mean and variance of the chi-square distribution is k and $2k$ respectively. For different degrees of freedom, the shapes of the chi-square distribution are presented in Figure 6.8.

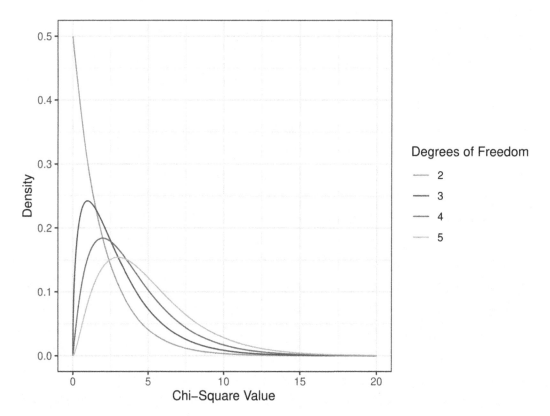

FIGURE 6.8
Plot of PDF of chi-square distribution for different degrees of freedom.

6.4.3 *F*-distribution

The *F*-distribution, also known as the Fisher-Snedecor distribution, is a probability distribution that arises in statistics and is commonly used in various statistical tests, particularly in the analysis of variance (ANOVA), regression analysis, and experimental design. It is used to compare variances or test hypotheses related to variances.

The *F*-distribution with n_1 and n_2 degrees of freedom is the distribution of

$$F = \frac{S_1 / n_1}{S_2 / n_2}$$

where S_1 and S_2 are independent random variables with chi-square distributions with respective degrees of freedom n_1 and n_2.

The probability density function for *F* can be demonstrated as follows:

$$f\left(F; n_1, n_2\right) = \frac{1}{B\left(\dfrac{n_1}{2}, \dfrac{n_2}{2}\right)} \left(\frac{n_1}{n_2}\right)^{\frac{n_1}{2}} F^{\frac{n_1}{2} - 1} \left(1 + \frac{n_1}{n_2} F\right)^{-\frac{(n_1 + n_2)}{2}} ; 0 \le F \le \infty.$$

Here, *B* is the beta function.

For real-life applications of the above-discussed three sampling distributions, the readers are recommended to read Chapter 7: Statistical Inferences.

6.5 Conclusion and Exercises

This section presents the summary of the chapter and includes problem-solving exercises.

6.5.1 Concluding Remarks

In order to properly understand and forecast uncertain events, probability and probability distributions are essential. We may measure the possibility of various outcomes via the concept of probability, which serves as a basis for decision-making across a range of domains. In this chapter, we discuss different concepts and rules of probability with examples.

The authors also discuss some widely used probability distributions including Binomial, Poisson, Normal, Exponential distributions with real-life examples. Moreover, we also present the concept of chi-square, Student's *t*, and *F* distributions that are widely used in hypothesis testing.

By studying probability and distributions, we gain valuable tools for statistical inference, risk assessment, and making informed choices in the face of uncertainty. As we move forward, a solid grasp of these concepts will continue to be essential in different fields for everyday decision-making.

6.5.2 Practice Questions

Exercise 6.1   Researchers aim to determine the prevalence of nutritional status among children under five who reside in a specific location within a specific country. Suppose they conducted a sample survey and got the following information.

Family Wealth Status	Number of Children			
	Stunted	Wasted	Underweight	Total
Poorest	230	185	220	635
Poorer	158	175	185	518
Middle	120	146	143	409
Richer	82	97	77	256
Richest	50	47	35	132
Total	640	650	660	1950

Now, find the followings

i.   A child is selected randomly, what is the probability that the selected children is (a) stunted? (b) wasted? (c) underweight?

ii. A child is selected randomly, what is the probability that the selected children is from (a) poorest family? (b) poorer family? (c) middle family? (d) richer family? (e) richest family?

iii. A child is selected randomly, what is the probability that the selected children is stunted and from poorer family? Also, compute all other conditional probabilities.

Exercise 6.2   Within a specific group of people, 15% of individuals have a high risk of heart attack. From this population, 3 are chosen at random. Find the probability that exactly 1 of the 3 selected people have high risk of heart attack? **Hint**: independent events.

Exercise 6.3   Suppose that in a certain population, one in every 500 individuals has a rare disease. And let's say there is a decent, if not perfect, test for this illness. The test is positive 99% of the time if the patient has the illness. However, the test also results in occasional false positives, meaning that 1% of individuals who are not infected also test positive. Someone recently had a positive test. What is the probability that he has this illness? **Hint**: use Bayes' theorem.

Exercise 6.4   Suppose there are three boxes with apples: A, B, and C. Someone wants to buy an apple box. There is a 0.5, 0.3, and 0.2 chance of selecting boxes A, B, and C, respectively. Each box contains a few rotting apples. It is assumed that there are 5%, 10%, and 15% of rotten apples in Boxes A, B, and C, respectively. Determine the probability that an apple will be selected from boxes A, B, and C given it is rotten? **Hint**: use Bayes' theorem.

Exercise 6.5   A salesman is employed by a computer shop to sell PC's. The salary (USD) he gets in a day is calculated by the formula $g(x) = 50 + 10x$, where

x is the number of PC's he sells in a day. The following table shows the number of PCs he sold in a day with their corresponding probabilities:

Number of PC's (x)	0	1	2	3	4
Probability $P(x)$	0.05	0.2	0.4	0.2	0.15

Find the expected daily salary of the salesman. Also, find the standard deviation. **Hint:** $E(x) = \Sigma\, g(x) \times P(x)$.

Exercise 6.6 The gap width is an important property of a magnetic recording head. In coded units, the probability density function of the gap width of a magnetic recording head is

$$f(x) = kx; 0 \le x \le 2.$$

a. Find the value of k. **Hint:** Total area under the graph of $f(x)$ is 1.
b. What is the probability that the gap width is at least 1 unit.
c. What is the probability that the gap width lies between 0.5 units and 1.5 units?

Exercise 6.7 It is observed that 25% of the time, phone lines to airline reservation systems are busy. It is presumed that the lines are occupied independently of one another during subsequent calls. Let's say ten people contact the airline. Now, compute the followings

a. Probability of exactly three calls, the lines are occupied.
b. Probability of at least one call, the lines are not occupied.
c. What is the expected number of calls in which the lines are all occupied?

Hint: Binomial distribution.

Exercise 6.8 Suppose you are investigating the relationship between smoking and lung cancer using a case-control study. If the probability of lung cancer cases among smokers is 0.47. Let us select ten cases of cancer, find the followings:

a. Probability of less than 3 smokers among the selected cases.
b. Probability of more than 7 smokers among the selected cases.
c. Probability between 2 and 5 smokers among the selected cases.
d. What is the expected the number of smokers?

Hint: Binomial distribution

Exercise 6.9 A website's content modifications are distributed according to a Poisson distribution, with an average of 0.15 changes every day. Compute the following probabilities:

a. 2 or more changes in a day.
b. No content changes in five days.
c. 3 or fewer changes in five days.

Hint: For more than one day use $n\lambda$ instead of λ.

Exercise 6.10 Suppose the file transfer speed on a weekday evening from a university server to a student's home computer follows a normal distribution with a mean of 60 kilobits per second and a standard deviation of 4 kilobits per second. Find the probabilities of:

    a. The file will transfer at a speed of 65 kilobits per second or more.

    b. The file transfer speed will be between 50 and 80 kilobits per second or more.

    c. The file will transfer at a speed of less than 58 kilobits per second.

Exercise 6.11 The survival time, in weeks, of an animal exposed to a specific amount of gamma radiation was found to have an exponential distribution with a mean of 20 weeks in a biomedical research project. Compute the following probabilities:

    a. An animal survives more than 30 weeks.

    b. An animal survives between 25 and 40 weeks.

    c. An animal survives less than 25 weeks.

Bibliography

Ahsanullah, M., Kibria, B. G., & Shakil, M. (2014). *Normal and student's t distributions and their applications* (Vol. 4). Atlantis Press.

Badge, J., & Badge, R. (2023). *Introduction to probability and statistics: Probability and statistics*. Shashwat Publication.

Balakrishnan, N., & Nevzorov, V. B. (2004). *A primer on statistical distributions*. John Wiley & Sons.

Balakrishnan, N., Voinov, V., & Nikulin, M. S. (2013). *Chi-squared goodness of fit tests with applications*. Academic Press.

Bandyopadhyay, S., & Bhattacharya, R. (2014). *Discrete and continuous simulation: Theory and practice*. CRC Press.

Daniel, W. W., & Cross, C. L. (2018). *Biostatistics: a foundation for analysis in the health sciences*. Wiley.

Forbes, C., Evans, M., Hastings, N., & Peacock, B. (2011). *Statistical distributions*. John Wiley & Sons.

Grimmett, G., & Welsh, D. J. (2014). *Probability: An introduction*. Oxford University Press.

Kaptein, M., & Van Den Heuvel, E. (2022). *Statistics for data scientists: An introduction to probability, statistics, and data analysis*. Springer.

Krishnamoorthy, K. (2006). *Handbook of statistical distributions with applications*. Chapman and Hall/CRC.

Matloff, N. (2019). *Probability and statistics for data science: Math+ R+ data*. CRC Press.

Montgomery, D. C., & Runger, G. C. (2020). *Applied statistics and probability for engineers*. John Wiley & Sons.

Pishro-Nik, H. (2014). *Introduction to probability, statistics, and random processes* (p. 732). Kappa Research, LLC.

Proschan, M. A., & Shaw, P. A. (2018). *Essentials of probability theory for statisticians*. CRC Press.

Rahman, A. (2008). *Bayesian predictive inference for some linear models under student-t errors* (1st ed.). VDM Verlag.

Rahman, A., & Upadhyay, S. K. (2015). A Bayesian reweighting technique for small area estimation. In U. Singh, A. Loganathan, S. K. Upadhyay, & D. K. Dey (Eds.), *Current trends in Bayesian methodology with applications* (1st ed., pp. 503–519). CRC Press.

Ross, S. (2009). *Probability and statistics for engineers and scientists*. Elsevier.

Ross, S. M. (2019). *First course in probability, A, global edition*. Pearson.

Ugarte, M. D., Militino, A. F., & Arnholt, A. T. (2008). *Probability and statistics with R*. CRC Press.

7

Statistical Inferences

7.1 Statistical Estimation

The estimation process is a powerful tool that enables us to draw meaningful conclusions about the population of interest based on sample observations or data. Statistical estimation involves using sample data from biostatistics to make educated guesses about population parameters. These parameters could include means, proportions, variances, and more. It allows us to make data-centric informed decisions and predictions in situations where the complete information set is not readily available.

7.1.1 Key Concepts

Population: A population refers to the entire group or collection of individuals, objects, events, or units of interest in a particular study or analysis.

Population parameters: Population parameters are specific numerical values that describe various characteristics of a population.

Sample: A sample refers to a subset of individuals, objects, events, or units selected from a larger group or population.

Sampling: Sampling refers to the process of selecting a subset of individuals, items, events, or units from a larger population for the purpose of conducting a study or analysis.

Estimate: An estimate refers to the calculated value that approximates an unknown population parameter based on sample data. It is also called a statistic.

Estimator: An estimator is a rule, or a formula used to calculate a statistic or prediction of a specific population parameter based on sample data. Estimators are functions that take the data from a sample and estimate an unknown population parameter.

Standard error (SE): Standard error (SE) is a measure of the variability or uncertainty associated with a sample statistic, such as the sample mean or sample proportion. It quantifies how much the statistic likely deviates from the actual parameter value due to random sampling variability. A smaller standard error indicates that the sample statistic is more precise and likely to be closer to the true value of the parameter.

Confidence interval (CI): A confidence interval is a range of values constructed around a statistic, such as a sample mean or a sample proportion, to estimate the likely range within which the respective true population parameter lies.

DOI: 10.1201/9781003426189-7

It quantifies the uncertainty associated with the estimate by specifying a confidence level, often expressed as a percentage, that the true parameter falls within the interval (Rahman et al., 2013; Rahman, 2017). A wider confidence level reflecting greater uncertainty, i.e., the estimate is not entirely informative enough to make a valid decision.

Example 7.1

Suppose we consider a health study that investigates the average cholesterol levels among adults aged 45 to 65 years in a particular city.

In this example, the entire group of adults aged 45 to 65 years living in the specific city is the 'population'. It includes every eligible individual within the defined age range in that city.

The average cholesterol levels of all adults aged 45 to 65 years in the city is a 'population parameter'. It is generally denoted by the Greek letter μ(mu).

Due to practical constraints, measuring the cholesterol levels of every single adult in the city might not be feasible. Instead, a subset of this population, known as the '**sample**', is selected. For instance, a sample of 100 individuals was randomly selected from this age group for the study.

The sample mean (average) cholesterol levels of the 100 selected adults is an 'estimator' for the population parameter of average cholesterol levels. Mathematically, this 'estimator' is computed by summing up all the cholesterol levels and dividing by the sample size (100).

If the calculated sample mean cholesterol level is 180 mg/dL, then 180 mg/dL is the 'estimate' for the average cholesterol level of all adults aged 45 to 65 years in the city.

The 'standard error' of the sample mean (SE of the mean) is determined as dividing the standard deviation by the square root of the sample size.

7.1.2 Types of Statistical Estimation

Statistical estimation is mainly classified into two categories: point estimation and interval estimation.

Point estimation: It provides a single numerical value as an estimate for the unknown population parameter. This value is often determined from the sample data and serves as the best guess for the parameter. Common point estimators include sample mean, sample proportion, and sample variance. The point estimate can be illustrated in Figure 7.1.

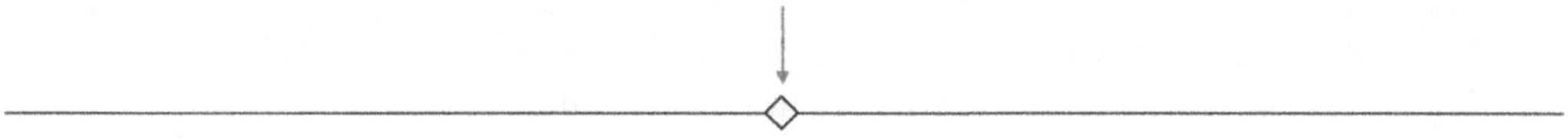

FIGURE 7.1
Diagram for the concept of point estimation.

Example 7.2

Let us consider a dataset named 'Body Fat Prediction Dataset' which is available on the Kaggle database and consists of the following variables:

1. Density determined from underwater weighing
2. Percent body fat from Siri's (1956) equation
3. Age (years)

4. Weight (lbs)
5. Height (inches)
6. Neck circumference (cm)
7. Chest circumference (cm)
8. Abdomen 2 circumference (cm)
9. Hip circumference (cm)
10. Thigh circumference (cm)
11. Knee circumference (cm)
12. Ankle circumference (cm)
13. Biceps (extended) circumference (cm)
14. Forearm circumference (cm)
15. Wrist circumference (cm)

First, download the dataset from your Kaggle account into your computer (assume that it is stored in the folder 'Book>Chapter 7' on the 'E' drive) and read the dataset from the specified location using the following R-code:

R-code:

```
data_bfat <- read.csv('E:/Book/Chapter 7/bodyfat.csv', header = TRUE)
attach(data_bfat)
head(data_bfat)
```

Output:

```
Console  Terminal ×  Background Jobs ×
R  R4.3.2 · ~/
> data_bfat <- read.csv("E:/Book/Chapter 7/bodyfat.csv", header = TRUE)
> attach(data_bfat)
> head(data_bfat)
  Density BodyFat Age Weight Height Neck Chest Abdomen   Hip Thigh Knee Ankle Biceps Forearm Wrist
1  1.0708    12.3  23 154.25  67.75 36.2  93.1    85.2  94.5  59.0 37.3  21.9   32.0    27.4  17.1
2  1.0853     6.1  22 173.25  72.25 38.5  93.6    83.0  98.7  58.7 37.3  23.4   30.5    28.9  18.2
3  1.0414    25.3  22 154.00  66.25 34.0  95.8    87.9  99.2  59.6 38.9  24.0   28.8    25.2  16.6
4  1.0751    10.4  26 184.75  72.25 37.4 101.8    86.4 101.2  60.1 37.3  22.8   32.4    29.4  18.2
5  1.0340    28.7  24 184.25  71.25 34.4  97.3   100.0 101.9  63.2 42.2  24.0   32.2    27.7  17.7
6  1.0502    20.9  24 210.25  74.75 39.0 104.5    94.4 107.8  66.0 42.0  25.6   35.7    30.6  18.8
>
```

Now, we want to compute the point estimate of average body fat and its standard error. You may use the following code in R:

R-code:

```
#Mean of body fat
mean(data_bfat$BodyFat)
#standard error of mean
sd(data_bfat$BodyFat)/sqrt(length(data_bfat$BodyFat))
```

Output:

```
Console  Terminal ×  Background Jobs ×
R  R4.3.2 · ~/
> #Mean of body fat
> mean(data_bfat$BodyFat)
[1] 19.15079
> #standard error of mean
> sd(data_bfat$BodyFat)/sqrt(length(data_bfat$BodyFat))
[1] 0.5271811
>
```

Again, suppose we want to find the proportion of how many people have a hip circumference larger than 100 cm and its standard error.

A proportion represents the fraction or percentage of occurrences that fall into a particular category. The following formula is used to calculate a proportion:

$$\text{Proportion}\,(p) = \frac{(\text{Count of specific outcome})}{(\text{Total count})}.$$

The standard error of proportion is obtained using the following formula:

$$\text{Standard Error of Proportion}\,(\text{SE proportion}) = \sqrt{\frac{p(1-p)}{n}}$$

where p is the sample proportion and n is the sample size.

In R, we use the following commands:

R-code:

```
#proportion
sum(data_bfat$Hip>100)/length(data_bfat$Hip)
#SE
p<- sum(data_bfat$Hip>100)/length(data_bfat$Hip)
SE_p <- sqrt(p*(1-p)/length(data_bfat$Hip))
SE_p
```

Output:

```
Console   Terminal ×   Background Jobs ×

R   R 4.3.2 · ~/
> #proportion
> sum(data_bfat$Hip>100)/length(data_bfat$Hip)
[1] 0.4365079
> #SE
> p<- sum(data_bfat$Hip>100)/length(data_bfat$Hip)
> SE_p <- sqrt(p*(1-p)/length(data_bfat$Hip))
> SE_p
[1] 0.03124206
>
```

Here, to find how many hip circumferences are larger than 100 cm, we use `sum(data_bfat$Hip>100)` command. Hence, about 43.65% of people included in the dataset had hip circumferences larger than 100 cm.

Interval estimation: Interval estimation provides a range of values (confidence interval) within which the true population parameter is likely to fall. The interval estimate considers the uncertainty associated with the estimate and is accompanied by a specified confidence level. Confidence intervals provide a more complete picture of the possible range of values for the parameter. The interval estimate, or a 'confidence interval', consists of upper and lower confidence limits. We can present the interval estimation in the following way. In many statistical modeling studies, it is also called uncertainty intervals (e.g., see GBD 2019 Tobacco Collaborators, 2021). We can present the interval estimation in the following way (Figure 7.2).

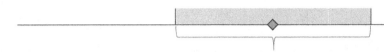

FIGURE 7.2
Diagram for the concept of interval estimation.

Interval estimators are statistical techniques that generate a range of estimates for an unknown parameter. The sampling distribution of a point estimator typically serves as the basis for interval estimators. For instance, the interval estimator uses the point estimate and its bound to the estimation error. It can be written as:

$$\text{point estimate} \pm \text{bound on the error of estimation.}$$

Example 7.3

Recall the data used in **Example 7.2**, we now want to estimate the 90%, 95%, and 99% interval estimation for the average Weight (lbs).

In particular, an interval estimate for μ may be written as follows when sampling from a normal distribution with known variance:

$$\bar{x} \pm Z_{(1-\alpha/2)} \sigma_{\bar{x}}$$

where $Z_{(1-\alpha/2)}$ is the value of Z lies $(1-\alpha/2)$ and $\alpha/2$ to the left and right the area under the normal curve respectively.

The widely used values are 90%, 95%, and 99%, which are associated with the value of Z of 1.645, 1.96, and 2.58 respectively.

In R, we use the following codes to get the desired outputs.

R-code:

```
#define variable
weight<-data_bfat$Weight
#mean and SE
mean(weight)
SE_mean<-sd(weight)/sqrt(length(weight))
# 90% CI
lower_90=mean(weight)-1.645*SE_mean
lower_90
upper_90=mean(weight)+1.645*SE_mean
upper_90
# 95% CI
lower_95=mean(weight)-1.96*SE_mean
lower_95
upper_95=mean(weight)+1.96*SE_mean
upper_95
# 99% CI
lower_99=mean(weight)-2.58*SE_mean
lower_99
upper_99=mean(weight)+2.58*SE_mean
upper_99
```

Output:

```
Console  Terminal ×  Background Jobs ×

R  R 4.3.2 · ~/
> #define variable
> weight<-data_bfat$weight
> #mean and SE
> mean(weight)
[1] 178.9244
> SE_mean<-sd(weight)/sqrt(length(weight))
> # 90% CI
> lower_90=mean(weight)-1.645*SE_mean
> lower_90
[1] 175.8789
> upper_90=mean(weight)+1.645*SE_mean
> upper_90
[1] 181.9699
> # 95% CI
> lower_95=mean(weight)-1.96*SE_mean
> lower_95
[1] 175.2958
> upper_95=mean(weight)+1.96*SE_mean
> upper_95
[1] 182.553
> # 99% CI
> lower_99=mean(weight)-2.58*SE_mean
> lower_99
[1] 174.1479
> upper_99=mean(weight)+2.58*SE_mean
> upper_99
[1] 183.7009
> |
```

Here, we interpret that we are 90% confident that the population mean is between 175.8789 and 181.9699 lbs.

Example 7.4

Recall the data used in **Example 7.2**. We now want to estimate the 90%, 95%, and 99% interval estimations for the proportion of chest circumferences larger than 100 cm.

When both np and $n(1-p)$ are greater than 5, the sampling distribution of the estimated proportion ($\hat{p}$) is approximately close to the normal distribution. Hence, the $100(1-\alpha)$ percent confidence interval for proportion can be expressed as

$$\hat{p} \pm Z_{(1-\alpha/2)}\sigma_{\hat{p}}$$

where

$$\sigma_{\hat{p}} = \sqrt{\frac{\hat{p}(1-\hat{p})}{n}}.$$

Therefore, in R, we use the following commands to calculate the 90%, 95%, and 99% interval estimation of the proportion that the Chest circumferences are larger than 100 cm. Output is also presented along with R commands.

R-code:

```
# define variable
chest<-data_bfat$Chest
```

```
n<-length(chest)
#estimate of the proportion
p_100<- sum(chest>100)/n
p_100
#SE
SE_p_100 <- sqrt(p_100*(1-p_100)/n)
SE_p_100
# 90% CI
lower_p90=p_100-1.645*SE_p_100
lower_p90
upper_p90=p_100+1.645*SE_p_100
upper_p90
# 95% CI
lower_p95=p_100-1.96*SE_p_100
lower_p95
upper_p95=p_100+1.96*SE_p_100
upper_p95
# 99% CI
lower_p99=p_100-2.58*SE_p_100
lower_p99
upper_p99=p_100+2.58*SE_p_100
upper_p99
```

Output:

```
Console   Terminal ×   Background Jobs ×

R 4.3.2 · ~/
> # define variable
> chest<-data_bfat$Chest
> n<-length(chest)
> #estimate of the proportion
> p_100<- sum(chest>100)/n
> p_100
[1] 0.484127
> #SE
> SE_p_100 <- sqrt(p_100*(1-p_100)/n)
> SE_p_100
[1] 0.03148116
> # 90% CI
> lower_p90=p_100-1.645*SE_p_100
> lower_p90
[1] 0.4323405
> upper_p90=p_100+1.645*SE_p_100
> upper_p90
[1] 0.5359135
> # 95% CI
> lower_p95=p_100-1.96*SE_p_100
> lower_p95
[1] 0.4224239
> upper_p95=p_100+1.96*SE_p_100
> upper_p95
[1] 0.5458301
> # 99% CI
> lower_p99=p_100-2.58*SE_p_100
> lower_p99
[1] 0.4029056
> upper_p99=p_100+2.58*SE_p_100
> upper_p99
[1] 0.5653484
> |
```

Thus, it can be said that we are 95% confident that the population proportion is between 0.422 and 0.546. The results mean that we are 95% confident that the true proportion of individuals with Chest circumferences larger than 100 cm in the population lies between 0.422 cm and 0.546 cm.

7.2 Hypothesis Testing

Hypothesis testing is a technique used to make decisions about population parameters based on sample data. It involves comparing a null hypothesis (a default assumption) with an alternative hypothesis (a claim we are testing). The goal is to determine whether there is enough evidence to reject the null hypothesis or not. The statistical conclusion should not be taken as final; instead, it should be considered along with all other relevant evidence.

7.2.1 Key Concepts of Hypothesis Testing

The following concepts are important to perform statistical hypothesis testing.

Hypothesis: A hypothesis is a statement or assumption about a population parameter that you want to test using data.

Null hypothesis (H_0): This is the default hypothesis that there is no effect, no difference, or no relationship in the population. It represents the status quo or the absence of an effect.

Alternative hypothesis (H_a or H_1): This is the hypothesis that contradicts the null hypothesis. It declares that there is an effect, a difference, or a relationship in the population.

Test statistic: A test statistic is calculated from sample data during hypothesis testing. It serves as a basis for making decisions about whether to reject or fail to reject the null hypothesis. The general form of the test statistic is defined below.

$$\text{test statistic} = \frac{\text{relevant statistic} - \text{value of the hypothesized parameter}}{\text{standard error of the relevant statistic}}.$$

Level of significance: The level of significance, denoted by α, is a critical parameter in hypothesis testing that determines the threshold for deciding about whether to reject or fail to reject the null hypothesis. Typically, researchers choose a significance level before conducting the hypothesis test. Common choices for the significance level include 0.05 (5%), 0.01 (1%), and 0.10 (10%).

Critical value: A critical value is a specific threshold or cut-off point used to determine whether to reject the null hypothesis. It is derived from the chosen level of significance (α) and the characteristics of the statistical distribution relevant to the test being performed. Figure 7.3 shows the boundary between the acceptance and rejection areas as well as the value of the standard statistic at which we reject the null hypothesis.

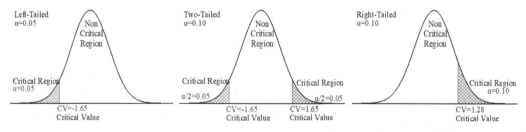

FIGURE 7.3
Boundary between the acceptance and rejection areas.

Types of error: Two kinds of errors can happen in hypothesis testing: Type I error and Type II error. These errors are associated with the decisions made when testing a hypothesis based on sample data.

Type I error (false positive): When we reject the null hypothesis when it is true, we are making a Type I error. In other words, we may conclude that there is an effect or difference when no such effect or difference exists in the population. The level of significance is the probability of committing a Type I error. For example, in a medical context, a Type I error would correspond to diagnosing a healthy patient as having a disease.

Type II error (false negative): A Type II error happens when we fail to reject the null hypothesis when it is actually false. In this case, you conclude that there is no effect or difference when there is a real effect or difference in the population. For instance, in a medical context, a Type II error would correspond to failing to diagnose a patient with a disease when they actually have the disease.

Both types of error can be summarized and presented below:

Decision	Null Hypothesis	
	True	False
Reject	Type I error (false positive)	Correct decision
Fail to reject	Correct decision	Type II error (false negative)

Power of the test: The probability of Type II error is denoted by the symbol β and $(1 - \beta)$ is termed the power of the test, which represents the probability of accurately rejecting the null hypothesis when it is false.

One-sided test: A one-sided test, also known as a one-tailed test, is a type of hypothesis test where the alternative hypothesis identifies a direction for the effect or difference being tested. In other words, it focuses on whether a parameter is significantly larger or smaller than a certain value, rather than simply detecting a difference without specifying the direction. One-sided tests are used when the research question or hypothesis is specifically concerned with an effect in a particular direction. The sampling distribution of the Z-statistic, left- and right-tailed test, with 0.05 level of significance are presented in Figure 7.4.

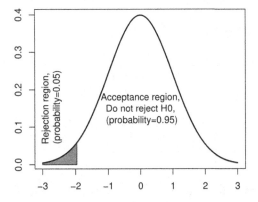 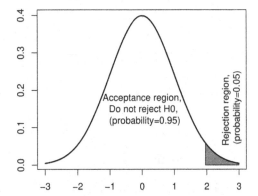

FIGURE 7.4
Boundary between rejection and acceptance region for left- and right-tailed test with 0.05 level of significance.

Two-sided test: A two-sided test is also known as a two-tailed test, is a kind of hypothesis test when the alternative hypothesis does not define a specific direction for the effect or difference under the test. It is used to determine whether a parameter is significantly different from a certain value, regardless of whether it is larger or smaller. For Z-statistics, the critical value and regions of non-rejection and rejection of two-tailed test with 0.05 significance level are illustrated in Figure 7.5.

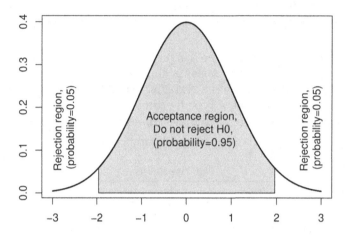

FIGURE 7.5
Boundary between rejection and acceptance region for two-tailed test with 0.05 level of significance.

P-value: The p-value or 'probability value' is a measure that quantifies the strength of evidence against the null hypothesis. It expresses the probability that, under the assumption that the null hypothesis is true, a test statistic will be observed that is more extreme than the value computed from the sample data.

- A small p-value ($\leq \alpha$) implies that the observed data is unlikely to have occurred under the null hypothesis, providing evidence against the null hypothesis. This may lead to rejecting the null hypothesis.

- A large p-value ($> \alpha$) indicates that the observed data is consistent with what would be expected under the null hypothesis. This does not provide strong evidence to reject the null hypothesis.

Criteria for hypothesis construction: We consider the following four points to formulate a hypothesis:

a. It should be empirically testable, whether it is right or wrong;

b. It should be specific and precise;

c. The statements in the hypothesis should not be contradictory; and

d. It should describe one issue only.

7.2.2 Steps of Hypothesis Testing

A typical hypothesis testing includes the following steps:

1. **Formulating hypotheses**: Define the null and alternative hypotheses based on the research question.

2. **Selecting a significance level**: Choose a significance level.

3. **Calculating test statistic**: Calculate the appropriate test statistic.

4. **Determining critical value/*p*-value**: Calculate the critical value of the test statistic or *p*-value corresponding to the calculated value of the test statistic.

5. **Comparing critical value/*p*-value with significance level**: If the calculated value of the test statistic is larger than the critical value or the *p*-value is less than the chosen significance level (α) suggesting we have evidence to reject the null hypothesis.

6. **Drawing conclusion**: Make a decision based on the critical value or *p*-value. If the null hypothesis is rejected, we may consider the alternative hypothesis or may be decided further investigation.

7.2.3 Mean Test

For testing the mean of a population with a dataset, we consider the following specific cases:

7.2.3.1 Comparison of a Sample Mean with an Assigned Population Mean

Testing of a hypothesis about a population mean, we need to keep in mind two things, i.e., sample size (small or large) and population variance (known or unknown).

Case 1: Population variance is known or the sample size is large

Suppose sampling is from a population which is normally distributed and the population variance is known or the sample size is large (≥30), for testing $H_0 : \mu = \mu_0$ the following test statistic is used

$$Z = \frac{\bar{x} - \mu_0}{\sigma / \sqrt{n}} \sim N(0,1)$$

where

$\bar{x}$ is the sample mean

σ is the known standard deviation

n is the sample size, and under the null hypothesis

Z follows a standard normal distribution.

Example 7.5

The mean lifetime of 50 hospital electric equipment produced by a company is found to be 14 years with a standard deviation of 3 years. Test the hypothesis that the mean lifetime of the equipment produced by the company is 15 years a 5% significance level.
To justify the hypothesis, we use the following steps:

Step 1: To verify the claim, we consider the following hypothesis,

$$H_0 : \mu = 15 \text{ against } H_1 : \mu \neq 15.$$

Step 2: Significant level is $\alpha = 0.05$.

Step 3: We assumed that the sample comes from a population whose lifetimes are normally distributed. Hence, the test statistics is $Z = \dfrac{\bar{x} - \mu_0}{\sigma / \sqrt{n}}$.

Now, in R, we use the following command for computing the test statistic:

R-code:

```
# Population parameters (known)
population_mean <- 15
population_sd <- 3
# Define the significance level
alpha <- 0.05
# Calculate sample statistics
sample_mean <- 14
sample_size <- 50
# Calculate z-score
z_score <- (sample_mean - population_mean) / (population_sd /
sqrt(sample_size))
z_score
```

Output:

```
Console   Terminal ×   Background Jobs ×                                                       ─ □
R  R 4.3.2 · ~/
> # Population parameters (known)
> population_mean <- 15
> population_sd <- 3
> # Define the significance level
> alpha <- 0.05
> # Calculate sample statistics
> sample_mean <- 14
> sample_size <- 50
> # Calculate z-score
> z_score <- (sample_mean - population_mean) / (population_sd / sqrt(sample_size))
> z_score
[1] -2.357023
>
```

Step 4:   Hence, the test is two-sided, to calculate the p-value, we have to divide α by 2 and the rejection region of the test statistic $\alpha/2 = 0.025$ is related with small values and $\alpha/2 = 0.025$ is associated with large values. In R, we use the following command:

R-code:

```
# Calculate the critical value or p-value for a two-tailed test
Critical_value <- qnorm(1-alpha/2)
Critical_value
#P-value
p_value <- pnorm(z_score)
p_value
```

Output:

```
Console   Terminal ×   Background Jobs ×                                                       ─ □
R  R 4.3.2 · ~/
> # Calculate the critical value or p-value for a two-tailed test
> Critical_value <- qnorm(1-alpha/2)
> Critical_value
[1] 1.959964
> #P-value
> p_value <- pnorm(z_score)
> p_value
[1] 0.009211063
>
```

Step 5: We know that for 5% level of significance, reject H_0 if the calculated value of the test statistic is either ≥ 1.96 or ≤ -1.96. Otherwise, failed to reject H_0. Again, if the p-value is less than or equal to the significance level, we reject the null hypothesis; however, if the p-value is greater than the significance level, we failed to reject the null hypothesis. We use the following code in R:

R-code:

```
# Perform the z-test based on test statistic
if (abs(z_score) > Critical_value) {
  cat('Reject the null hypothesis')
} else {
  cat('Fail to reject the null hypothesis')
}
#Perform the z-test based on p-value
if (p_value  <=  alpha/2) {
  cat('Reject the null hypothesis')
} else {
  cat('Fail to reject the null hypothesis')
}
```

Output:

```
Console   Terminal ×   Background Jobs ×

R  R 4.3.2 · ~/
> # Perform the z-test based on test statistic
> if (abs(z_score) > Critical_value) {
+     cat("Reject the null hypothesis")
+ } else {
+     cat("Fail to reject the null hypothesis")
+ }
Reject the null hypothesis> #Perform the z-test based on p-value
> if (p_value  <=  alpha/2) {
+     cat("Reject the null hypothesis")
+ } else {
+     cat("Fail to reject the null hypothesis")
+ }
Reject the null hypothesis
> |
```

Step 6: Hence, the according to the decision rule, we are able to reject the null hypothesis since the computed vale of the test statistic is smaller than the critical value ($-2.357023 < -1.96$) or p-value = 0.009 which is smaller than $\alpha/2 = 0.025$ indicating the null hypothesis is rejected at the 0.05 level of significance. Therefore, it can be concluded that the mean lifetime of the equipment is not 15 years.

Example 7.6

Recall the dataset used in **Example 7.2**, suppose we want to test the hypothesis that the body fat is 19% with a standard deviation 2 at 5% level of significance.

Step 1: Let us consider the hypothesis $H_0 : \mu = 19$ against $H_1 : \mu \neq 19$.

Step 2: Level of significance is 0.05.

Steps 3 & 4: In this dataset, we have 252 observations. Therefore, we can use the Z-test here. In R, we can use the command 'z.test'; however, we need to install 'BSDA' Package. The code along with output is given below:

R-code:

```
install.packages('BSDA')
library(BSDA)
body_fat<-data_bfat$BodyFat
z.test(body_fat, mu=19, sigma.x=2)
```

Output:

```
Console   Terminal ×   Background Jobs ×

R  R 4.3.2 · ~/
> library(BSDA)
> body_fat<-data_bfat$BodyFat
> z.test(body_fat, mu=19, sigma.x=2)

        One-sample z-Test

data:  body_fat
z = 1.1969, p-value = 0.2314
alternative hypothesis: true mean is not equal to 19
95 percent confidence interval:
 18.90386 19.39773
sample estimates:
mean of x
 19.15079

>
```

Steps 5 & 6:   The test statistic for the one sample Z-test is 1.1969 and the corresponding p-value is 0.2314. Since this p-value is larger than 0.025, we do not have sufficient evidence to reject the null hypothesis. Therefore, it can be said that the population mean of body fat is 19%.

Case 2: Population variance is unknown or the sample size is small

In real-world scenarios, the population variance is generally unknown. If the sampling is from an approximately normal population with an unknown variance or the sample size ($n < 30$) is small, then for testing the mean to an assigned value, i.e., $H_0 : \mu = \mu_0$, we use the following test statistic

$$t = \frac{\bar{x} - \mu_0}{s / \sqrt{n}} \sim t_{n-1,\alpha}.$$

Under H_0, the test statistic is distributed as Student's t with $n - 1$ degrees of freedom, s is the sample standard deviation.

Example 7.7

A researcher studying the blood pressure of a group of patients at a hospital. He/she want to determine whether the average blood pressure of a sample is significantly different from the population mean blood pressure 120 mm Hg. The sample consists of the following ten blood pressure measurements:

$$125, 130, 122, 128, 123, 126, 119, 131, 124, 129.$$

Step 1:   Let us consider the hypothesis,

$$H_0 : \mu = 120 \text{ against } H_1 : \mu \neq 120.$$

Step 2: We also consider 5% level of significance.

Steps 3 & 4: Now, in R, we use the following command to perform the test and got the following output:

R-code:

```
install.packages('BSDA')
library(BSDA)
# Create a vector of blood pressure measurements
blood_pressure <- c(125, 130, 122, 128, 123, 126, 119, 131, 124, 129)
# Specify the population mean for comparison
population_mean <- 120
# Perform a one-sample t-test
t.test(blood_pressure, mu = 120)
```

Output:

```
Console   Terminal ×   Background Jobs ×                                    ─ ⬚

R  R 4.3.2 · ~/
> library(BSDA)
> # Create a vector of blood pressure measurements
> blood_pressure <- c(125, 130, 122, 128, 123, 126, 119, 131, 124, 129)
> # Specify the population mean for comparison
> population_mean <- 120
> # Perform a one-sample t-test
> t.test(blood_pressure, mu = 120)

        One Sample t-test

data:  blood_pressure
t = 4.7048, df = 9, p-value = 0.001113
alternative hypothesis: true mean is not equal to 120
95 percent confidence interval:
 122.9594 128.4406
sample estimates:
mean of x
    125.7

>
```

Steps 5 & 6: It is seen that the *p*-value = 0.001113 which is smaller than 0.025. Therefore, we have sufficient evidence to reject the null hypothesis at 5% level of significance indicating the population mean blood pressure is not 120 mm Hg; i.e., there is evidence to suggest a significant difference between the sample and population means.

7.2.3.2 Comparison of Two Independent Populations Means

Suppose we want to test equality of two independent populations means, i.e., we have to test the hypothesis $H_0 : \mu_1 = \mu_2$ against $H_1 : \mu_1 \neq \mu_2$ which is equivalent to test $H_0 : \mu_1 - \mu_2 = 0$ against $H_1 : \mu_1 - \mu_2 \neq 0$. Here, μ_1 and μ_2 represents the means of sample 1 and sample 2 respectively.

Now, we have the following three cases:

Case 1: Variances are known or samples are large ($n_1 > 30$, $n_2 > 30$).

Case 2: Small samples and variances are unknown however assumed they are equal.

Case 3: Small samples and variances are unknown, however assumed they are not equal.

Case 1: Variances are known or samples are large ($n_1 > 30$, $n_2 > 30$)

It is assumed that two independent samples are taken from a normally distributed population with a known variance or both of the sample sizes are large enough; the test statistic for testing the null hypothesis of equal population means is expressed as

$$Z = \frac{\bar{x}_1 - \bar{x}_2}{\sqrt{\frac{\sigma_1^2}{n_1} + \frac{\sigma_2^2}{n_2}}} \sim N(0,1)$$

where $\bar{x}_1$ and $\bar{x}_2$ are the sample means of the two groups, σ_1 and σ_2 are the population standard deviations (known), n_1 and n_2 are the sample sizes of the two groups. When the null hypothesis is true, Z follows the standard normal distribution.

Example 7.8

A researcher wants to check whether there is a difference between average uric acid levels for males and females or not. He/she collected samples from 50 males and 45 females and found the average uric acid levels are 4.5 mg/100 ml and 4.25 mg/100 ml, with standard deviations of 1.5 mg/100 ml and 1.2 mg/100 ml for males and females respectively. At a 10% level of significance, the researcher wants to test the hypothesis that the average uric acid levels for males and females are equal.

Step 1:   Let us consider the hypothesis,

$$H_0 : \mu_1 = \mu_2, \text{i.e.,} H_0 : \mu_1 - \mu_2 = 0 \text{ against } H_1 : \mu_1 - \mu_2 \neq 0.$$

Step 2:   The level of significance is $\alpha = 0.10$.

Steps 3 & 4:   Here both the standard deviations are known and the sample size is large, so we use the following test statistic to test the hypothesis given in Step-1.

$$Z = \frac{\bar{x}_1 - \bar{x}_2}{\sqrt{\frac{\sigma_1^2}{n_1} + \frac{\sigma_2^2}{n_2}}}$$

In R, we use the following commands and get the desired outputs:

R-code:

```
# Define your sample means, population standard deviations, and sample
sizes
mean1 <- 4.5
mean2 <- 4.25
sd1 <- 1.5
sd2 <- 1.2
n1 <- 50
n2 <- 45
# Calculate the Z-statistic
Z <- (mean1-mean2)/sqrt((sd1^2/n1)+(sd2^2/ n2))
# Find the p-value
p_value <- 2*(1-pnorm(abs(Z)))
# Print the Z-statistic and p-value
cat('Z-statistic:', Z, '\n')
cat('p-value:', p_value, '\n')
```

Output:

```
Console   Terminal    Background Jobs

R R 4.3.2 · ~/
> # Define your sample means, population standard deviations, and sample sizes
> mean1 <- 4.5
> mean2 <- 4.25
> sd1 <- 1.5
> sd2 <- 1.2
> n1 <- 50
> n2 <- 45
> # Calculate the Z-statistic
> Z <- (mean1-mean2)/sqrt((sd1^2/n1)+(sd2^2/ n2))
> # Find the p-value
> p_value <- 2*(1-pnorm(abs(Z)))
> # Print the Z-statistic and p-value
> cat("Z-statistic:", Z, "\n")
Z-statistic: 0.9009375
> cat("p-value:", p_value, "\n")
p-value: 0.3676216
>
```

Steps 5 & 6: It is observed that the *p*-value of the corresponding Z-value is 0.368 which is larger than the significance level 0.10. Thus, we have not sufficient evidence to reject the null hypothesis at 10% level of significance suggesting that there are no significant differences between the uric acid levels of males and females.

Case 2: Small samples and variances are unknown but assumed they are equal

For conducting a hypothesis test for the equality of two population means ($H_0 : \mu_1 = \mu_2$) when samples are small and variances are unknown however assumed to be equal, we use the following test statistic:

$$t = \frac{\overline{X}_1 - \overline{X}_2}{S_p \sqrt{\left(\dfrac{1}{n_1} + \dfrac{1}{n_2} \right)}}$$

where $S_p^2 = \dfrac{(n_1 - 1)S_1^2 + (n_2 - 1)S_2^2}{n_1 + n_2 - 2}$ and when the null hypothesis is true, the test statistic is distributed as Student's t with $n_1 + n_2 - 2$ degrees of freedom.

Example 7.9

Suppose we have samples consisting of cholesterol levels of two independent groups of patients and presented in Table 7.1:

TABLE 7.1

Cholesterol Levels of Two Independent Groups of Patients

Group 1	180, 185, 190, 195, 200, 205, 210, 215, 220, 225
Group 2	170, 175, 180, 185, 190, 195, 200, 205, 210, 215

In this example, we want to test the equality of the cholesterol levels of two different groups of patients: Group 1 and Group 2.

Step 1: Let us consider the hypothesis $H_0 : \mu_1 = \mu_2$, i.e., $H_0 : \mu_1 - \mu_2 = 0$ against $H_1 : \mu_1 - \mu_2 \neq 0$.

Step 2: Suppose the significance level is $\alpha = 0.05$.

Steps 3 & 4: To calculate the test statistic and p-value we use the following command in R:

R-code:

```
# Create two independent sample vectors (cholesterol levels)
group1 <- c(180, 185, 190, 195, 200, 205, 210, 215, 220, 225)
group2 <- c(170, 175, 180, 185, 190, 195, 200, 205, 210, 215)
# sample size
n1 <-length(group1)
n2 <-length(group2)
#means
m1 <- mean(group1)
m2 <- mean(group2)
#standard deviations
s1 <- sd(group1)
s2 <- sd(group2)
sp <- sqrt(((n1-1)*s1^2+(n2-1)*s2^2)/(n1+n2-2))
sp
# Your calculated t-statistic
t_statistic <-(m1-m2)/(sp*sqrt((1/n1)+(1/n2)))
t_statistic
# Degrees of freedom (typically based on your sample sizes and
assumptions)
df <- n1+n2-2   # Replace with your actual degrees of freedom
# Calculate the two-tailed p-value (for a two-sample test)
p_value <- 2 * (1 - pt(abs(t_statistic), df))
# Print the p-value
cat('P-value:', p_value, '\n')
```

Output:

```
Console   Terminal ×   Background Jobs ×
R  R 4.3.2 · ~/
> # Create two independent sample vectors (cholesterol levels)
> group1 <- c(180, 185, 190, 195, 200, 205, 210, 215, 220, 225)
> group2 <- c(170, 175, 180, 185, 190, 195, 200, 205, 210, 215)
> # sample size
> n1 <-length(group1)
> n2 <-length(group2)
> #means
> m1 <- mean(group1)
> m2 <- mean(group2)
> #standard deviations
> s1 <- sd(group1)
> s2 <- sd(group2)
> sp <- sqrt(((n1-1)*s1^2+(n2-1)*s2^2)/(n1+n2-2))
> sp
[1] 15.13825
> # Your calculated t-statistic
> t_statistic <-(m1-m2)/(sp*sqrt((1/n1)+(1/n2)))
> t_statistic
[1] 1.477098
> # Degrees of freedom (typically based on your sample sizes and assumptions)
> df <- n1+n2-2   # Replace with your actual degrees of freedom
> # Calculate the two-tailed p-value (for a two-sample test)
> p_value <- 2 * (1 - pt(abs(t_statistic), df))
> # Print the p-value
> cat("P-value:", p_value, "\n")
P-value: 0.1569322
>
```

Or, simply we use the 't.test()' function to perform the above test of hypothesis as like below:

R-code:

```
# Perform a two-sample t-test with equal variances assumed
result <- t.test(group1, group2, var.equal = TRUE)
# View the results
print(result)
```

Output:

```
Console   Terminal ×   Background Jobs ×                                    ⊟⬜
R  R 4.3.2 · ~/
> # Perform a two-sample t-test with equal variances assumed
> result <- t.test(group1, group2, var.equal = TRUE)
> # View the results
> print(result)

        Two Sample t-test

data:  group1 and group2
t = 1.4771, df = 18, p-value = 0.1569
alternative hypothesis: true difference in means is not equal to 0
95 percent confidence interval:
 -4.223309 24.223309
sample estimates:
mean of x mean of y
    202.5     192.5

>
```

Steps 5 & 6: Here, the *p*-value is larger than the chosen significance level 0.05, suggesting that we have not sufficient evidence to reject the null hypothesis at 5% level of significance. Therefore, it is said that there is evidence of not a significant difference between the two group means in terms of cholesterol levels; i.e., the mean cholesterol levels of two groups are equal.

Case 3: Small samples and variances are unknown but assumed they are not equal

Suppose two independent random samples are drawn from normally distributed populations with unknown as well as unequal variances. In this circumstance, for testing the hypothesis $H_0 : \mu_1 = \mu_2$, we use the following test statistic:

$$t = \frac{\bar{X}_1 - \bar{X}_2}{\sqrt{\left(\dfrac{S_1^2}{n_1} + \dfrac{S_2^2}{n_2}\right)}}.$$

Here, the test statistic follows *t*-distribution with $\dfrac{\left(\dfrac{S_1^2}{n_1} + \dfrac{S_2^2}{n_2}\right)^2}{\dfrac{\left(\dfrac{S_1^2}{n_1}\right)^2}{n_1 - 1} + \dfrac{\left(\dfrac{S_2^2}{n_2}\right)^2}{n_2 - 1}}$ degrees of freedom. It is also known as Welch's *t*-test.

Example 7.10

Suppose a study is aimed to investigate the effect of smoking on lung destruction. We have the data in Table 7.2 for two groups: non-smoker and smoker:

TABLE 7.2

Lung Destructive Index Scores for Two Groups: Non-smoker and Smoker

Group	Lung destructive index scores
Non-smokers	15.1, 7.0, 12.8, 11.0, 8.7, 17.9, 6.5, 13.2, 18.9, 8.5
Smokers	15.5, 15.9, 14.3, 26.5, 18.4, 25.3, 18.8, 16.3, 21.8, 23.5

Test a hypothesis that the average score of Lung destructive index are equal between smokers and non-smokers at 5% level of significance.

Step 1: Let us consider the hypothesis $H_0 : \mu_1 = \mu_2$, i.e., $H_0 : \mu_1 - \mu_2 = 0$ against $H_1 : \mu_1 - \mu_2 \neq 0$.

Step 2: It is given that the significance level is $\alpha = 0.05$.

Steps 3 & 4: To calculate the test statistic and corresponding *p*-value, we use the following command in R:

R-code:

```
#Data
non_smokers  <- c(15.1,  7.0,  12.8,  11.0,  8.7,  17.9,  6.5,  13.2,
18.9,  8.5)
smokers      <- c(15.5,  15.9,  14.3,  26.5,  18.4,  25.3,  18.8,
16.3,  21.8,  23.5)
#sample size
n_NS <-length(non_smokers)
n_S <-length(smokers)
#means
m_NS <- mean(non_smokers)
m_S <- mean(smokers)
#standard deviations
s_NS <- sd(non_smokers)
s_S <- sd(smokers)
# Your calculated Welch's t-test
t_statistic_W <-(m_NS-m_S)/sqrt((s_NS^2/n_NS)+(s_S^2/n_S))
t_statistic_W
# Degrees of freedom (typically based on your sample sizes and
assumptions)
df_W <- ((s_NS^2/n_NS)+(s_S^2/n_NS))^2/(((s_NS^2/n_NS)^2)/(n_NS-1) +
((s_S^2/n_S)^2)/(n_S-1)) # Replace with your actual degrees of freedom
# Calculate the two-tailed p-value (for a two-sample test)
p_value_W <- 2 * (1 - pt(abs(t_statistic_W), df_W))
# Print the p-value
cat('P-value:', p_value_W, '\n')
```

Output:

```
Console   Terminal ×   Background Jobs ×
R R 4.3.2 · ~/
> #Data
> non_smokers   <- c(15.1,   7.0,   12.8,   11.0,   8.7,   17.9,   6.5,   13.2,   18.9,   8.5)
> smokers       <- c(15.5,   15.9,   14.3,   26.5,   18.4,   25.3,   18.8,   16.3,   21.8,   23.5)
> #sample size
> n_NS <-length(non_smokers)
> n_S <-length(smokers)
> #means
> m_NS <- mean(non_smokers)
> m_S <- mean(smokers)
> #standard deviations
> s_NS <- sd(non_smokers)
> s_S <- sd(smokers)
> # Your calculated Welch's t-test
> t_statistic_W <-(m_NS-m_S)/sqrt((s_NS^2/n_NS)+(s_S^2/n_S))
> t_statistic_W
[1] -3.916433
> # Degrees of freedom (typically based on your sample sizes and assumptions)
> df_W <- ((s_NS^2/n_NS)+(s_S^2/n_NS))^2/(((s_NS^2/n_NS)^2)/(n_NS-1) + ((s_S^2/n_S)^2)/(n_S-1))
# Replace with your actual degrees of freedom
> # Calculate the two-tailed p-value (for a two-sample test)
> p_value_W <- 2 * (1 - pt(abs(t_statistic_W), df_W))
> # Print the p-value
> cat("P-value:", p_value_W, "\n")
P-value: 0.001011728
> |
```

Or, simply, we use the 't.test' function and supply 'var.equal = FALSE' to get the desired results. The R-codes are given below:

R-code:

```
# Perform Welch's t-test
result2 <- t.test(non_smokers, smokers, var.equal = FALSE)
# View the results
print(result2)
```

Output:

```
Console   Terminal ×   Background Jobs ×
R R 4.3.2 · ~/
> # Perform Welch's t-test
> result2 <- t.test(non_smokers, smokers, var.equal = FALSE)
> # View the results
> print(result2)

        Welch Two Sample t-test

data:  non_smokers and smokers
t = -3.9164, df = 18, p-value = 0.001012
alternative hypothesis: true difference in means is not equal to 0
95 percent confidence interval:
 -11.784485  -3.555515
sample estimates:
mean of x mean of y
    11.96     19.63

> |
```

Steps 5 & 6:   It is observed that the *p*-value is less than the significance level, which indicates that we have sufficient evidence to reject the null hypothesis. It concludes that there is a significant difference between the means of lung destructive index among smokers and non-smokers at the 5% level of significance.

7.2.3.3 Comparison of Two Correlated Samples Means

In many situations, individuals or subjects in a study exhibit inherent variability. By comparing paired observations (e.g., before and after measurements on the same subjects), paired tests account for this individual variability more effectively than comparing independent samples. The use of paired comparison tests helps to increase the precision and validity of statistical analyses, making them particularly valuable in situations where individual variability, confounding factors, or the impact of interventions or treatments need to be carefully considered.

We use the difference between pairs of observations to perform the analysis instead of using individual observations. Suppose we have n sample differences calculated from the n pairs of measurements (x_{Bi} and x_{Ai} represents before and after measurements on the same subjects) and follows a normal distribution. In such situations, to test the hypothesis $H_0 : \mu_B = \mu_A$, i.e., $H_0 : \mu_B - \mu_A = 0$ vs $H_A : \mu_B - \mu_A \neq 0$, we use the following test statistic,

$$t = \frac{\bar{d}}{s_d \big/ \sqrt{n}}$$

where $d_i = x_{Bi} - x_{Ai}$, $s_d = \sqrt{\dfrac{1}{n-1} \sum_{i=1}^{n} \left(d_i - \bar{d}\right)^2}$ and t is distributed as Student's t with $(n - 1)$

degrees of freedom.

Example 7.11

In a proteomics study, you have data on protein abundance levels measured in two different conditions (Table 7.3).

Suppose we want to check whether there is a significant difference or not in protein abundance between these conditions at 5% level of significance.

Step 1:   Let us consider the hypothesis $H_0 : \mu_B = \mu_A$, i.e., $H_0 : \mu_B - \mu_A = 0$ against $H_A :$ $\mu_B - \mu_A \neq 0$.

Step 2:   It is given that the significance level is $\alpha = 0.05$.

Steps 3 & 4:   To calculate the test statistic and its *p*-value, we use the following command in R:

TABLE 7.3

Data on Protein Abundance Levels Measured in Two Different Conditions

Condition 1	11, 12, 10, 13, 14, 11, 12
Condition 2	9, 11, 10, 11, 12, 12, 13

R-code:

```
# Create vectors for protein abundance levels in two conditions
condition_1 <- c(11, 12, 10, 13, 14, 11, 12)
condition_2 <- c(9, 11, 10, 11, 12, 12, 13)
# Calculate the differences between measurements in the two conditions
differences <- condition_2 - condition_1
# Perform a paired samples t-test
result_p <- t.test(differences)
# View the results
print(result_p)
```

Output:

```
Console   Terminal ×   Background Jobs ×                                          ━ □

R  R 4.3.2 · ~/
> # Create vectors for protein abundance levels in two conditions
> condition_1 <- c(11, 12, 10, 13, 14, 11, 12)
> condition_2 <- c(9, 11, 10, 11, 12, 12, 13)
> # Calculate the differences between measurements in the two conditions
> differences <- condition_2 - condition_1
> # Perform a paired samples t-test
> result_p <- t.test(differences)
> # View the results
> print(result_p)

        One Sample t-test

data:  differences
t = -1.3693, df = 6, p-value = 0.2199
alternative hypothesis: true mean is not equal to 0
95 percent confidence interval:
 -1.9906941  0.5621227
sample estimates:
 mean of x
-0.7142857

> |
```

Steps 5 & 6:   Result depicts that the p-value is greater than the significance level suggesting that we have not sufficient evidence to reject the null hypothesis at 5% level of significance. That is there is no significant difference between means of protein abundance levels in two different conditions.

7.2.3.4 Comparison of k (k > 2) Independent Sample Means

To verify whether there are significant differences among the means of three or more independent groups, i.e., we test the following hypothesis:

$$H_0 : \mu_1 = \mu_2 = \ldots = \mu_k \text{ against } H_1 : \text{not all } \mu_j \text{ are equal}, j = 1, 2, \ldots, k.$$

For testing the hypothesis, we use the F-statistic and defined as follows:

$$F = \frac{\text{Between} - \text{Group Variability (SSB)}}{\text{Within} - \text{Group Variability (SSW)}}.$$

Here, 'Between-Group Variability (SSB)' also known as 'treatment' or 'factor' variation that measures the variation between the means of the different groups. It is computed as the sum of squares between groups, i.e., $\text{SSB} = \sum_{i=1}^{k} n_i (X_i - \bar{X})^2$.

Again, 'Within-Group Variability (SSW)' also known as 'error' or 'residual' variation and measures the variation within each group, or the variation of individual data points within their respective groups. It is determined as the sum of squares within groups, i.e.,

$$SSW = \sum_{i=1}^{k} \sum_{j=1}^{n_i} \left(X_{ij} - \bar{X}_i \right)^2.$$

Here, k is the number of groups, n_i is the sample size of the i-th group, X_{ij} is the j-th observation in the i-th group, $\bar{X}_i$ is the mean of the i-th group, and $\bar{X}$ is the overall mean of all data points.

The test statistic follows the F distribution with $(k - 1)$ and $(N - k)$ degrees of freedoms, where N is the total number of observations.

Generally, all the information is presented in Table 7.4 called the one-way analysis of variance (ANOVA) table.

TABLE 7.4

One-Way Analysis of Variance

Source of Variation	Degrees of Freedom (DF)	Sum of Squares (SS)	Mean Square (MS)	F-Statistic (F)
Between Groups (Treatment)	$k - 1$	SSB	$MSB = SSB/(k-1)$	$F = MSB/MSW$
Within Groups (Error or Residual)	$N - k$	SSW	$MSW = SSW/(N-k)$	

TABLE 7.5

Mean Cholesterol Levels among Four Different Drug Treatments

Treatment 1	180, 182, 175, 178, 185, 190, 188, 195, 197, 200
Treatment 2	170, 172, 180, 175, 178, 181, 174, 182, 185, 194
Treatment 3	164, 165, 171, 162, 167, 162, 180, 167, 169, 175
Treatment 4	152, 165, 156, 155, 172, 158, 162, 159, 160, 163

Example 7.12

Suppose we want to compare the mean cholesterol levels among four different drug treatments data are given (Table 7.5).

Hence, we consider the following steps.

Step 1: Let us consider the hypothesis $H_0 : \mu_1 = \mu_2 = \mu_3 = \mu_4$ against H_1 : not all μ_j are equal, $j = 1, 2, ..., 4$.

Step 2: We assume that the significance level is $\alpha = 0.05$.

Steps 3 & 4: To calculate the test statistic and p-value, we write the following commands in R:

R-code:

```
# Create four independent sample vectors (cholesterol levels for each
treatment)
treatment1 <- c(180, 182, 175, 178, 185, 190, 188, 195, 197, 200)
treatment2 <- c( 170, 172, 180, 175, 178, 181, 174, 182, 185, 194)
treatment3 <- c(164, 165, 171, 162, 167, 162, 180, 167, 169, 175)
treatment4 <- c(152, 165, 156, 155, 172, 158, 162, 159, 160, 163)
```

```
# Combine the data into a single data frame
data <- data.frame(Treatment = rep(c('Treatment1', 'Treatment2',
'Treatment3', 'Treatment4'), each = 5),
                   Cholesterol = c(treatment1, treatment2, treatment3,
treatment4))
# Perform one-way ANOVA
result <- aov(Cholesterol ~ Treatment, data = data)
# View the ANOVA summary
summary(result)
```

Output:

```
Console  Terminal ×  Background Jobs ×                                    ▬☐

R R 4.3.2 · ~/ ⇌

> # Create four independent sample vectors (cholesterol levels for each treatment)
> treatment1 <- c(180, 182, 175, 178, 185, 190, 188, 195, 197, 200)
> treatment2 <- c( 170, 172, 180, 175, 178, 181, 174, 182, 185, 194)
> treatment3 <- c(164, 165, 171, 162, 167, 162, 180, 167, 169, 175)
> treatment4 <- c(152, 165, 156, 155, 172, 158, 162, 159, 160, 163)
> # Combine the data into a single data frame
> data <- data.frame(Treatment = rep(c("Treatment1", "Treatment2", "Treatment3", "Treatment4"),
each = 5),
+                    Cholesterol = c(treatment1, treatment2, treatment3, treatment4))
> # Perform one-way ANOVA
> result <- aov(Cholesterol ~ Treatment, data = data)
> # View the ANOVA summary
> summary(result)
            Df Sum Sq Mean Sq F value Pr(>F)
Treatment    3   1166   388.8   2.974 0.0444 *
Residuals   36   4705   130.7
---
Signif. codes:  0 '***' 0.001 '**' 0.01 '*' 0.05 '.' 0.1 ' ' 1
>
```

Steps 5 & 6: Here, the p-value is less than the considered significance level indicating the null hypothesis is rejected at 5% level of significance. Thus, we conclude that not all population means of cholesterol levels among four different drug treatments are equal.

7.2.4 Proportion Test

In many circumstances, we are concerned only with the presence or absence of some specified attribute or characteristic by the individuals selected. To test the proportion of a particular event, we consider the following cases:

a. Comparison of p with population parameter p_0.

b. Comparison of two proportions.

7.2.4.1 *Comparison of p with Population Parameter p_0*

For testing a proportion of an event with a given population proportion, we consider the hypothesis $H_0 : p = p_0$ against $H_A : p \neq p_0$. Hence, we consider the following test statistic

$$Z = \frac{\hat{p} - p_0}{\sqrt{\dfrac{p_0(1-p_0)}{n}}}$$

where under the null hypothesis and for large sample and with the help of the central limit theorem, the test statistics is distributed approximately as the standard normal.

Example 7.13

Recall the 'Body Fat Prediction Dataset' from **Example 7.2**, and we wish to test the hypothesis that the 45% of the population has hip circumference more than 100 cm.

Step 1: Here, we consider the following hypothesis,

$$H_0 : p = 0.45 \text{ against } H_A : p \neq 0.45.$$

Step 2: We assume that the significance level is $\alpha = 0.05$.

Steps 3 & 4: To test the hypothesis, we use the following test statistic

$$Z = \frac{\hat{p} - p_0}{\sqrt{\dfrac{p_0(1 - p_0)}{n}}}.$$

To calculate the test statistic and p-value, we write the following commands in R:

R-code:

```
data_bfat <- read.csv('E:/Book/Chapter 7/bodyfat.csv', header = TRUE)
attach(data_bfat)
#estimate of the proportion
hip<-data_bfat$Hip
#mean(hip)
n<-length(hip)
p_hat<- sum(hip>100)/n
p_hat
# Hypothesized population proportion
p0 <- 0.45  # Hypothesized proportion (45% have hip circumference more
than 100 cm)
# Calculate the test statistic using the formula
z <- (p_hat-p0)/sqrt((p0*(1-p0))/n)
# Calculate the two-tailed p-value
p_value <- 2*(1-pnorm(abs(z)))
# View the test statistic and p-value
cat('Test Statistic (Z):', z, '\n')
cat('Two-Tailed p-Value:', p_value, '\n')
```

Output:

```
Console  Terminal ×   Background Jobs ×                                                    ─ □
R  R 4.3.2 · ~/
> data_bfat <- read.csv("E:/Book/Chapter 7/bodyfat.csv", header = TRUE)
> attach(data_bfat)
> #estimate of the proportion
> hip<-data_bfat$Hip
> #mean(hip)
> n<-length(hip)
> p_hat<- sum(hip>100)/n
> p_hat
[1] 0.4365079
> # Hypothesized population proportion
> p0 <- 0.45  # Hypothesized proportion (45% have hip circumference more than 100 cm)
> # Calculate the test statistic using the formula
> z <- (p_hat-p0)/sqrt((p0*(1-p0))/n)
> # Calculate the two-tailed p-value
> p_value <- 2*(1-pnorm(abs(z)))
> # View the test statistic and p-value
> cat("Test Statistic (Z):", z, "\n")
Test Statistic (Z): -0.4305177
> cat("Two-Tailed p-Value:", p_value, "\n")
Two-Tailed p-Value: 0.6668191
> |
```

Steps 5 & 6:   The *p*-value of the test is 0.6668191, which is greater than the significance level (0.05) indicate that we have no sufficient evidence to reject the null hypothesis. Thus, we may conclude that the proportion of hip circumference more than 100 cm is not different from 0.45 at 5% level of significance.

7.2.4.2 Comparison of Two Proportions

Sometimes we test the hypothesis that two population proportions are equal or the difference between two population proportions is zero, i.e., $H_0 : p_1 = p_2$ or $H_0 : p_1 - p_2 = 0$ against $H_A : p_1 - p_2 \neq 0$.

To test the hypothesis, we use the following test statistic,

$$Z = \frac{\hat{p}_1 - \hat{p}_2}{\sqrt{\frac{\bar{p}(1-\bar{p})}{n_1} + \frac{\bar{p}(1-\bar{p})}{n_2}}}$$

where $\bar{p} = \frac{x_1 + x_2}{n_1 + n_2}$, here x_1 and x_2 are the numbers possessing the characteristic of interest in the first and second samples respectively; n_1 and n_2 are the total number of observations belongs to the first and second samples respectively.

If the null hypothesis is true, the test statistic is distributed approximately as the standard normal distribution.

Example 7.14

Recall the 'Body Fat Prediction Dataset' from **Example 7.2**, let us split the dataset into two groups based on 'age' is more than or equal to 45 years and more than 45 years. Now, we test the hypothesis that the proportion of abdomen 2 circumferences are more than 90 cm of group one is equal to the proportion of abdomen 2 circumferences are more than 90 cm of group two. Consider 5% level of significance.

Step 1:   First, we state the following hypothesis

$$H_0 : p_1 - p_2 = 0 \text{ against } H_A : p_1 - p_2 \neq 0.$$

Step 2:   It is given that the significance level is $\alpha = 0.05$.

Steps 3 & 4:   To test the hypothesis, we use the following test statistic

$$Z = \frac{\hat{p}_1 - \hat{p}_2}{\sqrt{\frac{\bar{p}(1-\bar{p})}{n_1} + \frac{\bar{p}(1-\bar{p})}{n_2}}}.$$

To compute the value of the test statistic and its *p*-value, we write the following commands in R:

R-code:

```
data_bfat <- read.csv('E:/Book/Chapter 7/bodyfat.csv', header = TRUE)
attach(data_bfat)
#split the dataset into two groups
g1 <- data_bfat[data_bfat$Age<=45, ] #group1
g2 <- data_bfat[data_bfat$Age>45, ] #group2
```

```
#select variable
abdomen_g1<-g1$Abdomen
abdomen_g2<-g2$Abdomen
#calculate proportion
n1<-length(abdomen_g1)
x1<-sum(abdomen_g1>90)
p1_hat<-x1/n1
n2<-length(abdomen_g2)
x2<-sum(abdomen_g2>90)
p2_hat<-x2/n2
#pooled proportion
p_hat<-(x1+x2)/(n1+n2)
p_hat
# Calculate the test statistic
z<-(p1_hat-p2_hat)/sqrt((((p_hat*(1-p_hat))/n1)+((p_hat*(1-p_hat))/n2))
# Calculate the two-tailed p-value
p_value <- 2*(1-pnorm(abs(z)))
# View the test statistic and p-value
cat('Test Statistic (Z):', z, '\n')
cat('Two-Tailed p-Value:', p_value, '\n')
```

Output:

```
Console   Terminal ×   Background Jobs ×
R R 4.3.2 · ~/
> data_bfat <- read.csv("E:/Book/Chapter 7/bodyfat.csv", header = TRUE)
> attach(data_bfat)
> #split the dataset into two groups
> g1 <- data_bfat[data_bfat$Age<=45, ] #group1
> g2 <- data_bfat[data_bfat$Age>45, ] #group2
> #select variable
> abdomen_g1<-g1$Abdomen
> abdomen_g2<-g2$Abdomen
> #calculate proportion
> n1<-length(abdomen_g1)
> x1<-sum(abdomen_g1>90)
> p1_hat<-x1/n1
> n2<-length(abdomen_g2)
> x2<-sum(abdomen_g2>90)
> p2_hat<-x2/n2
> #pooled proportion
> p_hat<-(x1+x2)/(n1+n2)
> p_hat
[1] 0.547619
> # Calculate the test statistic
> z<-(p1_hat-p2_hat)/sqrt(((p_hat*(1-p_hat))/n1)+((p_hat*(1-p_hat))/n2))
> # Calculate the two-tailed p-value
> p_value <- 2*(1-pnorm(abs(z)))
> # View the test statistic and p-value
> cat("Test Statistic (Z):", z, "\n")
Test Statistic (Z): -1.162435
> cat("Two-Tailed p-value:", p_value, "\n")
Two-Tailed p-value: 0.2450587
> |
```

Steps 5 & 6: It is observed that the *p*-value is greater than the given significance level suggesting that we have not sufficient evidence to reject the null hypothesis at 5% level of significance. Therefore, we conclude that the proportion of abdomen 2 circumferences are more than 90 cm of two groups are equal.

7.2.5 Variance Test

There are two possible cases for testing the variance:

  a. Testing a single variance.
  b. Equality of two variances.

7.2.5.1 Testing a Single Variance

It is assumed that the sample are drawn from a normally distributed population, then for testing a hypothesis about a population variance, i.e., $H_0 : \sigma^2 = \sigma_0^2$ against $H_A : \sigma^2 \neq \sigma_0^2$, where σ_0^2 is a specified value, we use the following test statistic:

$$\chi^2 = \frac{(n-1)s^2}{\sigma^2}$$

where n is the sample size, s^2 is the sample variance, σ^2 is the hypothesized population variance and the test statistic follows a chi-squared distribution with $(n - 1)$ degrees of freedom under the true null hypothesis.

Example 7.15

Recall the 'Body Fat Prediction Dataset' from **Example 7.2**. From this dataset, suppose we want to test the hypothesis that variance of 'Height' is 10, i.e., $H_0 : \sigma^2 = 10$ considering 5% level of significance.

Step 1: First, we state the following hypothesis,

$$H_0 : \sigma^2 = 10 \text{ against } H_0 : \sigma^2 \neq 10.$$

Step 2: It is given that the significance level is $\alpha = 0.05$.

Steps 3 & 4: To test the hypothesis, we use the following test statistic,

$$\chi^2 = \frac{(n-1)s^2}{\sigma^2}.$$

To calculate the value of the test statistic and the corresponding p-value, we use the following commands in R:

R-code:

```
# sample data
height<-data_bfat$Height
# Hypothesized population variance
var_pop <- 10
# Sample size
n <- length(height)
# Calculate the sample variance (S^2)
sample_var <- var(height)
# Calculate the chi-squared test statistic
chi_squared <- ((n - 1) * sample_var)/var_pop
# Calculate the p-value using the chi-squared distribution (two-sided)
p_value <- 2*(1 - pchisq(chi_squared, df = n - 1))
# View the test statistic and p-value
cat('Chi-Squared Test Statistic:', chi_squared, '\n')
cat('P-Value:', p_value, '\n')
```

Output:

```
Console   Terminal   Background Jobs                                              ▭ ☐
 R  R 4.3.2 · ~/
> # sample data
> height<-data_bfat$Height
> # Hypothesized population variance
> var_pop <- 10
> # Sample size
> n <- length(height)
> # Calculate the sample variance (S^2)
> sample_var <- var(height)
> # Calculate the chi-squared test statistic
> chi_squared <- ((n - 1) * sample_var)/var_pop
> # Calculate the p-value using the chi-squared distribution (two-sided)
> p_value <- 2*(1 - pchisq(chi_squared, df = n - 1))
> # View the test statistic and p-value
> cat("Chi-Squared Test Statistic:", chi_squared, "\n")
Chi-Squared Test Statistic: 336.7545
> cat("P-Value:", p_value, "\n")
P-Value: 0.0004812731
> |
```

Or, simply we use R-package as below:

R-code:

```
install.packages('EnvStats')
library(EnvStats)
varTest(height, sigma.squared = var_pop)
```

Output:

```
Console   Terminal   Background Jobs                                              ▭ ☐
 R  R 4.3.2 · ~/
> library(EnvStats)
> varTest(height, sigma.squared = var_pop)

Results of Hypothesis Test
--------------------------

Null Hypothesis:              variance = 10

Alternative Hypothesis:       True variance is not equal to 10

Test Name:                    Chi-Squared Test on variance

Estimated Parameter(s):       variance = 13.41651

Data:                         height

Test Statistic:               Chi-Squared = 336.7545

Test Statistic Parameter:     df = 251

P-value:                      0.0004812731

95% Confidence Interval:      LCL = 11.34708
                              UCL = 16.11187

> |
```

Steps 5 & 6: From the findings, it is observed that the *p*-value of the test statistic is less than the significance level suggesting that we have sufficient evidence to reject the null hypothesis at 5% level of significance. Thus, the variance of 'Height' is 10.

7.2.5.2 *Testing the Equality of Two Variances*

Testing of the null hypothesis that two population variances are equal is also known as the variance ratio test. When we attempt to test our hypothesis that two population variances are the same, we are actually testing the hypothesis that their ratio is equal to 1. Therefore, to test the null hypothesis $H_0 : \sigma_1^2 = \sigma_2^2$ or $H_0 : \dfrac{\sigma_1^2}{\sigma_2^2} = 1$ vs $H_A : \sigma_1^2 \neq \sigma_2^2$, we use the following test statistic:

$$F = \frac{s_1^2}{s_2^2}.$$

When the null hypothesis is true and under the assumption that the samples are drawn from the normal population, the test statistic follows F distribution with $(n_1 - 1)$ and $(n_2 - 1)$ degrees of freedom.

Example 7.16

Recall the 'Body Fat Prediction Dataset' from **Example 7.2**. From this dataset, we split the dataset into two groups based on the age variable, i.e., group 1: age ≤45 years; group 2: age >45 years. Now, we want to test the hypothesis that variance of 'Weight' of these two groups are equal, i.e., $H_0 : \sigma_1^2 = \sigma_2^2$ and choose 5% level of significance.

Step 1: First, we state the following hypothesis,

$$H_0 : \sigma_1^2 = \sigma_2^2 \text{ against } H_A : \sigma_1^2 \neq \sigma_2^2.$$

Step 2: It is given that the significance level is $\alpha = 0.05$.

Steps 3 & 4: To test the hypothesis, we use the following test statistic,

$$F = \frac{s_1^2}{s_2^2}.$$

To calculate the value of the F statistic and its p-value, one can use the following commands in R:

R-code:

```
# Split data into two groups
Sample1<-data_bfat[data_bfat$Age<=45, ]
Sample2<-data_bfat[data_bfat$Age>45, ]
# Select variable for two groups
weight_g1<-Sample1$Weight
weight_g2<-Sample2$Weight
# Calculate the variances of both groups
var_1 <- var(weight_g1)
var_2 <- var(weight_g2)
# Calculate the F-statistic
F_statistic <- var_1/var_2
# Calculate the degrees of freedom
df1 <- length(weight_g1) - 1
df2 <- length(weight_g2) - 1
# Calculate the p-value
p_value <- 2*(1 - pf(F_statistic, df1, df2, lower.tail = FALSE))
# View the results
cat('F-Statistic:', F_statistic, '\n')
cat('P-Value:', p_value, '\n')
```

Output:

```
Console   Terminal ×   Background Jobs ×                                              ▭ ☐
 R  R 4.3.2 · ~/ ⏏
> # Split data into two groups
> Sample1<-data_bfat[data_bfat$Age<=45, ]
> Sample2<-data_bfat[data_bfat$Age>45, ]
> # Select variable for two groups
> weight_g1<-Sample1$weight
> weight_g2<-Sample2$weight
> # Calculate the variances of both groups
> var_1 <- var(weight_g1)
> var_2 <- var(weight_g2)
> # Calculate the F-statistic
> F_statistic <- var_1/var_2
> # Calculate the degrees of freedom
> df1 <- length(weight_g1) - 1
> df2 <- length(weight_g2) - 1
> # Calculate the p-value
> p_value <- 2*(1 - pf(F_statistic, df1, df2, lower.tail = FALSE))
> # View the results
> cat("F-Statistic:", F_statistic, "\n")
F-Statistic: 0.9015899
> cat("P-Value:", p_value, "\n")
P-Value: 0.5606544
> |
```

Or, simple use var.test() function as below:

R-code:

```
var.test(weight_g1, weight_g2) # var1 and var2 have the same variance
```

Output:

```
Console   Terminal ×   Background Jobs ×                                              ▭ ☐
 R  R 4.3.2 · ~/ ⏏
> var.test(weight_g1, weight_g2) # var1 and var2 have the same variance

        F test to compare two variances

data:  weight_g1 and weight_g2
F = 0.90159, num df = 137, denom df = 113, p-value = 0.5607
alternative hypothesis: true ratio of variances is not equal to 1
95 percent confidence interval:
 0.6309608 1.2804803
sample estimates:
ratio of variances
         0.9015899

> |
```

Steps 5 & 6:   Results reveal that the *p*-value of the *F* statistic is larger than the chosen significance level. Therefore, it may suggest that we have no sufficient evidence to reject the null hypothesis indicating the variance of the weight of two groups is equal at 5% level of significance.

7.2.6 Correlation Coefficient Test

For testing the correlation coefficient, specifically the Pearson correlation coefficient (denoted by r), we consider the hypothesis $H_0 : r = 0$ that is there is no correlation between the two variables against $H_A : r \neq 0$; i.e., there is a correlation between two variables. To test the hypothesis, use the following test statistic:

$$t = \frac{r\sqrt{n-2}}{\sqrt{1-r^2}}$$

where r is the sample correlation coefficient, n is the number of data pairs (sample size) and the test statistic follows a t-distribution with $(n-2)$ degrees of freedom.

Example 7.17

Again, recall the 'Body Fat Prediction Dataset' from **Example 7.2**. Now, we want to test the hypothesis that the correlation between Height and Weight is zero, i.e., $H_0 : r = 0$ against $H_A : r \neq 0$ at 5% level of significance.

Step 1: We define the hypothesis as follows,

$$H_0 : r = 0 \text{ against } H_A : r \neq 0.$$

Step 2: It is given that the significance level is $\alpha = 0.05$.
Steps 3 & 4: We use the following test statistic:

$$t = \frac{r\sqrt{n-2}}{\sqrt{1-r^2}}.$$

To compute the t statistic and its p-value, one can use the following commands to get the desired output in R:

R-code:

```
# Sample data for two variables
height<-data_bfat$Height
weight<-data_bfat$Weight
# Calculate the sample correlation coefficient (r)
r <- cor(height, weight)
r
# Sample size
n <- length(height)
# Calculate the test statistic (t)
t_statistic <- (r*sqrt(n-2))/sqrt(1-r^2)
# Degrees of freedom
df <- n-2
# Calculate the p-value
p_value <- 2 * (1 - pt(abs(t_statistic), df))
# View the results
cat('Test Statistic (t):', t_statistic, '\n')
cat('Two-Tailed p-Value:', p_value, '\n')
```

Output:

```
Console   Terminal ×   Background Jobs ×                                                      ▭ ▢
R  R 4.3.2 · ~/
> # Sample data for two variables
> height<-data_bfat$Height
> weight<-data_bfat$Weight
> # Calculate the sample correlation coefficient (r)
> r <- cor(height, weight)
> r
[1] 0.3082785
> # Sample size
> n <- length(height)
> # Calculate the test statistic (t)
> t_statistic <- (r*sqrt(n-2))/sqrt(1-r^2)
> # Degrees of freedom
> df <- n-2
> # Calculate the p-value
> p_value <- 2 * (1 - pt(abs(t_statistic), df))
> # View the results
> cat("Test Statistic (t):", t_statistic, "\n")
Test Statistic (t): 5.123864
> cat("Two-Tailed p-value:", p_value, "\n")
Two-Tailed p-value: 5.989982e-07
>
```

Or, simply we use `cor.test()` function as below:

R-code:

```
# Perform the correlation coefficient test
cor_test <- cor.test(height, weight)
# View the test results
print(cor_test)
```

Output:

```
Console   Terminal ×   Background Jobs ×                                                      ▭ ▢
R  R 4.3.2 · ~/
> # Perform the correlation coefficient test
> cor_test <- cor.test(height, weight)
> # View the test results
> print(cor_test)

        Pearson's product-moment correlation

data:  height and weight
t = 5.1239, df = 250, p-value = 5.99e-07
alternative hypothesis: true correlation is not equal to 0
95 percent confidence interval:
 0.1920207 0.4160038
sample estimates:
      cor
0.3082785

>
```

Steps 5 & 6:   From the results, we see that the *p*-value of the test statistic is less than the chosen significance level, suggesting that we have sufficient evidence to reject the null hypothesis. Thus, there is a significant correlation between height and weight of the respondents 5% level of significance.

7.2.7 Test of Regression Coefficient

We assume that X and Y values are normally distributed and after estimating the regression coefficients, we need to test the hypothesis whether the coefficients are statistically significant or not before prediction. For these purposes, we consider the null hypothesis $H_0 : \beta_i = 0; i = 1, 2, \ldots, k$ where k represents the number of covariates used in the regression model. For testing the hypothesis $H_0 : \beta_i = 0$ against $H_0 : \beta_i \neq 0; i = 1, 2, \ldots, k$, we use the following test statistic:

$$t_i = \frac{\hat{\beta}_i}{SE\left(\hat{\beta}_i\right)}$$

where $SE\left(\hat{\beta}_i\right)$ is the estimate of the standard error of $\hat{\beta}_i$, and under the null hypothesis, the test statistic (t_i) follows Student's t with $(n - 2)$ degrees of freedom.

Example 7.18

We recall the 'Body Fat Prediction Dataset' from **Example 7.2**. Using this dataset, we first fit a multiple regression model considering 'Percent body fat' as the dependent variable and following independent variables.

- a. Age (years)
- b. Weight (lbs)
- c. Height (inches)
- d. Neck circumference (cm)
- e. Chest circumference (cm)
- f. Abdomen 2 circumference (cm)
- g. Hip circumference (cm)
- h. Thigh circumference (cm)

Now, we want to test the hypothesis that the regression coefficients of 'age' and 'Abdomen 2 circumference' are zero, i.e., $H_{01} : \beta_{age} = 0$ against $H_{A1} : \beta_{age} \neq 0$, and $H_{02} : \beta_{abdomen} = 0$ against $H_{A2} : \beta_{abdomen} \neq 0$ at 5% level of significance.

Step 1: We define the hypothesis as follows,

    i. $H_{01} : \beta_{age} = 0$ against $H_{A1} : \beta_{age} \neq 0$

    ii. $H_{02} : \beta_{abdomen} = 0$ against $H_{A2} : \beta_{abdomen} \neq 0$.

Step 2: It is given that the significance level is $\alpha = 0.05$.

Steps 3 & 4: To test the hypothesis, we use the following test statistic:

$$t_i = \frac{\hat{\beta}_i}{SE\left(\hat{\beta}_i\right)}, i = 1, 2.$$

Now, to calculate the value of the test statistic and their corresponding p-value, we use the following commands in R and get the following outputs. Here, '`lm()`' function is used to fit the linear regression model. We use '`summary()`' function to view all outputs of the regression model.

R-code:

```
# Define variable
bodyfat <- data_bfat$BodyFat
age <- data_bfat$Age
height <- data_bfat$Height
weight <- data_bfat$Weight
neck <- data_bfat$Neck
```

```
chest <- data_bfat$Chest
abdomen <- data_bfat$Abdomen
hip <- data_bfat$Hip
thigh <- data_bfat$Thigh
# Fit the multiple linear regression model
model2 <- lm(bodyfat ~ age + height + weight + neck + chest + abdomen
+ hip + thigh )
coef1 <- coef(model2)['age']
coef2 <- coef(model2)['abdomen']
se1 <- summary(model2)$coefficients['age', 'Std. Error']
se2 <- summary(model2)$coefficients['abdomen', 'Std. Error']
# Sample size
n <- length(bodyfat)
# Calculate the test statistic
t_statistic1 <- coef1 / se1
t_statistic2 <- coef2 / se2
# Degrees of freedom
df <- n - 2
# Calculate the two-tailed p-value
p_value1 <- 2*(1-pt(abs(t_statistic1), df))
p_value2 <- 2*(1-pt(abs(t_statistic2), df))
# View the results
cat('Test Statistic (t):', t_statistic1, '\n')
cat('Two-Tailed p-Value:', p_value1, '\n')
cat('Test Statistic (t):', t_statistic2, '\n')
cat('Two-Tailed p-Value:', p_value2, '\n')
```

Output:

```
Console   Terminal ×   Background Jobs ×
R R 4.3.2 · ~/
> # Define variable
> bodyfat <- data_bfat$BodyFat
> age <- data_bfat$Age
> height <- data_bfat$Height
> weight <- data_bfat$Weight
> neck <- data_bfat$Neck
> chest <- data_bfat$Chest
> abdomen <- data_bfat$Abdomen
> hip <- data_bfat$Hip
> thigh <- data_bfat$Thigh
> # Fit the multiple linear regression model
> model2 <- lm(bodyfat ~ age + height + weight + neck + chest + abdomen + hip + thigh )
> coef1 <- coef(model2)["age"]
> coef2 <- coef(model2)["abdomen"]
> se1 <- summary(model2)$coefficients["age", "Std. Error"]
> se2 <- summary(model2)$coefficients["abdomen", "Std. Error"]
> # Sample size
> n <- length(bodyfat)
> # Calculate the test statistic
> t_statistic1 <- coef1 / se1
> t_statistic2 <- coef2 / se2
> # Degrees of freedom
> df <- n - 2
> # Calculate the two-tailed p-value
> p_value1 <- 2*(1-pt(abs(t_statistic1), df))
> p_value2 <- 2*(1-pt(abs(t_statistic2), df))
> # View the results
> cat("Test Statistic (t):", t_statistic1, "\n")
Test Statistic (t): 0.8585272
> cat("Two-Tailed p-Value:", p_value1, "\n")
Two-Tailed p-Value: 0.3914238
> cat("Test Statistic (t):", t_statistic2, "\n")
Test Statistic (t): 11.36538
> cat("Two-Tailed p-Value:", p_value2, "\n")
Two-Tailed p-Value: 0
> |
```

Or, we can use the following R command after defining the variables:

R-code:

```
model2 <- lm(bodyfat ~ age + height + weight + neck + chest + abdomen
+ hip + thigh )
summary(model2)
```

Output:

```
Console  Terminal ×  Background Jobs ×

R  R 4.3.2 · ~/
> model2 <- lm(bodyfat ~ age + height + weight + neck + chest + abdomen + hip + thigh )
> summary(model2)

Call:
lm(formula = bodyfat ~ age + height + weight + neck + chest +
    abdomen + hip + thigh)

Residuals:
    Min      1Q  Median      3Q     Max
-10.3609 -3.2386 -0.2673  2.8993 11.6395

Coefficients:
            Estimate Std. Error t value Pr(>|t|)
(Intercept) -20.94124   15.76756  -1.328   0.1854
age           0.02532    0.02950   0.859   0.3914
height       -0.10687    0.09683  -1.104   0.2708
weight       -0.09095    0.04853  -1.874   0.0621 .
neck         -0.55000    0.21819  -2.521   0.0124 *
chest         0.02073    0.09918   0.209   0.8346
abdomen       0.95988    0.08446  11.365   <2e-16 ***
hip          -0.26632    0.14693  -1.813   0.0711 .
thigh         0.32480    0.13560   2.395   0.0174 *
---
Signif. codes:  0 '***' 0.001 '**' 0.01 '*' 0.05 '.' 0.1 ' ' 1

Residual standard error: 4.391 on 243 degrees of freedom
Multiple R-squared:  0.7335,    Adjusted R-squared:  0.7248
F-statistic: 83.61 on 8 and 243 DF,  p-value: < 2.2e-16

> |
```

Steps 5 & 6: From the results, we observed that the p-value corresponding the coefficient of age is greater than the significance level suggesting we have not sufficient evidence to reject the null hypothesis $H_{01} : \beta_{age} = 0$ at 5% level of significance. It indicates that the age has no significant impact on the percentage of body fat for this dataset. However, the p-value of the coefficient of abdomen is less than the significance level indicating that we have sufficient evidence to reject the null hypothesis at 5% level of significance. Therefore, abdomen 2 circumference (cm) has a significant influence on the percentage of body fat.

7.3 Conclusion and Exercises

This section presents the summary of the chapter and includes problem-solving exercises.

7.3.1 Concluding Remarks

Estimation helps us to make educated guesses about population parameters based on sample data, while hypothesis testing allows us to assess the validity of our assumptions. In this chapter, we discuss the concept of statistical estimation along with examples. Moreover, we present different parametric tests commonly used in several fields including biostatistics, public health, finance, economics, and data science. With a solid understanding of estimation and hypothesis testing, you are well prepared to handle the statistical analysis in any fields.

7.3.2 Practice Questions

Exercise 7.1   Download a dataset named 'Medical Cost Personal Datasets' from 'Kaggle' database into your own computer. The dataset includes the following information:

   age: age of primary beneficiary

   sex: insurance contractor gender, female, male

   bmi: Body mass index

   children: Number of children covered by health insurance / Number of dependents

   smoker: Smoking

   region: the beneficiary's residential area

   charges: Individual medical costs billed by health insurance

Using this dataset,

   a.   Read the CSV data file from your computer in R and assign the name of your data file as `data_insurance`.

   b.   Find the mean of `age` and `bmi`.

   c.   Obtain the 95% confidence interval for both mean `age` and mean `bmi`.

   d.   Test a hypothesis that the mean `bmi` of the respondents is 30 considering 5% level of significance.

   Test a hypothesis that 45% of respondents have no child. Let $\alpha = 0.05$.

   Hints: For counting the number of respondents has no child use `sum(data_insurance$children=='0')` in R.

   e.   Split the dataset into male and female. Now, test the hypothesis that the average bmi of males is equal the average bmi of females. $\alpha = 0.05$.

   Hints: Assign the sex variable as a factor by using `sex<-as.factor(data_insurance$sex)` command in R. To split the data file into male and female you may use `male <- data_insurance[data_insurance$sex %in% 'male', ]` and `female <- data_insurance[data_insurance$sex %in% 'female', ]` in R.

   f.   Test the hypothesis that the variances of age for males and females are equal. Let $\alpha = 0.05$.

Bibliography

Ahsanullah, M., Kibria, B. G., & Shakil, M. (2014). *Normal and student's t distributions and their applications* (Vol. 4). Atlantis Press.

Balakrishnan, N., Voinov, V., & Nikulin, M. S. (2013). *Chi-squared goodness of fit tests with applications.* Academic Press.

Casella, G., & Berger, R. L. (2021). *Statistical inference.* Cengage Learning.

Conover, W. J., Johnson, M. E., & Johnson, M. M. (1981). A comparative study of tests for homogeneity of variances, with applications to the outer continental shelf bidding data. *Technometrics,* 23(4), 351–361.

Cox, D. R. (2006). *Principles of statistical inference.* Cambridge University Press.

Daniel, W. W., & Cross, C. L. (2018). *Biostatistics: A foundation for analysis in the health sciences.* Wiley.

Davis, C. B. (1994). Environmental Regulatory Statistics. In G. P. Patil, and C.R. Rao (Eds.), *Handbook of statistics, Vol. 12: Environmental statistics.* North-Holland, Amsterdam, a division of Elsevier, New York, NY, Chapter 26, 817–865.

Fowler, J., Jarvis, P., & Chevannes, M. (2021). *Practical statistics for nursing and health care.* John Wiley & Sons.

Frost, J. (2020). *Hypothesis testing: An intuitive guide for making data driven decisions.* Statistics by Jim Publishing.

Garthwaite, P. H., Jolliffe, I. T., & Jones, B. (2002). *Statistical inference.* Oxford Science Publications.

GBD 2019 Tobacco Collaborators. (2021). Spatial, temporal, and demographic patterns in prevalence of smoking tobacco use and attributable disease burden in 204 countries and territories, 1990–2019: A systematic analysis from the Global Burden of Disease Study 2019. *The Lancet,* 397(10292), 2337–2360. https://doi.org/10.1016/S0140-6736(21)01169-7

Gillard, J. (2020). *A first course in statistical inference* (p. 164). Springer.

Hossain, M. M. (2022). Practical statistics for nursing and health care. *Journal of the Royal Statistical Society Series A: Statistics in Society,* 185(3), 1461.

Lehmann, E. L., & Casella, G. (2006). *Theory of point estimation.* Springer Science & Business Media.

Li, B., & Babu, G. J. (2019). *A graduate course on statistical inference.* Springer.

Millard, S. P., & Neerchal, N. K. (2001). *Environmental Statistics with S-PLUS.* CRC Press.

Moulin, P., & Veeravalli, V. V. (2018). *Statistical inference for engineers and data scientists.* Cambridge University Press.

Rahman, A. (2017). Small area housing stress estimation in Australia: Calculating confidence intervals for a spatial microsimulation model. *Communications in Statistics Part B: Simulation and Computation,* 46(9), 7466–7484. https://doi.org/10.1080/03610918.2016.1241406

Rahman, A., Harding, A., Tanton, R., & Liu, S. (2013). Simulating the characteristics of populations at the small area level: New validation techniques for a spatial microsimulation model in Australia. *Computational Statistics and Data Analysis,* 57(1), 149–165. https://doi.org/10.1016/j.csda.2012.06.018

Siri, W. E. (1956). Gross composition of the body. In J. H. Lawrence and C. A. Tobias (Eds.), *Advances in biological and medical physics* (Vol. IV). Academic Press, Inc.

Srivastava, M. K., Khan, A. H., & Srivastava, N. (2014). *Statistical inference: Theory of estimation.* PHI Learning Pvt. Ltd.

van Belle, G., Fisher, L. D., Heagerty, P. J., & Lumley, T. (2004). *Biostatistics: A methodology for the health sciences* (2nd ed.). John Wiley & Sons.

Welch, B. L. (1947). The generalization of 'Student's' problem when several different population variances are involved. *Biometrika,* 34(1–2), 28–35. https://doi.org/10.1093/biomet/34.1-2.28

Young, G. A., & Smith, R. L. (2005). *Essentials of statistical inference* (Vol. 16). Cambridge University Press.

Zar, J. H. (2010). *Biostatistical analysis* (5th ed.). Prentice-Hall.

8

Normality Testing

8.1 What Is Normality?

In statistics, normality refers to the adherence of a dataset or a distribution to the characteristics of a normal distribution, also known as a Gaussian distribution or bell curve. A normal distribution is characterized by (Rahman & Harding, 2017; Mishra et al., 2019):

- Symmetric;
- Bell-shaped curve;
- Unimodal – it has one 'peak';
- Mean, median, and mode are approximately equal;
- About 68% of data falls within one standard deviation (σ) of the mean;
- About 95% of data falls within two standard deviations (2σ) of the mean; and
- About 99.7% of data falls within three standard deviations (3σ) of the mean.

8.2 Why Is Testing Normality Important?

To achieve the research objective, researchers must employ a range of statistical methods and tests, including *t*-tests, analysis of variance (ANOVA), and linear regression modeling, to analyze their data. Different statistical tools for analysis have different assumptions regarding the underlying distribution of the data researchers want to explore. Among these techniques, some have the assumption of normality of the data. For example, the t-test and ANOVA assume the data is normally distributed. Linear regression analysis considers that the underlying distribution of the residuals is normal (Brinkman et al., 2012; Mishra et al., 2019; Abdulla et al., 2023a, b; Hossain et al., 2023a, 2023b; Rahman et al., 2024).

When the data does not follow a normal distribution, lots of problems can arise in performing statistical analysis of the data. For instance, results can be misleading or inaccurate; type I and type II errors could increase. Moreover, non-normal data can make it challenging to interpret the results of statistical analyses, as the usual interpretations and assumptions may not hold. In addition, results obtained from non-normal data might not generalize well to the larger population or different contexts.

DOI: 10.1201/9781003426189-8

However, data often deviates from a perfect normal distribution in real-world scenarios due to various factors. Assessing normality allows researchers to make informed choices about appropriate methods and interpretations of their analyses. Therefore, prior to employing any statistical techniques or tests that rely on the assumption of normality, researchers and analysts should examine whether their data conforms to a normal distribution. If the data significantly deviates from normality, it might impact the validity of individual statistical analyses and require special considerations in the modeling process.

8.3 Graphical Methods for Investigating Normality

Graphical methods provide an intuitive and insightful way to visually inspect the distribution of data. These graphical techniques empower researchers and analysts to explore the symmetry, skewness, kurtosis, and overall shape of the dataset, shedding light on whether it adheres to the characteristics of a normal distribution. Three popular and commonly used graphical methods, such as probability histogram with the normal probability curve, probability-probability (P-P) plot, and quantile-quantile (Q-Q) plot for assessing the distribution shape of data are described below with real-life public health examples in R-programming languages.

8.3.1 Probability Histogram with Normal Probability Curve

The probability histogram is a great method to quickly visualize the distribution of a single variable by showing the probability of each outcome. Therefore, creating a probability histogram with a normal curve overlaid on it is a common way to visually assess that the distribution curve of the sample data approximately matches the normal distribution curve. This approach is valuable that can help us to determine whether our data follows a normal distribution or deviates significantly from it. Nevertheless, it may not be as effective in pinpointing normality specifically (in contrast to other symmetric distributions) unless the sample size is relatively large.

Example 8.1

Recall the 'Heart Disease Dataset' dataset from **Exercise 3.1**. We can import this data file into R using the following R-code:

R-code:

```
data <- read.csv('E:/Book/Chapter 8/Data/heart.csv', header = TRUE)
attach(data)
head(data)
```

Output:

```
Console   Terminal ×   Background Jobs ×                                    ▄ ☐
  R R 4.3.2 · ~/
> data <- read.csv("E:/Book/Chapter 8/Data/heart.csv", header = TRUE)
> attach(data)
> head(data)
  age sex cp trestbps chol fbs restecg thalach exang oldpeak slope ca thal target
1  52   1  0      125  212   0       1     168     0     1.0     2  2    3      0
2  53   1  0      140  203   1       0     155     1     3.1     0  0    3      0
3  70   1  0      145  174   0       1     125     1     2.6     0  0    3      0
4  61   1  0      148  203   0       1     161     0     0.0     2  1    3      0
5  62   0  0      138  294   1       1     106     0     1.9     1  3    2      0
6  58   0  0      100  248   0       0     122     0     1.0     1  0    2      1
> |
```

Suppose we want to check whether serum cholesterol measured in mg/dl follows normal distribution. We can check the normality of serum cholesterol by probability histogram with a normal probability curve using the following R-code:

R-code:

```
library('ggplot2')
phnc <- ggplot(data, aes(x = chol)) +
  geom_histogram(aes(y = ..density..), colour = 'blue', fill = 'gray')
+
  stat_function(fun = dnorm,
                args = list(mean = mean(chol),
                            sd = sd(chol)),
                col = 'red',
                size = 2)+
  labs(x='Serum Cholesterol in mg/dl', y='Density') +
  theme_bw()
phnc
```

The output of probability histogram with a normal probability curve is shown in Figure 8.1, which indicates that the distribution of serum cholesterol is positively skewed and peaked; therefore, we can conclude that it deviated from the normal symmetric curve.

8.3.2 Probability-Probability (P-P) Plot

The P-P plot (probability-probability plot or percent-percent plot or *p*-value plot) is a graphical method for assessing how closely two datasets agree or how closely a dataset fits a particular distribution. It compares the empirical cumulative distribution function (ECDF) and cumulative distribution function (CDF) of the hypothetical distribution by plotting the two distributions against each other; if they are similar, the data will appear nearly a straight line.

Example 8.2

Recall the dataset from **Example 8.1**. We can check the normality of the serum cholesterol measured in mg/dl by P-P plot using the following R-code:

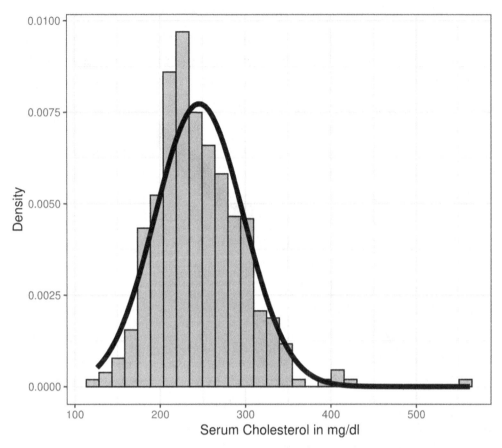

FIGURE 8.1
Probability histogram with a normal probability curve of serum cholesterol measured in mg/dl.

R-code:

```
install.packages('qqplotr')
library(qqplotr)
ppplot <- ggplot(data = data, mapping = aes(sample = chol)) +
    stat_pp_band(color = 'blue', fill = 'blue') +
    stat_pp_line(color = 'black') +
    stat_pp_point(color = 'red') +
    labs(x = 'Probability Points', y = 'Cumulative Probability') +
    theme_bw()
ppplot
```

The output of the P-P plot is illustrated in Figure 8.2. According to this P-P plot, the distribution of serum cholesterol measured in mg/dl is approximately normal because the P-P line almost looks like a straight line.

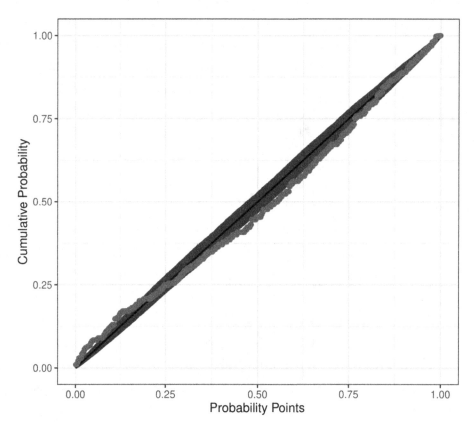

FIGURE 8.2
Probability-probability plot of serum cholesterol measured in mg/dl.

8.3.3 Quantile-Quantile (Q-Q) Plot

A Quantile-Quantile (Q-Q) plot is another powerful and well-known graphical method for assessing the normality of a dataset. It compares the quantiles of the sample data against the quantiles of a theoretical normal distribution. If the data points follow a straight line on the Q-Q plot, it suggests that the data closely follows a normal distribution.

Example 8.3

Recall the dataset from **Example 8.1**. We can check the normality of the serum cholesterol measured in mg/dl by Q-Q plot using the following R-code:

R-code:

```
qqplot <- ggplot(data = data, mapping = aes(sample = chol)) +
  stat_qq_band(color = 'blue', fill = 'blue') +
  stat_qq_line(color = 'black') +
  stat_qq_point(color = 'red') +
  labs(x = 'Theoretical Quantiles', y = 'Sample Quantiles') +
  theme_bw()
qqplot
```

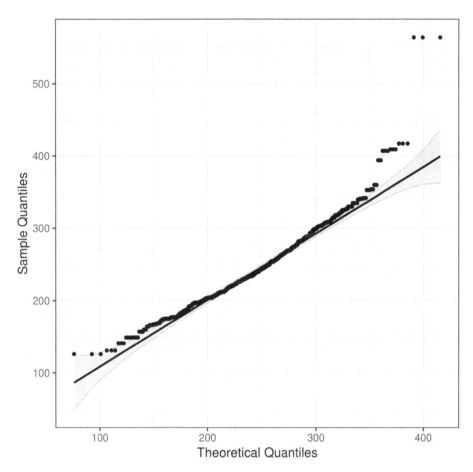

FIGURE 8.3
Quantile-quantile plot of serum cholesterol measured in mg/dl.

The output of the Q-Q plot is presented in Figure 8.3. We observed from this quantile-quantile plot that the Q-Q line does not follow a straight line and there are some outlier points; therefore, the distribution of serum cholesterol deviates from the normal distribution.

8.4 Frequentist Tests for Normality

Though the graphical methods for normality testing provide an intuitive and visual inspection to explore the distribution's shape, skewness, and potential outliers, they are not enough to finally decide whether a dataset follows normal distribution. Because graphical methods can help the researchers to assess only the distribution shape of the given dataset but cannot provide any quantitative measurement for testing the hypothesis that a given sample dataset follows normal distribution. In addition, visual inspection might be challenging in some cases, especially for large datasets. Therefore, frequentist tests are required to be conducted before making a final decision regarding whether a dataset follows normal distribution. Frequentist tests are structured frameworks for decision-making that provide quantitative measures and *p*-values that rigorously quantify the degree of departure from normality, allowing for straightforward statistical comparisons and hypothesis testing (Das et al., 2020; Abdulla et al., 2023a, b; Hossain et al., 2023a, 2023b; Ahmed et al., 2023; Imran et al., 2023).

In conclusion, while graphical methods help to quickly identify deviations from normality and guide further analyses, frequentist tests provide quantitative evidence. Therefore, researchers should not rely on only one approach to assess the normality of their dataset. Instead, they should be understood visually, through descriptive statistics and formal testing procedures, to conclude whether a given sample dataset meets the normality assumption because making the integration of both approaches is a powerful strategy in thorough normality exploration and assessment for a given dataset.

Some commonly used frequentist tests for testing the normality assumption of data are described below with real-life public health examples using R-programming languages.

8.4.1 D'Agostino's K-squared Test

D'Agostino's K-squared test, also known as the D'Agostino-Pearson omnibus test, is a statistical test used to assess whether a given dataset follows a normal distribution. It is based on the transformation of the sample skewness and sample kurtosis of the data. However, researchers should keep in mind that D'Agostino's K-squared test may not be appropriate for small sample sizes or if the data significantly deviates from normality in non-skewness and non-kurtosis ways. The D'Agostino's K-squared test statistic, denoted by K^2, is defined as:

$$K^2 = Z_1(g_1)^2 + Z_2(g_2)^2$$

where

$$Z_1(s) = \delta a \sinh\left(\frac{s}{\alpha\sqrt{\mu_2(s)}}\right);$$

$$\delta = 1/\sqrt{\ln W}; \alpha^2 = 2/(W^2 - 1); W^2 = \sqrt{2\gamma_2(s) + 4} - 1;$$

$$\mu_2(s) = \frac{6(n-2)}{(n+1)(n+3)};$$

$$\gamma_2(s) = \frac{36(n-7)(n^2 + 2n - 5)}{(n-2)(n+5)(n+7)(n+9)},$$

$$Z_2(k) = \sqrt{\frac{9A}{2}}\left\{1 - \frac{2}{9A} - \left(\frac{1 - 2/A}{1 + \frac{g_2 - \mu_1(k)}{\sqrt{\mu_2(k)}}\sqrt{2/(A-4)}}\right)^{1/3}\right\};$$

$$A = 6 + \frac{8}{\gamma_1(k)}\left(\frac{2}{\gamma_1(k)} + \sqrt{1 + 4/\gamma_1(k)}\right);$$

$$\mu_2(k) = \frac{24n(n-2)(n-3)}{(n+1)^2(n+3)(n+5)};$$

$$\gamma_1(k) = \frac{6(n^2 - 5n + 2)}{(n+7)(n+9)}\sqrt{\frac{6(n+3)(n+5)}{n(n-2)(n-3)}}, s = \text{sample skewness}; k = \text{kurtosis}.$$

The test statistic K^2 follows a chi-squared distribution with 2 degrees of freedom.

Example 8.4

Recall the dataset from **Example 8.1**. We can test the hypothesis that the serum cholesterol measured in mg/dl follows a normal distribution by D'Agostino's K-squared test using the following R-code:

R-code:

```
install.packages('fBasics')
library(fBasics)
dat = dagoTest(chol)
dat
```

Output:

```
Console   Terminal ×   Background Jobs ×

R  R 4.3.1 · ~/
> dat = dagoTest(chol)
> dat

Title:
 D'Agostino Normality Test

Test Results:
  STATISTIC:
    Chi2 | Omnibus: 236.9045
    Z3   | Skewness: 11.7142
    Z4   | Kurtosis: 9.9841
  P VALUE:
    Omnibus  Test: < 2.2e-16
    Skewness Test: < 2.2e-16
    Kurtosis Test: < 2.2e-16

> |
```

From the outputs, the D'Agostino's K-squared test ensures that the distribution of serum cholesterol does not follow a normal distribution.

8.4.2 Jarque-Bera Test

The Jarque-Bera test is a goodness-of-fit test of whether a given sample data follows a normal distribution. It is based on the skewness and kurtosis of the data, which are measures of the asymmetry and peakedness of the distribution, respectively. In a normal distribution, the skewness should be close to zero, and the kurtosis should be close to 3. If the data significantly deviates from a normal distribution, the skewness and/or kurtosis values will be different from those expected for a normal distribution. The hypotheses for the Jarque-Bera test are given below:

H_0: The data follow normal distribution.

H_1: The data do not follow normal distribution.

The Jarque-Bera test calculates a test statistic, usually denoted as JB, and compares it to a critical value from the chi-squared distribution with 2 degrees of freedom. The test is defined as follows:

$$JB = \frac{n}{6}\left(\text{skewness}^2 + \frac{1}{4}\left(\text{kurtosis} - 3\right)^2\right)$$

where n is the sample size of the given dataset.

Example 8.5

Recall the dataset from **Example 8.1**. We can test the hypothesis that the serum cholesterol measured in mg/dl follows a normal distribution by the Jarque-Bera test using the following R-code:

R-code:

```
install.packages('tseries')
library(tseries)
jb = jarque.bera.test(chol)
jb
```

Output:

```
Console   Terminal ×   Background Jobs ×
R  R 4.3.1 · ~/
> jb = jarque.bera.test(chol)
> jb

        Jarque Bera Test

data:  chol
X-squared = 870.13, df = 2, p-value < 2.2e-16

>
```

The results of the Jarque-Bera test in the output indicate that the distribution of serum cholesterol does not follow a normal distribution because the *p*-value is less than 0.001.

8.4.3 Anderson-Darling Test

The Anderson-Darling test was developed in 1952 by Theodore Anderson and Donald Darling. It is an extension of the Kolmogorov-Smirnov test used to assess whether a given dataset follows a specific theoretical probability distribution, such as the normal distribution. It is particularly useful when dealing with smaller sample sizes. The hypotheses for the Anderson-Darling test for the normal distribution are given below:

H_0: The data follow a specified distribution, i.e., normal distribution.

H_1: The data do not follow the specified distribution, i.e., normal distribution.

In essence, the Anderson-Darling test evaluates how well the observed data fits the chosen theoretical distribution by comparing the expected CDF with the empirical CDF of the data. The test calculates a statistic, the Anderson-Darling statistic, which measures the difference between the empirical distribution of the data and the expected distribution. The Anderson-Darling test is defined as:

$$AD = -n - \frac{1}{n}\sum_{i=1}^{n}(2i-1)\left[\ln F(X_i) + \ln\left(1 - F(X_{n-i+1})\right)\right].$$

Here, n is the sample size, $F(x)$ is the CDF for a specific distribution, and i is the i th sample, calculated when the data is sorted in ascending order.

The critical values for the Anderson-Darling test depend on the specific distribution being tested. The test is one-sided, and the hypothesis that the distribution is of a particular form is rejected if the test statistic, AD, is greater than the critical value or the p-value is smaller than the significance level.

Example 8.6

Recall the dataset from **Example 8.1**. We can test the hypothesis that the serum cholesterol measured in mg/dl follows a normal distribution by the Anderson-Darling test using the following R-code:

R-code:

```
install.packages('nortest')
library(nortest)
ad = ad.test(chol)
ad
```

Output:

```
Console  Terminal ×  Background Jobs ×

R  R 4.3.1 · ~/
> ad = ad.test(chol)
> ad

        Anderson-Darling normality test

data:  chol
A = 5.5575, p-value = 1.053e-13

>
```

The findings of the Anderson-Darling test reveal that the distribution of serum cholesterol does not follow the normal distribution.

8.4.4 Cramér-von Mises Test

The Cramér-von Mises test is a nonparametric test used to test the composite hypothesis, which states that the given independently and identically distributed random sample comes from a specific distribution, such as the normal distribution. The Cramér-von Mises test is based on the ECDF of the data and the CDF of the theoretical distribution we are testing against. The test calculates a statistic that quantifies the difference between the two distributions. The test statistic, denoted by W, is defined as:

$$W = \frac{1}{12n} + \sum_{i=1}^{n}\left[\frac{2i-1}{2n} - F\left(x_{(i)}\right)\right]^2.$$

Here, n is the sample size of the given dataset and $F(x_{(i)})$ is the cumulative distribution function of the specified normal distribution.

Example 8.7

Recall the dataset from **Example 8.1**. We can test the hypothesis that the serum cholesterol measured in mg/dl follows a normal distribution by the Cramér-von Mises test using the following R-code:

R-code:

```
library(nortest)
cvm = cvm.test(chol)
cvm
```

Output:

```
Console   Terminal ×   Background Jobs ×
R  R 4.3.1 · ~/
> cvm = cvm.test(chol)
> cvm

        Cramer-von Mises normality test

data:  chol
W = 0.90139, p-value = 4.015e-09

>
```

We can conclude from the findings of the Cramér-von Mises test that the distribution of serum cholesterol is deviated from normal distribution at 5% level of significance.

8.4.5 Kolmogorov-Smirnov Test

The Kolmogorov-Smirnov (K-S) test is a nonparametric test for evaluating goodness-of-fit. It is used to test the null hypothesis of whether two distributions are same or an underlying probability distribution follows a hypothesized, i.e., normal (theoretical) distribution. It is applicable only if the mean and the variance of the normal distribution are assumed known under the null hypothesis. The test statistic, denoted as the D or KS statistic, represents the maximum difference between the CDFs of the sample and the reference distributions (theoretical). The larger the D value, the more dissimilar the two distributions are. The Kolmogorov-Smirnov test is defined as:

$$D = \sup_{x} |F_n(x_i) - F(x_i)| = \max_{1 \le i \le N} \left(F(x_i) - \frac{i-1}{N}, \frac{i}{N} - F(x_i) \right).$$

Here, $F_n(x_i)$ is the theoretical distribution function and $F(x_i)$ is the empirical cumulative distribution of the sample being tested. For one sample test, the theoretical distribution function is defined as below:

$$F_n(x_i) = \frac{\text{number of elements in the sample} \le x_{(i)}}{n}.$$

Here, $x_{(i)}$ is the order statistic.

Example 8.8

Recall the dataset from **Example 8.1**. We can test the hypothesis that the serum cholesterol measured in mg/dl follows a normal distribution by one sample Kolmogorov-Smirnov (K-S) test using the following R-code:

R-code:

```
install.packages('stats')
library(stats)
ks = ks.test(chol, 'pnorm')
ks
```

Output:

```
Console   Terminal ×   Background Jobs ×

R  R 4.3.1 · ~/
> ks = ks.test(chol, "pnorm")
warning message:
In ks.test.default(chol, "pnorm") :
  ties should not be present for the Kolmogorov-Smirnov test
> ks

        Asymptotic one-sample Kolmogorov-Smirnov test

data:  chol
D = 1, p-value < 2.2e-16
alternative hypothesis: two-sided

> |
```

As *p*-value is less than 0.001, the one sample Kolmogorov-Smirnov (K-S) test concludes that the distribution of serum cholesterol does not follow normal distribution at 5% level of significance.

Now, we can test the hypothesis that the distribution of serum cholesterol and resting blood pressure in mm Hg are same by two sample Kolmogorov-Smirnov (K-S) test using the following R-code:

R-code:

```
ks = ks.test(chol, trestbps)
ks
```

Output:

```
Console   Terminal ×   Background Jobs ×
 R  R 4.3.1 · ~/
> ks = ks.test(chol, trestbps)
warning message:
In ks.test.default(chol, trestbps) :
  p-value will be approximate in the presence of ties
> ks

        Asymptotic two-sample Kolmogorov-Smirnov test

data:  chol and trestbps
D = 0.92976, p-value < 2.2e-16
alternative hypothesis: two-sided

> |
```

According to the *p*-value against two sample Kolmogorov-Smirnov (K-S) test, we can conclude that the distribution of serum cholesterol and resting blood pressure are not uniform.

8.4.6 Lilliefors Test

The Lilliefors test is a modification of the Kolmogorov-Smirnov one-sample test. It is employed when the parameters of the theoretical distribution are derived from the data itself rather than being known beforehand. The test statistic remains consistent with the Kolmogorov-Smirnov test, specifically the maximum difference between the empirical distribution function and the theoretical cumulative distribution function. However, the critical values with which *D* is compared are different. The null hypothesis of the Lilliefors test is that the sample data come from a normal distribution.

Example 8.9

Recall the dataset from **Example 8.1**. We can test the hypothesis that the serum cholesterol measured in mg/dl follows a normal distribution by the Lilliefors test using the following R-code:

R-code:

```
library(nortest)
lt = lillie.test(chol)
lt
```

Output:

```
Console   Terminal ×   Background Jobs ×                                    ⊟ ▯
R  R 4.3.1 · ~/
> lt = lillie.test(chol)
> lt

        Lilliefors (Kolmogorov-Smirnov) normality test

data:  chol
D = 0.056351, p-value = 3.951e-08

>
```

The Lilliefors test also conclude with the *p*-value of <0.001 that the serum cholesterol does not follow a normal distribution.

8.4.7 Shapiro-Wilk Test

The Shapiro-Wilk test is a statistical test used to determine if a given dataset follows a normal distribution. The Shapiro-Wilk test is a more suitable approach when the sample is smaller specifically less than 50 (Mishra et al., 2019). The hypotheses for the Shapiro-Wilk test are given below:

H_0: The data follow normal distribution; and

H_1: The data do not follow normal distribution.

The Shapiro-Wilk test statistic, denoted by W, is defined as:

$$W = \frac{\left(\sum_{i=1}^{n} a_i x_{(i)}\right)^2}{\sum_{i=1}^{n} (x_i - \bar{x})^2}.$$

Here, the $x_{(i)}$ denotes the ordered sample values ($x_{(1)}$ is the smallest) and the a_i indicates the constants generated from the means, variances and covariances of the order statistics of a sample of size n from a normal distribution.

Example 8.10

Recall the dataset from **Example 8.1**. We can test the hypothesis that the serum cholesterol measured in mg/dl follows a normal distribution by Shapiro-Wilk test using the following R-code:

R-code:

```
install.packages('dplyr')
library(dplyr)
sw = shapiro.test(chol)
sw
```

Output:

```
Console   Terminal ×   Background Jobs ×                                                    ▭▢
R  R 4.3.1 · ~/
> sw = shapiro.test(chol)
> sw

        Shapiro-wilk normality test

data:  chol
W = 0.95022, p-value < 2.2e-16

>
```

We observed that the *p*-value against the Shapiro-Wilk test is <0.001; therefore, we can conclude that the serum cholesterol does not follow a normal distribution.

8.4.8 Shapiro-Francia Test

The Shapiro-Francia test is an improvement upon the Shapiro-Wilk test and is designed to provide more accurate results for large datasets. Like the Shapiro-Wilk test, the Shapiro-Francia test also assesses whether the data follows a normal distribution. This is actually a formalization of the Q-Q plot to compare two distributions. The test statistic for the Shapiro-Francia test is defined as the correlation between the ordered sample values and the (approximated) expected ordered quantiles from the standard normal distribution. The mathematical formula of this test is given below:

$$W' = \frac{\sum_{i=1}^{n}\left(x_{(i)} - \bar{x}\right)\left(m_i - \bar{m}\right)}{\sqrt{\sum_{i=1}^{n}\left(x_{(i)} - \bar{x}\right)^2 \sum_{i=1}^{n}\left(m_i - \bar{m}\right)^2}}.$$

Here, $x_{(i)}$ is the *i*th ordered value from the given sample of size *n* and m_i is the *i*th ordered value of a normal distribution.

Example 8.11

Recall the dataset from **Example 8.1**. We can test the hypothesis that the serum cholesterol measured in mg/dl follows a normal distribution by Shapiro-Francia test using the following R-code:

R-code:

```
library(nortest)
sf = sf.test(chol)
sf
```

Output:

```
Console    Terminal ×    Background Jobs ×                                   ▭☐

R  R 4.3.1 · ~/
> sf = sf.test(chol)
> sf

          Shapiro-Francia normality test

data:  chol
W = 0.94921, p-value < 2.2e-16

> |
```

The Shapiro-Francia test also confirms that the distribution of serum cholesterol is deviated from a normal distribution at 5% level of significance.

8.4.9 Pearson's Chi-squared Test

Pearson's chi-square test for normality allows checking whether a model or theory follows an approximately normal distribution. It is useful, especially when the given variable is categorical, the data is randomly sampled, and the expected value for the number of sample observations for each level is greater than 5. The null hypothesis is that the given data is sampled from a normal distribution. Given the mean and standard deviation obtained from the given sample, the expected values under the normal distribution for every data point can be calculated. The test statistic is defined as:

$$\chi^2 = \sum_{i=1}^{n} \frac{(O_i - E_i)^2}{E_i}.$$

Here, χ^2 follows chi-square distribution with $df = n - 1 - p$, O_i is the observed ith value, E_i is the expected ith value, n is the number of observations, and p is the number of estimated parameters.

Example 8.12

Recall the dataset from **Example 8.1**. We can test the hypothesis that the serum cholesterol measured in mg/dl follows a normal distribution by Pearson's chi-square test using the following R-code:

R-code:

```
library(nortest)
pct = pearson.test(chol)
pct
```

Output:

```
Console   Terminal ×   Background Jobs ×                                    ▭ ☐
ⓡ  R 4.3.1 · ~/
> pct = pearson.test(chol)
> pct

        Pearson chi-square normality test

data:  chol
P = 84.09, p-value = 5.019e-07

> |
```

According to the Pearson's chi-square test, we can also take decision that the distribution of serum cholesterol is deviated from a normal distribution at 5% level of significance since the p-value is less than 0.05.

8.5 Conclusion and Exercises

This section concludes the chapter with some practical problem-solving exercises. These concepts and measures are used in the subsequent chapters.

8.5.1 Concluding Remarks

In conclusion, this chapter has provided a comprehensive exploration of various statistical methods employed to assess the normal distribution of data in the field of biostatics. In practice, a combination of graphical methods and formal tests is often recommended for a robust assessment of normality. Therefore, throughout this chapter, various normality tests (both of graphical and frequentist tests) have been explored, each with its suitability and application in real-life data using R-programming. The understanding gained from this chapter equips researchers with the knowledge needed to make informed decisions about the assumptions underlying statistical analyses and enhances the reliability of study outcomes. The knowledge gained from this chapter is required as we move through the following chapters on the statistical methods that require to meet the normality assumption for the data.

8.5.2 Practice Questions

Exercise 8.1   Using appropriate approach, test the normality of the maximum heart rate (variable name: *thalach*) from the 'Heart Disease Dataset' that is used in this chapter.

Exercise 8.2   Recall the data 'Sleep Health and Lifestyle Dataset' from **Example 3.15**. Check the normality of the sleep duration (variable name: *Sleep.Duration*) of the individuals using different graphical methods and the frequentist tests.

Bibliography

Abdulla, F., Rahman, A., & Hossain, M. M. (2023a). Prevalence and risk predictors of childhood stunting in Bangladesh. *PLoS One*, *18*(1), e0279901. https://doi.org/10.1371/journal.pone.0279901

Ahmed, K. T., Karimuzzaman, M., Mahmud, S., Rahman, L., Hossain, M. M., & Rahman, A. (2023b). Influencing factors associated with maternal delivery at home in urban areas: A cross-sectional analysis of the Bangladesh Demographic and Health Survey 2017–2018 data. *Journal of Health, Population and Nutrition*, 42, 1–13. Article 83. https://doi.org/10.1186/s41043-023-00428-9

Anderson, T. W., & Darling, D. A. (1954). A test of goodness of fit. *Journal of the American Statistical Association*, 49(268), 765–769.

Arnold, B. C., Balakrishnan, N., & Nagaraja, H. N. (2008). *A first course in order statistics*. Society for Industrial and Applied Mathematics.

Bera, A. K., & Jarque, C. M. (1981). Efficient tests for normality, homoscedasticity and serial independence of regression residuals: Monte Carlo evidence. *Economics Letters*, 7(4), 313–318.

Brinkman, S. A., Gialamas, A., Rahman, A., Mittinty, M. N., Gregory, T. A., Silburn, S., Goldfeld, S., Zubrick, S. R., Carr, V., Janus, M., Hertzman, C., & Lynch, J. W. (2012). Jurisdictional, socioeconomic and gender inequalities in child health and development: Analysis of a national census of 5-year-olds in Australia. *BMJ Open*, 2(5), 1–14 (e001075). https://doi.org/10.1136/bmjopen-2012-001075

Cochran, W. G. (1952). The χ^2 test of goodness of fit. *The Annals of Mathematical Statistics*, 23(3), 315–345.

Conover, W. J. (1999). *Practical nonparametric statistics* (Vol. 350). John Wiley & Sons.

Corder, G. W., & Foreman, D. I. (2014). *Nonparametric statistics: A step-by-step approach*. John Wiley & Sons.

Cramér, H. (1928a). On the composition of elementary errors: First paper: Mathematical deductions. *Scandinavian Actuarial Journal*, 1928(1), 13–74.

Cramér, H. (1928b). On the composition of elementary errors: second paper: Statistical applications. *Scandinavian Actuarial Journal*, 1928(1), 141–180.

D'Agostino, R. B. (2017). Tests for the normal distribution. In D'Agostino, R. B. & Stephens, M. A. (Eds.), *Goodness-of-fit-techniques* (pp. 367–420). Routledge.

D'Agostino, R. B., Belanger, A., & D'Agostino Jr, R. B. (1990). A suggestion for using powerful and informative tests of normality. *The American Statistician*, 44(4), 316–321.

Das, S., Kumar, B., Hossain, M. Z., Rahman, S. T., & Rahman, A. (2020). Estimation of child undernutrition at disaggregated administrative tiers of a north-eastern district of Bangladesh: An application of small area estimation method. In A. Rahman (Ed.), *Statistics for data science and policy analysis* (pp. 267–281). Springer. https://doi.org/10.1007/978-981-15-1735-8_20

Gan, F. F., & Koehler, K. J. (1990). Goodness-of-fit tests based on P-P probability plots. *Technometrics*, 32(3), 289–303.

Geary, R. C. (1947). Testing for normality. *Biometrika*, 34(3/4), 209–242.

Ghasemi, A., & Zahediasl, S. (2012). Normality tests for statistical analysis: A guide for non-statisticians. *International Journal of Endocrinology and Metabolism*, 10(2), 486.

Hollander, M., Wolfe, D. A., & Chicken, E. (2013). *Nonparametric statistical methods*. John Wiley & Sons.

Hossain, M. M., Abdulla, F., Hai, A., Khan, M. T. F., Rahman, A., & Rahman, A. (2023a). Exploring the prevalence, duration and determinants of participation in household chores among children aged 5–17 years in Bangladesh. *Child Indicators Research*, 16(5), 2107–2124. https://doi.org/10.1007/s12187-023-10051-z

Hossain, M. M., Abdulla, F., & Rahman, A. (2023b). Prevalence and risk factors of underweight among under-5 children in Bangladesh: Evidence from a countrywide cross-sectional study. *PLoS One*, 18(4), Article e0284797. https://doi.org/10.1371/journal.pone.0284797

Imran, A., Khanom, N., & Rahman, A. (2023). Parental perception of children's online behaviour: A study on ethnic communities in Australia. *International Journal of Environmental Research and Public Health*, 20(7), 1–22. Article 5342. https://doi.org/10.3390/ijerph20075342

Jarque, C. M., & Bera, A. K. (1980). Efficient tests for normality, homoscedasticity and serial independence of regression residuals. *Economics Letters*, 6(3), 255–259.

Jarque, C. M., & Bera, A. K. (1987). A test for normality of observations and regression residuals. *International Statistical Review/Revue Internationale de Statistique*, 55(2), 163–172.

Kolassa, J. E. (2020). *An introduction to nonparametric statistics*. CRC Press.

Kvam, P., Vidakovic, B., & Kim, S. J. (2022). *Nonparametric statistics with applications to science and engineering with R* (Vol. 1). John Wiley & Sons.

Lilliefors, H. W. (1967). On the Kolmogorov-Smirnov test for normality with mean and variance unknown. *Journal of the American Statistical Association*, 62(318), 399–402.

Lilliefors, H. W. (1969). On the Kolmogorov-Smirnov test for the exponential distribution with mean unknown. *Journal of the American Statistical Association*, 64(325), 387–389.

Marden, J. I. (2004). Positions and QQ plots. *Statistical Science*, 19(4), 606–614.

Massey Jr, F. J. (1951). The Kolmogorov-Smirnov test for goodness of fit. *Journal of the American statistical Association*, 46(253), 68–78.

Mishra, P., Pandey, C. K., Singh, U., Gupta, A., Sahu, C., & Keshri, A. (2019). Descriptive statistics and normality tests for statistical data. *Annals of Cardiac Anaesthesia*, 22(1), 67. https://doi.org/10.4103/aca.aca_157_18

Rahman, A., & Harding, A. (2017). *Small area estimation and microsimulation modeling*. CRC Press. https://doi.org/10.1201/9781315372143

Rahman, A., Othman, N., Kuddus, M. A., & Hasan, M. Z. (2024). Impact of the COVID-19 pandemic on child malnutrition in Selangor, Malaysia: A pilot study. *Journal of Infection and Public Health*, 17(5), 833–842. https://doi.org/10.1016/j.jiph.2024.02.019

Razali, N. M., & Wah, Y. B. (2011). Power comparisons of Shapiro-Wilk, Kolmogorov-Smirnov, Lilliefors and Anderson-Darling tests. *Journal of statistical modeling and analytics*, 2(1), 21–33.

Shapiro, S. S., & Francia, R. S. (1972). An approximate analysis of variance test for normality. *Journal of the American Statistical Association*, 67(337), 215–216.

Shapiro, S. S., & Wilk, M. B. (1965). An analysis of variance test for normality (complete samples). *Biometrika*, 52(3/4), 591–611.

Slakter, M. J. (1965). A comparison of the Pearson chi-square and Kolmogorov goodness-of-fit tests with respect to validity. *Journal of the American Statistical Association*, 60(311), 854–858.

Sprent, P., & Smeeton, N. C. (2007). *Applied nonparametric statistical methods*. CRC Press.

Thode, H. C. (2002). *Testing for normality* (Vol. 164). CRC Press.

Tiku, M. L., & Akkaya, A. D. (2004). *Robust estimation and hypothesis testing*. New Age International.

Wyłomańska, A., Iskander, D. R., & Burnecki, K. (2020). Omnibus test for normality based on the Edgeworth expansion. *PLoS One*, 15(6), e0233901.

Yap, B. W., & Sim, C. H. (2011). Comparisons of various types of normality tests. *Journal of Statistical Computation and Simulation*, 81(12), 2141–2155.

9

Nonparametric Tests and Applications

9.1 Parametric Tests vs Nonparametric Tests

Parametric and nonparametric tests are two categories of statistical tests used for different types of data and under different assumptions. The key differences between them are given below:

a. **Assumptions**:
  - *Parametric tests*: Parametric tests make specific assumptions about the underlying distribution of the data. The most common assumption is that the data are normally distributed, and they often assume equal variances between groups (homoscedasticity). These assumptions are essential for the validity of parametric tests.
  - *Nonparametric tests*: Nonparametric tests do not rely on specific distributional assumptions. They are considered distribution-free tests, which means they can be applied when the data do not meet the assumptions of parametric tests.

b. **Data type**:
  - *Parametric tests*: Parametric tests are primarily used for continuous data, specifically interval or ratio data. They require data that can be described in terms of specific parameters such as means and variances.
  - *Nonparametric tests*: Nonparametric tests are more versatile in terms of data type. They can be applied to nominal, ordinal, interval, and ratio data.

9.2 Common Nonparametric Tests

In real life, different nonparametric tests are used. Commonly used nonparametric tests are given below.

Chi-square test: Testing association between two categorical variables.

Anderson-Darling test: A goodness-of-fit test used to determine if a sample comes from a specific distribution, e.g., normal distribution or other.

DOI: 10.1201/9781003426189-9

Sign test: Tests whether the median of a sample is equal to a specified value, often used for small sample sizes.

Wilcoxon Rank-Sum test (Mann-Whitney U test): Compares two independent groups to determine if they come from the same population or have different central tendencies.

Wilcoxon Signed-Rank test: Compares two related (paired) groups to determine if there is a significant difference between them.

Kruskal-Wallis test: Compares the medians of three or more independent groups to determine if they are statistically different.

Runs test: Checks for randomness in a sequence of data by examining the number of runs (consecutive increasing or decreasing values).

9.2.1 Chi-Square Test of Independence

This test is used to assess whether there is a significant association between two nominal (categorical) variables. It compares observed frequencies to expected frequencies under the assumption of independence. We consider the following hypothesis to test the significance of the association between two variables.

H_0: There is no association between the two variables.

H_1: There is an association between the two variables.

To test this hypothesis, we choose the following formula for the chi-square statistic:

$$\chi^2 = \sum \frac{(O-E)^2}{E}$$

where χ^2(chi-square) is the test statistic, Σ represents the summation symbol, indicating that you should sum over all categories or cells in your contingency table, O is the observed frequency in a specific cell of the contingency table, and E is the expected frequency in the same cell under the assumption of independence.

Example 9.1

Let us consider a dataset named 'Diabetes, Hypertension and Stroke Prediction', available on the Kaggle database, and we consider the 'diabetes_data' file, which includes the variables presented in Table 9.1:

First, download the dataset from your Kaggle account into your computer (assume that it is stored in the folder 'Book>Chapter 9' on the 'E' drive) and read the dataset from the specified location.

Step 1: Now we want to test the hypothesis that
H_0: There is no association between smoking status and the incidence of Stroke.
H_1: There is an association between smoking status and the incidence of Stroke.

Step 2: The level of significance is $\alpha = 0.05$.

Step 3 & 4: For this purpose, we may use the following code in R:

TABLE 9.1

Description of the Variables Included in the 'Diabetes, Hypertension and Stroke Prediction' Dataset

Variable Name	Descriptions
Age	13-level age category 1 = 18–24/2 = 25–29/3 = 30–34/4 = 35–39/5 = 40–44/6 = 45–49/7 = 50–54/8 = 55–59/9 = 60–64/10 = 65–69/11 = 70–74/12 = 75–79/13 = 80 or older
Sex	Patient's gender (1: male; 0: female)
HighChol	0 = no high cholesterol 1 = high cholesterol
CholCheck	0 = no cholesterol check in 5 years 1 = yes cholesterol check in 5 years
BMI	Body mass index
Smoker	Have you smoked at least 100 cigarettes in your entire life? [Note: 5 packs = 100 cigarettes] 0 = no 1 = yes
HeartDiseaseorAttack	Coronary heart disease (CHD) or myocardial infarction (MI) 0 = no 1 = yes
PhysActivity	Physical activity in past 30 days - not including job 0 = no 1 = yes
Fruits	Consume fruit 1 or more times per day 0 = no 1 = yes
Veggies	Consume vegetables 1 or more times per day 0 = no 1 = yes
HvyAlcoholConsump	Adult men ≥ 14 drinks per week and adult women ≥ 7 drinks per week 0 = no 1 = yes
GenHlth	Would you say that in general your health is: scale 1–5 1 = excellent 2 = very good 3 = good 4 = fair 5 = poor
MentHlth	Days of poor mental health scale 1–30 days
PhysHlth	Physical illness or injury days in past 30 days scale 1–30
DiffWalk	Do you have serious difficulty walking or climbing stairs? 0 = no 1 = yes
Stroke	You ever had a stroke. 0 = no, 1 = yes
HighBP	0 = no high, BP 1 = high BP
Diabetes	0= no diabetes, 1 = diabetes

R-code:

```
data_diabetes <- read.csv('E:/Book/Chapter 9/diabetes_data.csv',
header = TRUE)
attach(data_diabetes)
head(data_diabetes)
```

Output:

```
Console   Terminal ×   Background Jobs ×
R  R 4.3.2 · ~/
> data_diabetes <- read.csv("E:/Book/Chapter 9/diabetes_data.csv", header = TRUE)
> attach(data_diabetes)
> head(data_diabetes)
  Age Sex HighChol CholCheck BMI Smoker HeartDiseaseorAttack PhysActivity Fruits Veggies
1   4   1        0         1  26      0                    0            1      0       1
2  12   1        1         1  26      1                    0            0      1       0
3  13   1        0         1  26      0                    0            1      1       1
4  11   1        1         1  28      1                    0            1      1       1
5   8   0        0         1  29      1                    0            1      1       1
6   1   0        0         1  18      0                    0            1      1       1
  HvyAlcoholConsump GenHlth MentHlth PhysHlth DiffWalk Stroke HighBP Diabetes
1                 0       3        5       30        0      0      1        0
2                 0       3        0        0        0      1      1        0
3                 0       1        0       10        0      0      0        0
4                 0       3        0        3        0      0      1        0
5                 0       2        0        0        0      0      0        0
6                 0       2        7        0        0      0      0        0
> |
```

R-code:

```
# To see the frequency of the variables
install.packages('epiDisplay')
library(epiDisplay)
# Smoking status
tab1(data_diabetes$Smoker, sort.group = 'decreasing', cum.percent =
TRUE)
# Stroke status
tab1(data_diabetes$Stroke, sort.group = 'decreasing', cum.percent =
TRUE)
```

Output:

```
Console  Terminal ×  Background Jobs ×
R R 4.3.2 · ~/
> library(epiDisplay)
> # Smoking status
> tab1(data_diabetes$Smoker, sort.group = "decreasing", cum.percent = TRUE)
data_diabetes$Smoker :
        Frequency Percent Cum. percent
0           37094    52.5         52.5
1           33598    47.5        100.0
  Total     70692   100.0        100.0
> # Stroke status
> tab1(data_diabetes$Stroke, sort.group = "decreasing", cum.percent = TRUE)
data_diabetes$Stroke :
        Frequency Percent Cum. percent
0           66297    93.8         93.8
1            4395     6.2        100.0
  Total     70692   100.0        100.0
>
```

R-code:

```
# Create a contingency table by cross-tabulating two categorical
variables e.g., Smoker and Stroke
contingency_table<-table(data_diabetes$Smoker, data_diabetes$Stroke)
colnames(contingency_table) <- c('Stroke (No)', 'Stroke (Yes)')
rownames(contingency_table) <- c('Smoker (No)', 'Smoker (Yes)')
# View the contingency table
print(contingency_table)
# Perform the Chi-Square Test for Independence
result<- chisq.test(contingency_table)
print(result)
# Decision making
# Determine if the result is statistically significant at a 0.05
significance level
if (result$p.value < 0.05) {
  cat('There is a significant association between variables.\n')
} else {
  cat('There is no significant association between variables.\n')
}
```

Output:

```
Console   Terminal ×   Background Jobs ×

R  R 4.3.2 · ~/
> # Create a contingency table by cross-tabulating two categorical variables e.g., Smoker and St
roke
> contingency_table<-table(data_diabetes$Smoker, data_diabetes$Stroke)
> colnames(contingency_table) <- c("Stroke (No)", "Stroke (Yes)")
> rownames(contingency_table) <- c("Smoker (No)", "Smoker (Yes)")
> # View the contingency table
> print(contingency_table)

              Stroke (No) Stroke (Yes)
  Smoker (No)       35339         1755
  Smoker (Yes)      30958         2640
> # Perform the Chi-Square Test for Independence
> result<- chisq.test(contingency_table)
> print(result)

        Pearson's Chi-squared test with Yates' continuity correction

data:  contingency_table
X-squared = 295.01, df = 1, p-value < 2.2e-16

> # Decision making
> # Determine if the result is statistically significant at a 0.05 significance level
> if (result$p.value < 0.05) {
+     cat("There is a significant association between variables.\n")
+ } else {
+     cat("There is no significant association between variables.\n")
+ }
There is a significant association between variables.
> |
```

Step 5 & 6: The *p*-value of the chi-square test statistic is less than the significance level of 0.05, suggesting the null hypothesis is rejected at the 5% level of significance. Therefore, it is considered that there is a significant relationship between smoking status and the incidence of stroke.

9.2.2 Anderson-Darling Test

The Anderson-Darling test is used to assess whether a sample of data follows a particular probability distribution, typically the normal distribution. We consider the following hypothesis.

H_0: The data follow a specified distribution, e.g., the normal distribution.

H_1: The data do not follow the specified distribution.

The Anderson-Darling test statistic is written as

$$A^2 = -n - \sum_{i=1}^{n} \frac{2i-1}{n} \left[\ln\left(F(X_i)\right) + \ln\left(1 - F(X_{n+1-i})\right) \right],$$

where n is the sample size; $F(X_i)$ is the empirical distribution function (EDF) of the sample at the ith ordered data point.

If the *p*-value is less than the chosen significance level (e.g., 0.05), we reject the null hypothesis. This indicates that the data does not follow a normal distribution. Moreover, if the *p*-value is greater than the significance level, we would fail to reject the null hypothesis, suggesting that the data may follow a normal distribution.

Example 9.2

Recall the data used in **Example 9.1**. Now, we want to test the hypothesis that BMI follows the normal distribution.

Step 1: We can formulate the following hypothesis.
         H_0: The BMI follows the normal distribution.
         H_1: The BMI does not follow the normal distribution.
Step 2: The level of significance is $\alpha = 0.05$.
Step 3 & 4: To perform the Anderson-Darling test, we use the following R-code to make the decision:

R-code:

```
# sample of data
BMI<- data_diabetes$BMI
install.packages('nortest')
library(nortest)
result <- ad.test(BMI)
# Display the results
print(result)
#Decision making
# Determine if the result is statistically significant at a 0.05
significance level
if (result$p.value < 0.05) {
  cat('The data does not follow the normal distribution.\n')
} else {
  cat('The data follows the normal distribution.\n')
}
```

Output:

```
Console   Terminal    Background Jobs

R  R 4.3.2 · ~/
> # sample of data
> BMI<- data_diabetes$BMI
> library(nortest)
> result <- ad.test(BMI)
> # Display the results
> print(result)

        Anderson-Darling normality test

data:  BMI
A = 1207.5, p-value < 2.2e-16

> #Decision making
> # Determine if the result is statistically significant at a 0.05 significance level
> if (result$p.value < 0.05) {
+     cat("The data does not follow the normal distribution.\n")
+ } else {
+     cat("The data follows the normal distribution.\n")
+ }
The data does not follow the normal distribution.
>
```

Steps 5 & 6: It is observed that the *p*-value is less than the chosen significance level of 0.05. So, we reject the null hypothesis at the 5% level of significance. This indicates that the data does not follow a normal distribution.

9.2.3 Sign Test

The Sign test is a nonparametric statistical test used to determine if the median of a sample is equal to a specified value, often denoted as 'M_0'. It is a simple and robust test, particularly useful when dealing with small sample sizes or when the data is not normally distributed. In the Sign test, we consider the following hypothesis.

H_0: The median of the population from which the sample is drawn is equal to M_0.

H_1: The median of the population is not equal to M_0.

For testing the hypothesis, the test statistic is often denoted by 'T' is calculated based on the signs of the differences between each observation in the sample and M_0. The formula for the Sign test statistic (T) is given below:

$$T = \min(p, n - p),$$

where p is the number of positive signs i.e., the number of observations in the sample greater than M_0, and n is the total number of observations in the sample (excluding those equal to M_0).

If the p-value is less than our chosen significance level, e.g., 0.05, we can reject the null hypothesis, indicating that there is a significant difference between the sample median and the specified value M_0. Again, if the p-value is greater than the significance level, we would fail to reject the null hypothesis.

Example 9.3

Recall the data used in **Example 9.1**. Now we want to test the hypothesis that 'BMI' has a median (M) of 25.

Step 1: We consider the following hypothesis.
H_0: $M = M_0$ against H_1 : $M \neq M_0$.

Step 2: The level of significance is $\alpha = 0.05$.

Steps 3 & 4: To test the hypothesis, we use the following codes in R:

R-code:

```
median(BMI)
# The specified value
M0 <- 25
# Count the number of positive signs
p <- sum(BMI > M0)
# Calculate the sample size
n <- length(BMI)
# Perform the Sign Test
result <- binom.test(p, n, p = 0.5, alternative = 'two.sided')
# Display the test result
print(result)
#Decision making
# Determine if the result is statistically significant at a 0.05
significance level
if (result$p.value < 0.05) {
  cat('The median of the population is not equal to M0.\n')
} else {
  cat('The median of the population from which the sample is drawn is
equal to M0.\n')
}
```

Output:

```
Console   Terminal    Background Jobs

R R 4.3.2 . ~/
> median(BMI)
[1] 29
> # The specified value
> M0 <- 25
> # Count the number of positive signs
> p <- sum(BMI > M0)
> # Calculate the sample size
> n <- length(BMI)
> # Perform the Sign Test
> result <- binom.test(p, n, p = 0.5, alternative = "two.sided")
> # Display the test result
> print(result)

        Exact binomial test

data:  p and n
number of successes = 51548, number of trials = 70692, p-value < 2.2e-16
alternative hypothesis: true probability of success is not equal to 0.5
95 percent confidence interval:
 0.7258992 0.7324648
sample estimates:
probability of success
              0.7291914

> #Decision making
> # Determine if the result is statistically significant at a 0.05 significance level
> if (result$p.value < 0.05) {
+     cat("The median of the population is not equal to M0.\n")
+ } else {
+     cat("The median of the population from which the sample is drawn is equal to M0.\n")
+ }
The median of the population is not equal to M0.
>
```

Step 4 & 5: Here, the *p*-value is less than our chosen significance level 0.05; thus, we can reject the null hypothesis at the 5% level of significance, indicating that there is a significant difference between the sample median and the specified value $M_0 = 25$.

9.2.4 Wilcoxon Rank-Sum Test (Mann-Whitney U Test)

The Wilcoxon Rank-Sum test, also known as the Mann-Whitney U test, is used to compare two independent samples and determine whether they come from the same population or if there is a significant difference between their population medians. This test is particularly useful when dealing with non-normally distributed data or when the assumption of equal variances in the two groups is violated. We generally consider the following hypothesis:

H_0: The two independent samples come from the same population, and there is no significant difference between their population medians, against

H_1: The two independent samples come from different populations, and there is a significant difference between their population medians.

Suppose $X_1, X_2, ..., X_{n1}$ are an independently and identically distributed (i.i.d.) sample from X, $Y_1, Y_2, ..., Y_{n2}$ are an i.i.d. sample from Y, and both samples independent of each other. Then the Wilcoxon Rank-Sum test or Mann–Whitney U statistic is defined as:

$$U = \min(U_1, U_2)$$

where $U_1 = n_1 n_2 + \dfrac{n_1(n_1+1)}{2} - R_1$, and $U_2 = n_1 n_2 + \dfrac{n_2(n_2+1)}{2} - R_2$, here R_1 and R_2 are the sum of the ranks in X and Y, respectively.

The p-value less than our chosen significance level, e.g., 0.05, suggesting that we would reject the null hypothesis, that suggest there is a significant difference in the population medians of the two groups. Moreover, if the p-value is greater, we fail to reject the null hypothesis, suggesting that there is no significant difference in the population medians.

Example 9.4

Recall the data used in **Example 9.1**. Suppose we split the dataset into two groups named 'male' and 'female' based on the 'Sex' of the respondent. Now, we want to test the hypothesis that the median BMI for males and females are equal.

Step 1: We consider the hypothesis below.
$H_0 : M_{\text{male}} = M_{\text{female}}$ against $H_1 : M_{\text{male}} \neq M_{\text{female}}$.
Step 2: The level of significance is $\alpha = 0.05$.
Step 3 & 4: In R, we use the following code to perform this test:

R-code:

```
# Data for two independent groups
#1: male; 0: female
group1 <- data_diabetes[data_diabetes$Sex=='1', ]
male<- group1$BMI
group2 <- data_diabetes[data_diabetes$Sex=='0', ]
female <- group2$BMI
# Perform the Wilcoxon Rank-Sum Test (Mann-Whitney U Test)
result <- wilcox.test(male, female)
# Display the test result
print(result)
#Decision making
# Determine if the result is statistically significant at a 0.05
significance level
if (result$p.value < 0.05) {
  cat('There is a significant difference between medians.\n')
} else {
  cat('There is no significant difference between medians.\n')
}
```

Output:

```
Console   Terminal ×   Background Jobs ×                                             ⊡ ⊡
 R  R 4.3.2 · ~/
> # Data for two independent groups
> #1: male; 0: female
> group1 <- data_diabetes[data_diabetes$Sex=="1", ]
> male<- group1$BMI
> group2 <- data_diabetes[data_diabetes$Sex=="0", ]
> female <- group2$BMI
> # Perform the wilcoxon Rank-sum Test (Mann-whitney U Test)
> result <- wilcox.test(male, female)
> # Display the test result
> print(result)

        wilcoxon rank sum test with continuity correction

data:  male and female
w = 644086879, p-value < 2.2e-16
alternative hypothesis: true location shift is not equal to 0

> #Decision making
> # Determine if the result is statistically significant at a 0.05 significance level
> if (result$p.value < 0.05) {
+     cat("There is a significant difference between medians.\n")
+ } else {
+     cat("There is no significant difference between medians.\n")
+ }
There is a significant difference between medians.
> |
```

Step 5 & 6: It is evident that the *p*-value of the Mann-Whitney U test is less than the significance level of 0.05, suggesting that we would reject the null hypothesis at the 5% level of significance. Therefore, there is a significant difference between median BMI of males and females.

9.2.5 Wilcoxon Signed-Rank Test

The Wilcoxon Signed-Rank test is used to assess whether the median of a paired sample differs significantly from a specified value, usually zero. It is designed for situations where we have paired observations or two related measurements on the same subjects, and we want to determine if there is a significant change or difference. In such situations, we consider the hypothesis as follows:

H_0: The median of the differences between paired observations is equal to zero (or a specified value), against

H_1: The median of the differences is not equal to zero (or a specified value).

The Wilcoxon Signed-Rank test statistic is denoted by W and computed in the following way:

1. Calculate the differences between paired observations and denote these differences as $d_1, d_2, ..., d_n$.

2. Discard or assign a rank of zero to any differences equal to zero, i.e., no change. For the non-zero differences, assign ranks based on their absolute values, starting with the smallest absolute difference as rank 1, the next smallest as rank 2, and so on.

3. Calculate the sum of the ranks of the positive differences and denote this sum as W^+.

4. Calculate the sum of the ranks of the negative differences and denote this sum as W^-.

5. Calculate the test statistic W as $W = \min (W^+, W^-)$.

If the p-value is less than our chosen significance level 0.05, we can reject the null hypothesis. This suggests that there is a significant difference between the paired observations, and the median difference is not equal to zero. Again, if the p-value is greater than the significance level, we would fail to reject the null hypothesis, indicating that there is no significant difference between the paired observations or that the median difference is equal to zero.

Example 9.5

Suppose we have a hypothetical scenario in a clinical trial where we want to assess whether a new drug treatment has a significant effect on patients' blood pressure levels or not. We have the following sample of data of systolic blood pressure (mm Hg) before and after treatment of ten people.

Step 1: We define the hypothesis in the following way.

H_0: The median of the differences between before and after treatment of systolic blood pressure (mm Hg) is equal to zero, against

H_2: The median of the differences is not equal to zero.

Step 2: The level of significance is $\alpha = 0.05$.

Step 3 & 4: To test the hypothesis, we use the following R-code in R:

R-code:

```
# dataset of systolic blood pressure (mm Hg) before and after
treatment
before_treatment <- c(130, 140,  138, 125, 128,  142, 132, 135,
137,  128)
after_treatment <- c(122,  128,  135, 117, 121,  138, 128, 131,
134,  125)
# Calculate the differences in blood pressure levels (after - before)
d <- after_treatment - before_treatment
# Perform the Wilcoxon Signed-Rank Test
result <- wilcox.test(d, mu = 0, alternative = 'two.sided', exact =
FALSE)
# Display the test result
print(result)
#Decision making
# Determine if the result is statistically significant at a 0.05
significance level
if (result$p.value < 0.05) {
  cat('The median of the differences is not equal to zero.\n')
} else {
  cat('The median of the differences between paired observations is
equal to zero.\n')
}
```

TABLE 9.2

Sample of Data of Systolic Blood Pressure (mm Hg) before and after
Treatment of Ten People

Before	130, 140, 138, 125, 128, 142, 132, 135, 137, 128
After	122, 128, 135, 117, 121, 138, 128, 131, 134, 125

Output:

```
Console   Terminal ×   Background Jobs ×                                              ⊟ ☐
ℝ  R 4.3.2 · ~/
> # dataset of systolic blood pressure (mm Hg) before and after treatment
> before_treatment <- c(130, 140,   138, 125,  128,   142,  132,  135,  137,  128)
> after_treatment <- c(122, 128,   135, 117,  121,   138,  128,  131,  134,   125)
> # Calculate the differences in blood pressure levels (after - before)
> d <- after_treatment - before_treatment
> # Perform the Wilcoxon Signed-Rank Test
> result <- wilcox.test(d, mu = 0, alternative = "two.sided", exact = FALSE)
> # Display the test result
> print(result)

        Wilcoxon signed rank test with continuity correction

data:  d
V = 0, p-value = 0.005635
alternative hypothesis: true location is not equal to 0

> #Decision making
> # Determine if the result is statistically significant at a 0.05 significance level
> if (result$p.value < 0.05) {
+    cat("The median of the differences is not equal to zero.\n")
+ } else {
+    cat("The median of the differences between paired observations is equal to zero.\n")
+ }
The median of the differences is not equal to zero.
>
```

Step 5 & 6: Here, the *p*-value is less than the chosen significance level 0.05; thus, we
can reject the null hypothesis at the 5% level of significance. This suggests that there is a significant difference between the paired measurements of before and after treatment of systolic blood pressure (mm Hg).

9.2.6 Kruskal-Wallis Test

The Kruskal-Wallis test is applied to determine whether there are significant differences
among three or more independent groups when the dependent variable is ordinal or continuous but not normally distributed. It is a robust alternative to one-way ANOVA when
we have non-normally distributed data and three or more groups. In this test, we consider
the following hypothesis:

H_0: There is no significant difference in the medians of the groups, against

H_1: At least one group differs significantly from the others in terms of medians.

The Kruskal-Wallis test statistic is denoted by H and can be obtained in the following way:

1) Rank all the data points from all groups combined, regardless of their group
   membership.
2) Calculate the sum of the ranks for each group.

3) Calculate the test statistic H based on the ranked data using the following formula:

$$H = \frac{12}{N(N+1)} \sum_{i=}^{g} \frac{R_i^2}{n_i} - 3(N+1),$$

where N is the total number of data points across all groups, g is the number of groups, R_i is the sum of the ranks for group i, and n_i is the number of observations in group i.

The Kruskal-Wallis test statistic H follows a chi-squared (χ^2) distribution with $(g-1)$ degrees of freedom.

If the p-value is less than the chosen significance level e.g., 0.05, we can reject the null hypothesis, indicating that there is a significant difference in medians among the groups. Also, if the p-value is greater than the significance level, we would fail to reject the null hypothesis, suggesting that there is no significant difference in medians among the groups.

Example 9.6

Recall the data used in **Example 9.1**. Now we split the dataset into five groups based on the general health status (1 = excellent, 2 = very good, 3 = good, 4 = fair, and 5 = poor). Now we compare the median BMI of these groups.

Step 1: We set the hypothesis as below:
H_0: There is no significant difference in the medians of the groups, against
H_1: At least one group differs significantly from the others in terms of medians.

Step 2: The level of significance is $\alpha = 0.05$.

Step 3 & 4: To test the hypothesis, we employ the following code in R:

R-code:

```
#data preparation
#1 = excellent, 2 = very good, 3 = good, 4 = fair, 5 = poor
#Split data into different groups
g1 <- data_diabetes[data_diabetes$GenHlth=='1', ]
g2 <- data_diabetes[data_diabetes$GenHlth=='2', ]
g3 <- data_diabetes[data_diabetes$GenHlth=='3', ]
g4 <- data_diabetes[data_diabetes$GenHlth=='4', ]
g5 <- data_diabetes[data_diabetes$GenHlth=='5', ]
# define variable
excellent <- g1$BMI
very_good <- g2$BMI
good <- g3$BMI
fair <- g4$BMI
poor <- g5$BMI
# Combine the data into a list
data_list <- list(excellent, very_good, good, fair, poor)
# Perform the Kruskal-Wallis Test
result <- kruskal.test(data_list)
# Display the test result
print(result)
#Decision making
# Determine if the result is statistically significant at a 0.05
significance level
if (result$p.value < 0.05) {
```

```
    cat('At least one group differs significantly from the others in
    terms of medians.\n')
    } else {
    cat('There is no significant difference in the medians of the
    groups.\n')
    }
```

Output:

```
Console   Terminal    Background Jobs
R  R 4.3.2 · ~/
> #data preparation
> #1 = excellent, 2 = very good, 3 = good, 4 = fair, 5 = poor
> #Split data into different groups
> g1 <- data_diabetes[data_diabetes$GenHlth=="1", ]
> g2 <- data_diabetes[data_diabetes$GenHlth=="2", ]
> g3 <- data_diabetes[data_diabetes$GenHlth=="3", ]
> g4 <- data_diabetes[data_diabetes$GenHlth=="4", ]
> g5 <- data_diabetes[data_diabetes$GenHlth=="5", ]
> # define variable
> excellent <- g1$BMI
> very_good <- g2$BMI
> good <- g3$BMI
> fair <- g4$BMI
> poor <- g5$BMI
> # Combine the data into a list
> data_list <- list(excellent, very_good, good, fair, poor)
> # Perform the Kruskal-wallis Test
> result <- kruskal.test(data_list)
> # Display the test result
> print(result)

        Kruskal-wallis rank sum test

data:  data_list
Kruskal-wallis chi-squared = 6253.6, df = 4, p-value < 2.2e-16

> #Decision making
> # Determine if the result is statistically significant at a 0.05 significance level
> if (result$p.value < 0.05) {
+    cat("At least one group differs significantly from the others in terms of medians.\n")
+ } else {
+    cat("There is no significant difference in the medians of the groups.\n")
+ }
At least one group differs significantly from the others in terms of medians.
> |
```

Step 5 & 6:   It is observed that the *p*-value is less than the significance level 0.05; thus, we may reject the null hypothesis at the 5% level of significance. Therefore, we can conclude that at least one group differs significantly from the others in terms of medians BMI.

9.2.7 Runs Test

The Runs test, also known as the Wald-Wolfowitz Runs test, is applied to determine if a sequence of data points is random or exhibits some form of systematic pattern. It is commonly used to analyze data in time series or to check for randomness in a sequence of events. The test is based on the concept of 'runs' which are consecutive sequences of data points that have the same characteristics. In a binary sequence, e.g., '0' and '1', a run is a continuous sequence of identical values. In the Runs test, we consider the following hypothesis:

H_0: The sequence is random, against

H_1: The sequence is not random; it exhibits some systematic pattern.

Firstly, we need to determine the expected number of runs and variance of the number of runs that are used to calculate the test statistics. The formula for calculating the expected number of runs in a binary sequence is given below:

$$R\_expected = 1 + \frac{2n_0n_1}{(n_0 + n_1)},$$

and the variance of the number of runs is determined by

$$\text{var}(R\_observed) = \frac{2n_0n_1(2n_0n_1 - n_0 - n_1)}{(n_0 + n_1)^2(n_0 + n_1 - 1)},$$

where n_0 is the number of zeros in the binary sequence, n_1 is the number of ones in the binary sequence.

The formula for the Z statistic in the Runs test is as follows:

$$Z = \frac{(R\_observed - R\_expected)}{\sqrt{\text{var}(R\_observed)}} \sim N(0,1).$$

If the p-value is less than the chosen significance level, we reject the null hypothesis, indicating a significant departure from randomness; otherwise, the sequence is consistent with randomness.

Example 9.7

Recall the data used in **Example 9.1**. We select the variable 'Diabetes (0 = no diabetes, 1 = diabetes)'.

Step 1: Now we test the hypothesis that the
H_0: The sequence is random, against
H_1: The sequence is not random.

Step 2: The level of significance is $\alpha = 0.05$.

Step 3 & 4: In R, we write the following commands to get the desired decision:

R-code:

```
diabetes <- data_diabetes$Diabetes
install.packages('lawstat')
library(lawstat)
# Perform the Runs Test and calculate the p-value
runs_test_result <- runs.test(diabetes)
# Display the test result
print(runs_test_result)
# Extract the p-value from the test result
p_value <- runs_test_result$p.value
# Define a significance level (alpha)
alpha <- 0.05
```

```
# Determine whether to reject the null hypothesis based on the p-value
if (p_value < alpha) {
  cat('Reject the null hypothesis. The sequence exhibits a significant
departure from randomness.\n')
} else {
  cat('Fail to reject the null hypothesis. The sequence is consistent
with randomness.\n')
}
```

Output:

```
Console   Terminal ×   Background Jobs ×                                          ⊖ ▢
R  R 4.3.2 · ~/
> diabetes <- data_diabetes$Diabetes
> library(lawstat)
> # Perform the Runs Test and calculate the p-value
> runs_test_result <- runs.test(diabetes)
> # Display the test result
> print(runs_test_result)

        Runs Test - Two sided

data: diabetes
Standardized Runs Statistic = -265.87, p-value < 2.2e-16

> # Extract the p-value from the test result
> p_value <- runs_test_result$p.value
> # Define a significance level (alpha)
> alpha <- 0.05
> # Determine whether to reject the null hypothesis based on the p-value
> if (p_value < alpha) {
+   cat("Reject the null hypothesis. The sequence exhibits a significant departure from randomne
ss.\n")
+ } else {
+   cat("Fail to reject the null hypothesis. The sequence is consistent with randomness.\n")
+ }
Reject the null hypothesis. The sequence exhibits a significant departure from randomness.
> |
```

Steps 5 & 6:   It is observed that the *p*-value is less than the significance level 0.05, which suggests rejecting the null hypothesis at the 5% level of significance. Therefore, the sequence of the status of diabetes indicates a statistically significant departure from randomness.

9.3 Conclusion and Exercises

This section presents the summary of the chapter with some real-life problem-solving exercises.

9.3.1 Concluding Remarks

The chapter explores a critical aspect of statistical analysis that is particularly useful when dealing with real-world data that may not conform to the stringent assumptions of parametric tests. This chapter scratches into the principles of nonparametric tests, highlighting their applicability. The chapter introduces key nonparametric tests, including the chi-square test for independence, Anderson-Darling test, Sign test, Wilcoxon Signed-Rank test, Mann-Whitney *U* test, Kruskal-Wallis test, and Runs test, along with various insights into their interpretation and practical use. It emphasizes the importance of choosing the right test based on the nature of the data and research questions.

This chapter furnished the reader with a solid understanding of nonparametric tests, their practical applications, and their significance in the broader context of biostatistical analysis and research. This chapter is a valuable resource for researchers, analysts, and statisticians, offering a comprehensive understanding of nonparametric tests and their pivotal role in modern data analysis. In the subsequent chapters, we will discuss different statistical methods for assessing association between variables and those methods are based on different assumptions. However, when researchers deal with some biostatistics-related real-life data that does not meet the assumptions needed to perform those methods, then the researchers face challenges. In that situation, the knowledge gained from this chapter will guide the researchers to deal with their data.

9.3.2 Practice Questions

Exercise 9.1   Every year, almost 17 million individuals worldwide pass away from cardiovascular disorders (CVDs), which primarily manifest as myocardial infarctions and heart failures. Download a dataset named 'Heart Failure Prediction' from the Kaggle database and store it on your computer. One of the main causes of heart failure is CVDs, and this dataset includes 12 variables that are useful in predicting heart failure–related mortality. The description of the attributes is given below.

> age: Age of the patient
> anaemia: Haemoglobin level of patient (0 = No; 1 = Yes)
> creatinine_phosphokinase: Level of the CPK enzyme in the blood (mcg/L)
> diabetes: If the patient has diabetes (0 = No; 1 = Yes)
> ejection_fraction: Percentage of blood leaving the heart at each contraction
> high_blood_pressure: If the patient has hypertension (0 = No; 1 = Yes)
> platelets: Platelet count of blood (kiloplatelets/mL)
> serum_creatinine: Level of serum creatinine in the blood (mg/dL)
> serum_sodium: Level of serum sodium in the blood (mEq/L)
> sex: Sex of the patient
> smoking: If the patient smokes or not (0 = No; 1 = Yes)
> time: Follow-up period (days)
> DEATH_EVENT: If the patient deceased during the follow-up period (0 = No; 1 = Yes)

Now solve the following problems:

a. Make a contingency table considering the variables *sex* and *DEATH_EVENT*. Then, test the hypothesis whether their association is significant or not at a 5% level of significance.

b. Test the hypothesis that the *age* of the respondent is normally distributed or not. Consider a 10% level of significance.

c. Consider the variable *platelets*. Now test the hypothesis that the median platelets are 260000. Choose $\alpha = 0.05$.

d. First split the dataset into two groups male and female using the variable *sex*. Now test the hypothesis that the medians of *creatinine_phosphokinase* for these two groups are equal or not at the 5% level of significance.

e. Now split the dataset based on the criteria of 'age' of the respondent as '40–50', '51–60', '61–70', '71–80', and '81+'. Now test the hypothesis that the medians of *serum_creatinine* for these groups are equal or not at the 5% level of significance.

f. Test the randomness of the variable *high_blood_pressure* at the 5% level of significance.

Exercise 9.2   Suppose you have gene expression measurements for a set of genes before and after a certain treatment. The hypothetical data are given below:

Before treatment	2.7, 3.2, 2.5, 2.8, 3.0, 3.1, 2.6
After treatment	2.2, 2.8, 2.3, 2.4, 2.6, 2.7, 2.5

Test the hypothesis that the median of the differences between before and after treatment are statistically significant or not at the 5% level of significance.

Bibliography

Bonnini, S., Corain, L., Marozzi, M., & Salmaso, L. (2014). *Nonparametric hypothesis testing: Rank and permutation methods with applications in R*. John Wiley & Sons.

Chicco, D., Jurman, G. (2020). Machine learning can predict survival of patients with heart failure from serum creatinine and ejection fraction alone. *BMC Medical Informatics and Decision Making* 20, 16. https://doi.org/10.1186/s12911-020-1023-5

Cline, G. (2019). *Nonparametric statistical methods using R*. Scientific e-Resources.

Daniel, W. W., & Cross, C. L. (2018). *Biostatistics: A foundation for analysis in the health sciences*. Wiley.

Deshpande, J. V., Naik-Nimbalkar, U., & Dewan, I. (2017). *Nonparametric statistics: Theory and methods*. World Scientific.

Dickhaus, T. (2018). *Theory of nonparametric tests*. Springer International Publishing.

Frost, J. (2020). *Hypothesis testing: An intuitive guide for making data driven decisions*. Statistics By Jim Publishing.

Gibbons, J. D., & Chakraborti, S. (2014). *Nonparametric statistical inference: Revised and expanded*. CRC Press.

Hartshorn, S. (2017). *Hypothesis testing: A visual introduction to statistical significance*. Independently Published.

Hollander, M., Wolfe, D. A., & Chicken, E. (2013). *Nonparametric statistical methods*. John Wiley & Sons.

Hossain, M. M. (2023). Applied statistics with R: A practical guide for the life sciences. *Journal of the Royal Statistical Society Series A: Statistics in Society*, 186(4), 900–901.

Kloke, J., & McKean, J. W. (2014). *Nonparametric statistical methods using R*. CRC Press.

Lehmann, E. L., & Romano, J. P. (2022). *Testing statistical hypotheses* (Vol. 1). Springer.

Lehmann, E. L., Romano, J. P., & Casella, G. (1986). *Testing statistical hypotheses* (Vol. 3). springer.

MacFarland, T. W., & Yates, J. M. (2016). *Introduction to nonparametric statistics for the biological sciences using R* (pp. 103–132). Springer.

Neuhauser, M. (2011). *Nonparametric statistical tests: A computational approach*. CRC Press.

Pons, O. (2013). *Statistical tests of nonparametric hypotheses: Asymptotic theory*. World Scientific.

Ruland, F. (2018). *The Wilcoxon-Mann-Whitney Test – An Introduction to Nonparametrics with Comments on the R Program wilcox.test*. Independently Published.

Sheskin, D. J. (2003). *Handbook of parametric and nonparametric statistical procedures*. Chapman and Hall/CRC.

Sprent, P., & Smeeton, N. C. (2016). *Applied nonparametric statistical methods*. CRC Press.

Taeger, D., & Kuhnt, S. (2014). *Statistical hypothesis testing with SAS and R*. John Wiley & Sons.

Touchon, J. C. (2021). *Applied statistics with R: A practical guide for the life sciences*. Oxford University Press.

10

Statistical Association and Correlation

10.1 Concept of Statistical Association and Causation

What does it mean to 'associate' or 'relate' two or more variables? Knowing the answer to this question is essential to comprehend a large portion of the statistical realm. In essence, the association is the observation that a pattern of data in one variable appears to be related to a pattern of data in one or more other variables in a certain way. It tells the researcher whether two variables are related. Therefore, the association is any statistical relationship between two or more variables. However, it does not explain the cause and effect between variables.

On the other hand, causation is the relationship in which one variable influences another. It is identified when a change in the value of one variable is due to a change in the value of another current variable. If an effect is caused by an exposure, it is said to be causative. Adverse exposure, such as higher risks from using illegal substances, working in a coal mine, or inhaling second-hand smoke, may be the cause. The absence of preventive exposure, such as not exercising or fasting, could be a causative factor. There needs to be a connection between the cause and the outcome, but the association alone is insufficient.

The presence of a statistical association between a risk factor and a disease does not necessarily imply a causal relationship. Similarly, no association does not necessarily mean the absence of a causal relationship. Determining whether a statistical association signifies a cause-and-effect relationship between exposure and disease requires considerations extending beyond the data of a single study. These considerations include evaluating the magnitude of the association, the consistency of findings in other studies, and the biological plausibility of the relationship (Hill, 2015). The following Bradford Hill criteria serve as a widely employed framework in epidemiology for evaluating whether an observed association is likely to be causal:

1. **Strength of association**: A higher magnitude of risk, indicating a stronger association between a risk factor and an outcome, increases the likelihood of considering the relationship as causal.
2. **Consistency**: The same findings across various populations, employing different study designs, and observed at different times contribute to the robustness of the association to be causal.
3. **Specificity**: This criterion implies a one-to-one relationship between the exposure and outcome, although such specificity is rare in reality.
4. **Temporal sequence**: It is essential for the exposure to precede the outcome, excluding the possibility of reverse causation.

DOI: 10.1201/9781003426189-10

5. **Biological gradient**: Changes in the intensity of the exposure results in a change in the severity or risk of the outcome, indicative of a dose-response relationship.

6. **Biological plausibility**: The presence of a potential biological mechanism that explains the observed association strengthens the credibility of the relationship to be causal.

7. **Coherence**: The observed relationship aligns with existing knowledge regarding the natural history or biology of the disease.

8. **Experiment**: Altering the frequency of the outcome by removing the exposure adds experimental support to the causal relationship.

9. **Analogy**: When the relationship is analogous to other established cause-and-effect relationships, it lends support to the relationship to be causal. For instance, if we know of the teratogenic effects of thalidomide, we might accept a cause-and-effect relationship for a similar agent based on less substantial evidence.

While extensively utilized, the criteria are not immune to criticism. Rothman posits that Hill did not intend these criteria to serve as a checklist for determining the interpretive causality of a reported association; however, they have been frequently employed in such a manner. Rothman argues that the Bradford Hill criteria have a shortfall in fulfilling the expectation of distinctly distinguishing between causal and noncausal relationships (Rothman, 2012). As an illustration, the initial criterion, 'strength of association', overlooks the fact that not every component cause will exhibit a strong association with the disease it produces. Additionally, the strength of the association is influenced by the prevalence of other factors (Rothman, 2012). Regarding the third criterion, 'specificity', which implies that a relationship is more likely to be causal if the exposure is linked to a single outcome, Rothman contends that this criterion is misleading. Causes can have multiple effects, as seen, for instance, with smoking (Rothman, 2012). The fifth criterion, 'biological gradient', proposes that the credibility of a causal association is increased when a dose-response curve can be illustrated (Lucas and McMichael, 2005). However, these associations might also arise from confounding or other biases (Rothman, 2012; Lucas & McMichael, 2005). According to Rothman, the sole criterion that can be regarded as causal is 'temporality', emphasizing that the cause must precede the effect. Nonetheless, establishing the precise time sequence for cause and effect can be challenging (Rothman, 2012). Causal inference is an intricate process, and arriving at a tentative inference about the causal or noncausal nature of an association involves subjectivity (Rothman, 2012).

If it seems a bit technical or hard to digest, think about the following simple examples:

Example 10.1

The summer season leads to a rise in ice cream sales. This statement has a causal association between the variables 'summer' and 'sales of ice cream'. As the summer season causes an increase in ice cream sales, we can infer that one variable has a causal effect on the other, and both variables are dependent on each other.

Example 10.2

A rise in the sales of air conditioners (AC) and ice cream is observed. This statement indicates an association, where the common cause for the increase is the 'summer season'. However, there is no causation, as the variables 'sales of AC' and 'sales of ice cream' lack a cause-and-effect relationship, and neither directly influences the other.

Example 10.3

Suppose there is a significant relationship between religion and breast cancer in a group of people where non-Muslim women are more likely to develop breast cancer, while Muslims exhibit a lower risk. Nonetheless, it is important to note that one's religion is not a causative factor for breast cancer. There are alternative explanations to consider.

Example 10.4

Extensive evidence has established a compelling link between individuals of lower socioeconomic status and an increased risk of lung cancer. However, this association does not necessarily imply that low socioeconomic status is a direct cause of lung cancer. A more plausible explanation could be that individuals with lower socioeconomic status are more prone to smoking and chronic exposure to air pollution. Such exposure to respiratory contaminants may lead to mutations in bronchial cells over time, ultimately contributing to the development of cancer.

Example 10.5

A statistical association exists between the number of individuals who drowned by falling into a pool and the number of films featuring Nicolas Cage in a given year. However, it is evident that there is no causal relationship between these two variables.

Example 10.6

A study may discover an association between the use of recreational drugs (exposure) and poor mental well-being (outcome), leading to the conclusion that drug use is likely to impact well-being negatively. An alternative explanation involving reverse causation suggests that individuals with poor mental well-being are more inclined to use recreational drugs, perhaps as a form of escapism.

Example 10.7

The relationship between lack of body exercise and heart disease explains both statistical association and causal relationship.

10.2 Importance of Measuring Statistical Association

The statistical association is concerned with how each variable is related to the other variable(s). By quantifying the strength and direction of these relationships, we gain insights into how changes in one variable may impact another. This understanding is fundamental for making informed, evidence-based decisions and drawing meaningful conclusions from data (Harjule et al., 2021; Rahman & Hossain, 2022; Ahmed et al., 2023; Ng et al., 2024). It also provides a basis for prediction (Sharif et al., 2022), invaluable in various fields such as infectious diseases (Marjan et al., 2021; Rahman et al., 2024), economics (Rahman & Harding, 2011), and epidemiology (Wong et al., 2023), where accurate forecasts can have significant implications. It also aids in identifying potential confounding factors that need to be controlled for in experimental design. In manufacturing and quality control, measuring associations between process variables can help identify which factors are responsible for variations in product quality. In fields like health insurance and finance, measuring associations helps assess and manage risk. For instance, understanding the association between certain factors and insurance claims can help insurance companies price policies appropriately and manage their risk portfolios. In business, including the biomedical business world, measuring associations can help companies identify factors influencing customer behavior, allowing for more effective marketing strategies and product development.

It is particularly crucial in the field of public health due to it guides the development of evidence-based policies, targeted interventions, and resource allocation, ultimately striving for a healthier and more equitable society. It allows public health professionals to identify and quantify risk factors for various health outcomes, disease transmission, and disparities in health outcomes. For example, studying the association between smoking and lung cancer has been instrumental in tobacco control efforts, leading to policies and campaigns aimed at reducing smoking rates. It also helps public health agencies target interventions and preventive measures more effectively. By understanding which factors are associated with the spread of diseases, such as vaccination rates and disease transmission, they can develop strategies to control outbreaks and prevent epidemics. It helps prioritize interventions, as resources can be directed toward factors with the strongest associations with adverse health outcomes. Public health policies and regulations are often informed by statistical associations. For instance, understanding the association between alcohol consumption and accidents has led to policies on blood alcohol limits for driving.

10.3 Describing Statistical Association

We can visually represent the statistical association through a scatterplot. When discussing statistical association, our focus lies in determining both the strength and direction of the association.

10.3.1 Strength

Examine the three scatterplots in Figure 10.1. The primary distinction among them pertains to the strength of the association between the variables. The strength of this association indicates how closely the distributions adhere to a pattern. Specifically in Figure 10.1, there is a notably strong association between X and Y_1, a moderate association between X

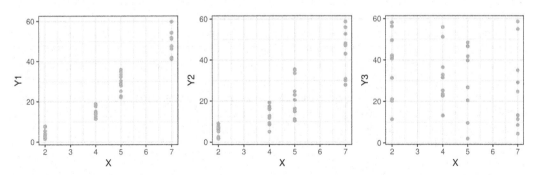

FIGURE 10.1
Scatterplots displaying varying degrees of strength of statistical association.

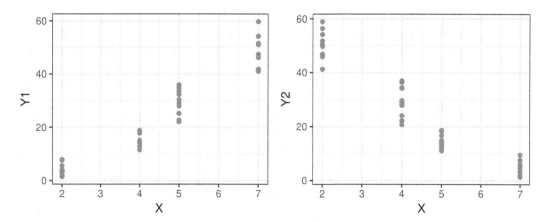

FIGURE 10.2
Scatterplots displaying different directions of statistical association.

and Y_2, and no discernible association between X and Y_3. In terms of measures of association, a value of zero denotes the absence of any relationship, while a value of one indicates a perfect relationship between the variables of interest.

10.3.2 Direction

Examine the two scatterplots in Figure 10.2. The primary distinction lies in the direction of the association. In Figure 10.2, a positive association is observed between X and Y_1, while a negative association is evident between X and Y_2. The direction of the association is consistently represented by a sign, either positive (+) or negative (−). Two variables exhibit a positive association when the values of one variable generally rise as the values of the other variable increase. Conversely, two variables show a negative association when the values of one variable tend to decrease as the values of the other variable increase.

10.4 Measures of Statistical Association

Measures of association comprise a variety of statistical techniques developed to quantify the strength and direction of relationships between two variables. These tools are important in comparing variables that may initially appear unrelated. It is essential for researchers to understand that measures of association differ from measures of statistical significance. A weak association can still be statistically significant; conversely, a strong one may lack statistical significance. In the realm of public health, measures of association are employed to assess disease occurrence in different groups, aiding in identifying risk factors associated with a particular disease of interest.

There are various methods for measuring statistical association. Some common approaches are listed below.

- Pearson's correlation coefficient.
- Spearman rank correlation coefficient.
- Kendall's rank correlation coefficient.

- Point-biserial correlation.
- Phi correlation coefficient.
- Cramer's *V*.
- Odds ratio.
- Relative risk.
- Regression analysis.

The approach employed to assess the strength and direction of an association relies on factors such as the study design, characteristics of the data for each variable, the research question, and assumptions about the data. Variables may be measured on an interval/ratio scale, an ordinal/rank scale, or a nominal/categorical scale.

10.4.1 Pearson's Correlation Coefficient

Pearson's correlation coefficient is a frequently employed statistic for computing the statistical association between two continuous variables. Regarded as the optimal method for measuring the association between variables of interest, it relies on the covariance method. This coefficient provides insights into both the magnitude and direction of the association. It is also referred to as Pearson's *r*, bivariate correlation, Pearson product-moment correlation coefficient (PPMCC), and simply, the correlation coefficient. The Pearson correlation coefficient is particularly suitable when all of the following conditions are met:

1. Both variables are quantitative;
2. The variables are normally distributed;
3. The data have no outliers;
4. The relationship is linear; and
5. The cases are independent.

Mathematically, the Pearson's correlation coefficient is defined as,

$$ r = \frac{n \sum_{i=1}^{n} x_i y_i - \left(\sum_{i=1}^{n} x_i \right) \left(\sum_{i=1}^{n} y_i \right)}{\sqrt{\left[n \sum_{i=1}^{n} x_i^2 - \left(\sum_{i=1}^{n} x_i \right)^2 \right] \left[n \sum_{i=1}^{n} y_i^2 - \left(\sum_{i=1}^{n} y_i \right)^2 \right]}} $$

The interpretation of Pearson's correlation coefficient, which ranges from −1 to +1, is as follows:

1. **Perfect**: A value near ±1 indicates a perfect correlation. In this scenario, both variables increase (if positive) or decrease (if negative) with the same increasing or decreasing rate.
2. **High degree**: When the observed coefficient value falls between ±0.50 and ±1, it is considered a strong correlation.
3. **Moderate degree**: When the value lies between ±0.30 and ±0.49, it is termed a moderate degree correlation.

4. **Low degree**: A value that falls between 0 and ±0.29 is indicative of a low degree correlation.

5. **No correlation**: A value of zero signifies no correlation between the variables.

Scatterplots of the above-mentioned degrees of Pearson's correlation coefficient are depicted in Figure 10.3.

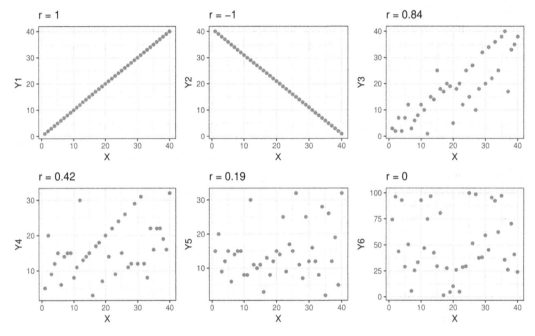

FIGURE 10.3
Scatterplots for different values of the measures of statistical association.

Figure 10.3 illustrates that X is perfect positively associated with Y_1, X is perfect negatively associated with Y_2, X and Y_3 have a high degree of statistical association, X and Y_4 have moderate degree of statistical association, and X is related to Y_5 with a low degree of statistical association. However, X and Y_6 have no statistical association.

Example 10.8

We have a dataset named 'Heart Disease Dataset' downloaded from Kaggle database that contains 4,238 observations under 16 variables. We can import this data file into R and see the first 5 observations using the following R-code:

R-code:

```
data <- read.csv('E:/Book/Chapter 10/heart_disease.csv', header =
TRUE)
attach(data)
head(data, 5)
```

Output:

```
Console   Terminal    Background Jobs
 R  R 4.3.2 · ~/
> data <- read.csv("E:/Book/Chapter 10/heart_disease.csv", header = TRUE)
> attach(data)
> head(data, 5)
  Gender age    education currentSmoker cigsPerDay BPMeds prevalentStroke prevalentHyp diabetes totChol
1   Male  39 postgraduate             0          0      0              no            0        0     195
2 Female  46 primaryschool            0          0      0              no            0        0     250
3   Male  48   uneducated             1         20      0              no            0        0     245
4 Female  61     graduate             1         30      0              no            1        0     225
5 Female  46     graduate             1         23      0              no            0        0     285
  sysBP diaBP   BMI heartRate glucose Heart_.stroke
1 106.0    70 26.97        80      77            No
2 121.0    81 28.73        95      76            No
3 127.5    80 25.34        75      70            No
4 150.0    95 28.58        65     103           yes
5 130.0    84 23.10        85      85            No
>
```

The description of the variables included in the 'Heart Disease Dataset' is presented in Table 10.1.

TABLE 10.1

Description of the Variables Included in the 'Heart Disease Dataset'

Variable Name	Description
Gender	Gender of the individual (Male and Female)
age	Age of the individual in years
education	Education level
currentSmoker	Smoking status (0 for no and 1 for yes)
cigsPerDay	The number of cigarettes consumed per day
BPMeds	Blood pressure medications (0 for no and 1 for yes)
prevalentStroke	Presence of prevalent stroke (0 for no and 1 for yes)
prevalentHyp	Presence of prevalent hypertension (0 for no and 1 for yes)
diabetes	Presence of diabetes (0 for no and 1 for yes)
totChol	Total cholesterol
sysBP	Systolic blood pressure
diaBP	Diastolic blood pressure
BMI	Body mass index
heartRate	The number of heart-beats per minute
glucose	Glucose level measured in mg/dL
Heart_ stroke	Presence of heart stroke (0 for no and 1 for yes)

Suppose we want to check whether age in years is associated with the number of cigarettes consumed per day. As both variables are continuous, we can calculate the Pearson correlation coefficient to check their association using the following R-code:

R-code:

```
p.correlation = cor(age, cigsPerDay, method = 'pearson')
p.correlation
```

Output:

```
Console   Terminal    Background Jobs
R 4.3.2 · ~/
> correlation = cor(age, cigsPerDay, method = "pearson")
> correlation
[1] NA
>
```

As we see in the output, the correlation return is missing. This is because there are some missing values in the dataset. These missing values can be estimated or omitted. For this example, we omit the missing values and calculate the correlation using the following R-code:

R-code:

```
data1 = data[complete.cases(data),]
attach(data1)
p.correlation = cor(age, cigsPerDay, method = 'pearson')
p.correlation
```

Output:

```
Console   Terminal    Background Jobs
R 4.3.2 · E:/Book - 05102023/Chapter 9 FA/
> data1 = data[complete.cases(data),]
> attach(data1)
The following objects are masked from data:

    age, BMI, BPMeds, cigsPerDay, currentSmoker, diabetes, diaBP, education, Gender, glucose,
    Heart_.stroke, heartRate, prevalentHyp, prevalentStroke, sysBP, totChol

> correlation = cor(age, cigsPerDay, method = "pearson")
> correlation
[1] -0.1890995
>
```

From the findings, we observed that age and the number of cigarettes consumed per day have a low degree of statistical association.

We can calculate the correlation among multiple variables using a correlation matrix using the following R-code:

R-code:

```
data2 = data1[, c(2, 5, 10, 11, 12, 13, 14, 15)]
p.correlation_matrix = cor(data2, method = 'pearson')
p.correlation_matrix
```

Output:

```
Console   Terminal    Background Jobs
R 4.3.2 · E:/Book - 05102023/Chapter 9 FA/
> data2 = data1[, c(2, 5, 10, 11, 12, 13, 14, 15)]
> correlation_matrix = cor(data2, method = "pearson")
> correlation_matrix
                    age   cigsPerDay      totChol        sysBP        diaBP          BMI    heartRate      glucose
age          1.000000000 -0.18909949   0.26776368   0.38855060   0.20888036   0.13717210 -0.002685426   0.11824473
cigsPerDay  -0.189099490  1.00000000  -0.03022238  -0.09476371  -0.05665012  -0.08688806  0.063549083  -0.05380272
totChol      0.267763684 -0.03022238   1.00000000   0.22012958   0.17498559   0.12079901  0.093057425   0.04974867
sysBP        0.388550599 -0.09476371   0.22012958   1.00000000   0.78672712   0.33100359  0.184901171   0.13470173
diaBP        0.208880362 -0.05665012   0.17498559   0.78672712   1.00000000   0.38561068  0.179008216   0.06370364
BMI          0.137172104 -0.08688806   0.12079901   0.33100359   0.38561068   1.00000000  0.074401235   0.08367110
heartRate   -0.002685426  0.06354908   0.09305743   0.18490117   0.17900822   0.07440124  1.000000000   0.09702585
glucose      0.118244733 -0.05380272   0.04974867   0.13470173   0.06370364   0.08367110  0.097025854   1.00000000
>
```

In the correlation matrix, the diagonal values indicate the correlation of each variable with itself and the values other than diagonal represent the bivariate association of each pair of the variables included in the dataset named 'data2'. We can visualize the correlation matrix as follows:

R-code:

```
install.packages('ggcorrplot')
library('ggcorrplot')
install.packages('stringi')
library('stringi')
corr.matrix.plot1 <- ggcorrplot(p.correlation_matrix)
corr.matrix.plot1
```

The output of matrix of scatter plot is illustrated in Figure 10.4.

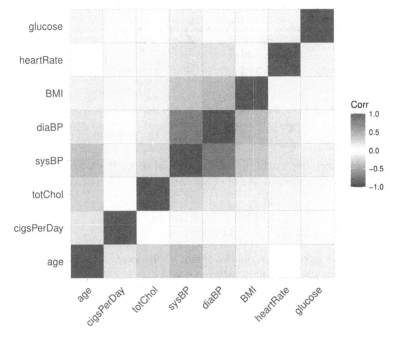

FIGURE 10.4
Heatmap displaying the bivariate association of each pair of the variables included in the dataset named 'data2'.

In the heatmap in Figure 10.4, the highest intensity of color represents the perfect statistical association and white color indicates no statistical association. Concentration of color is associated with the degree of statistical association.

10.4.2 Spearman Rank Correlation Coefficient

The Spearman rank correlation coefficient is a nonparametric measure of the strength and direction of the monotonic association between two variables measured on at least an ordinal scale or continuous (interval or ratio). The computation of this measurement requires paired observations of the variables X and Y. It is considered an alternative to Pearson's product-moment correlation when our data does not meet one or more statistical assumptions. The Spearman Rank Correlation can take a value from +1 to −1.

Only if all n ranks are distinct integers, Spearman rank correlation coefficient denoted by ρ can be computed using the following formula:

$$\rho = 1 - \frac{6 \sum_{i=1}^{n} d_i^2}{n(n^2 - 1)},$$

where ρ is the Spearman's rank correlation coefficient, d is the difference between the two ranks of each observation, and n is the number of observations.

In the presence of tied rank, the formula for examining the Spearman rank correlation coefficient is corrected as follows:

$$\rho = 1 - 6 \left[\frac{\sum_{i=1}^{n} d_i^2 + \sum_{j=1}^{n} T_{xj} + \sum_{k=1}^{r_2} T_{yk}}{n(n^2 - 1)} \right].$$

Here, $\sum_{j=1}^{r_1} T_{xj} = \dfrac{\sum_{j=1}^{r_1}\left(t_j^3 - t_j\right)}{12}$, $\sum_{k=1}^{r_2} T_{yk} = \dfrac{\sum_{k=1}^{r_2}\left(t_k^3 - t_k\right)}{12}$, t_j is the number of repetitions of a particular number in the X variable, and t_k is the number of repetitions of a particular number in the Y variable; j and k are the number of observation that has ties in the X and Y variables, respectively.

Example 10.9

Suppose we have a hypothetical dataset of size 40 observations of two variables, X and Y. We want to check the association between X and Y, where both variables are continuous. We can see the nature of the relationship between X and Y using the concept of scatterplot using the following R-code:

R-code:

```
data <- read.csv('E:/Book/Chapter 10/Spearman with No Ties.csv',
header = TRUE)
attach(data)
head(data, 5)
ggplot(data, aes(x=x, y=y)) +
  geom_point(color='orange', size = 4) +
  labs(x = 'X', y = 'Y') +
  geom_smooth(method=lm, se=FALSE, fullrange=TRUE, color='#2C3E50') +
theme_bw()
```

The output of scatter plot is shown in Figure 10.5.

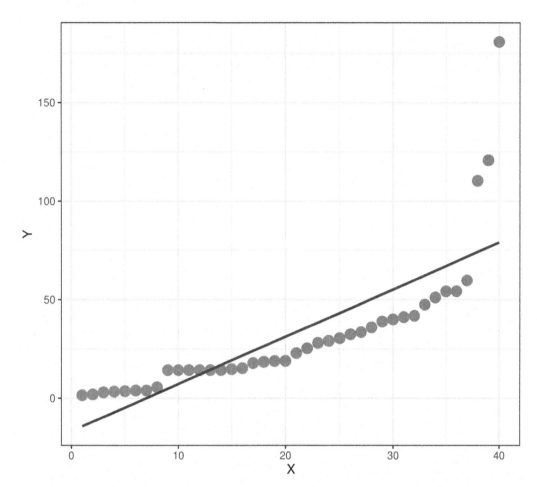

FIGURE 10.5
Scatterplot between the variables X and Y.

From the scatterplot in Figure 10.5, it is observed that the relationship between X and Y is a monotonic function, and there are some outliers in the dataset but no ties observation. Therefore, we can estimate this relationship by Spearman rank correlation coefficient using the following R-code:

R-code:

```
sp.correlation = cor(x, y, method = c('spearman'))
sp.correlation

#Or
ranked <- data.frame(cbind(rank(data$x, ties.method = 'average'),
                           rank(data$y, ties.method = 'average')))
colnames(ranked) <- c('X', 'Y')
sp.correlation <- cov(ranked) / (sd(ranked$X) * sd(ranked$Y))
sp.correlation[[2]]
```

Output:

```
Console   Terminal ×   Background Jobs ×

R  R 4.3.2 · ~/
> sp.correlation = cor(x, y, method = c("spearman"))
> sp.correlation
[1] 0.998357
>
> #or
>
> ranked <- data.frame(cbind(rank(data$x, ties.method = 'average'),
+                            rank(data$y, ties.method = 'average')))
> colnames(ranked) <- c('X', 'Y')
> sp.correlation <- cov(ranked) / (sd(ranked$X) * sd(ranked$Y))
> sp.correlation[[2]]
[1] 0.998357
>
```

We observed from the findings that there is a high degree of correlation of 0.998 between the variables X and Y.

10.4.3 Kendall Rank Correlation Coefficient

The Kendall rank correlation coefficient, also known as Kendall's τ coefficient, is a nonparametric measure of the strength and direction of the monotonic association between two variables, measured on at least an ordinal scale or continuous scale (interval or ratio). It is used as an alternative to the nonparametric Spearman rank-order correlation coefficient, especially in cases of small sample sizes and many tied ranks. To apply Kendall's coefficient, paired observations of two variables X and Y, such as the degree of deviation from diet guidelines and the degree of deviation from fluid guidelines for each patient in the sample, are required. Both variables need to be measured using at least an ordinal scale to calculate the correlation between them. Similar to Spearman's rank correlation, Kendall's τ coefficient is computed based on the ranks of the data and is denoted by τ. The formula for Kendall's τ is as follows:

$$\tau = \frac{C-D}{C+D}.$$

Here, C represents the number of concordant pairs which is equal to the number of larger ranks than a specific rank, and D represents the number of discordant pairs which is defined as the number of smaller ranks than a certain rank.

Similar to other correlation measures, Kendall's τ also ranges between −1 and +1. A positive correlation implies that the ranks of both variables increase together, while a negative correlation indicates that as the rank of one variable increases, the other variable decreases.

Example 10.10

Suppose we have a hypothetical dataset of size 40 observations of two variables: X and Y. We want to check the association between X and Y, where both variables are continuous. We can see the nature of the relationship between X and Y using the concept of scatterplot using the following R-code:

R-code:

```
data <- read.csv('E:/Book/Chapter 10/Kendall.csv', header = TRUE)
attach(data)
head(data, 5)
ggplot(data, aes(x=x, y=y)) +
  geom_point(color='orange', size = 4) +
  labs(x = 'X', y = 'Y') +
  geom_smooth(method=lm, se=FALSE, fullrange=TRUE, color='#2C3E50') +
  theme_bw()
```

The output of scatter plot is shown in Figure 10.6.

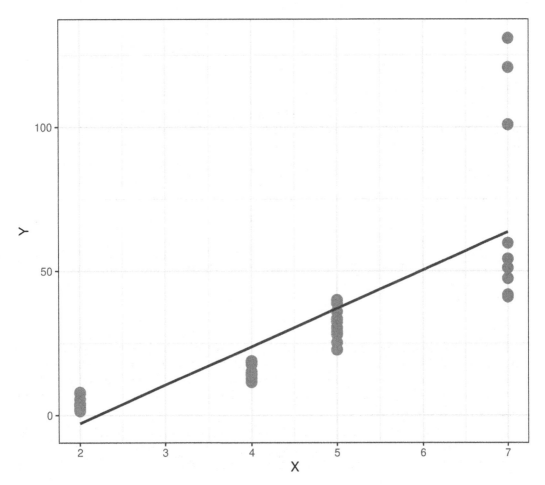

FIGURE 10.6
Scatterplot between the variables X and Y.

As we see in the scatterplot in Figure 10.6, there are some outliers in the dataset, the relationship is monotonic, and there are many tied observations in the X. Therefore, we can apply the Kendall rank correlation coefficient instead of the corrected Spearman rank correlation coefficient for estimating the correlation between X and Y using the following R-code:

R-code:

```
k.correlation = cor(x, y, method = 'kendall')
k.correlation
```

Output:

```
Console    Terminal ×    Background Jobs ×
R  R 4.3.2 · E:/Book - 05102023/Chapter 9 FA/
> k.correlation = cor(x, y, method = "kendall")
> k.correlation
[1] 0.877058
>
```

The variables X and Y are associated with a high degree of association of 0.877.

10.4.4 Point-biserial Correlation

The point-biserial correlation is employed to assess the relationship between a continuous variable and a dichotomous variable, such as a variable with two values (e.g., male/female, yes/no, true/false). It is applicable under the following assumptions:

1. The continuous variable has no outliers;
2. The continuous variable is approximately normally distributed; and
3. Homogeneity of variance for the continuous variable within both groups of the dichotomous variable.

Mathematically, the point-biserial correlation is defined as:

$$r_{pb} = \frac{M_1 - M_0}{s_n} \sqrt{\frac{n_1 n_0}{n^2}},$$

where M_1 represents the mean value of the continuous variable for all data points in group 1, and M_0 represents the mean value for all data points in group 2. Additionally, n_1 is the number of data points in group 1, n_0 is the number of data points in group 2, and n is the total sample size, s_n is the pool standard deviation for all data points in group 1 and group 2 and it is defined as:

$$s_n = \sqrt{\frac{1}{n} \sum_{i=1}^{n} (x_i - \bar{x})^2} \text{ or } s_n = \sqrt{\frac{1}{n-1} \sum_{i=1}^{n} (x_i - \bar{x})^2}.$$

Similar to Pearson's correlation coefficient, point-biserial correlation values range from −1.0 to +1.0.

Example 10.11

Recall the data from **Example 10.8**. We want to check the association between smoking status and systolic blood pressure, where smoking status is a binary variable and systolic blood pressure is a continuous variable. We can calculate the correlation between smoking status and systolic blood pressure by point-biserial correlation using the following R-code:

R-code:

```
install.packages('ltm')
library('ltm')
point.biserial = biserial.cor(sysBP, as.factor(currentSmoker), use =
c('all.obs'), level = 1)
point.biserial
```

Output:

```
Console   Terminal ×   Background Jobs ×
 R  R 4.3.2 · E:/Book - 05102023/Chapter 9 FA/
> point.biserial = biserial.cor(sysBP, as.factor(currentSmoker), use = c("all.obs"), level = 1)
> point.biserial
[1]  0.134371
>
```

The measure of point-biserial correlation indicates that the smoking status and systolic blood pressure of the individuals have a low degree of statistical association of 0.134.

10.4.5 Phi Correlation Coefficient

The Phi coefficient is a measure of the association between two binary variables (i.e., living/dead, male/female, success/failure, exposed/nonexposed). Sometimes, it is called a mean square contingency coefficient or Yule Phi coefficient. It is also referred to as the Matthews correlation coefficient (MCC) in machine learning, where it is used for measuring the quality of the binary classifiers.

Suppose a 2×2 table for two random variables x and y is distributed as follows:

	$y = 0$	$y = 1$
$x = 0$	A	B
$x = 1$	C	D

Mathematically, Phi coefficient is defined as follows:

$$\Phi = \frac{AD - BC}{\sqrt{(A+B)(C+D)(A+C)(B+D)}}.$$

Similar to Pearson correlation coefficient, the value of Phi coefficient ranges from -1 to 1. Interpretation is also similar as of Pearson correlation coefficient.

Example 10.12

Recall the data from **Example 10.8**. We want to measure the association between smoking status and heart stroke. As both of the variables have binary categories, we can calculate the correlation between smoking status and heart stroke by Phi correlation coefficient using the following R-code:

R-code:

```
install.packages('psych')
library('psych')
table <- table(data1$currentSmoker,data1$Heart_.stroke)
phi.correlation = phi(table, digits = 4)
phi.correlation
```

Output:

```
Console   Terminal ×   Background Jobs ×
R  R 4.3.2 · E:/Book - 05102023/Chapter 9 FA/
> table <- table(data1$currentSmoker,data1$Heart_.stroke)
> phi.correlation = phi(table, digits = 4)
> phi.correlation
[1] 0.0192
>
```

The estimated value of Phi coefficient shows that the association between smoking status and heart stroke is very low and it can be considered as approximately no statistical association between them.

10.4.6 Cramer's *V*

Cramer's *V* stands out as the preeminent among chi-square-based measures of nominal association. It is employed to investigate the association between two categorical variables in cases involving more than a 2 × 2 contingency table, such as 2 × 3. Notably, Cramer's *V* is a symmetrical measure, meaning that the designation of the independent (column) variable does not affect its calculation.

Let a sample of size n of the simultaneously distributed variables A and B for i = 1, 2, $\cdots$, r; j = 1, 2, $\cdots$, k be given by the frequencies. The chi-squared statistic then is:

$$\chi^2 = \sum_{i=1}^{r}\sum_{j=1}^{k}\frac{\left(n_{ij}-\dfrac{n_{i.}n_{.j}}{n}\right)^2}{\dfrac{n_{i.}n_{.j}}{n}}.$$

Here, $n_{i.} = \sum_{j=1}^{k} n_{ij}$ is the number of times the value A_i is observed and $n_{.j} = \sum_{i=1}^{r} n_{ij}$ is the number of times the value B_j is observed; and n_{ij} is the number of times the values (A_i, B_j) were observed. Then the Cramer's *V* is computed as follows:

$$V = \sqrt{\frac{\varphi^2}{\min(k-1,r-1)}} = \sqrt{\frac{\chi^2/n}{\min(k-1,r-1)}}.$$

Here, φ is the phi coefficient, χ^2 is derived from Pearson's chi-squared test, n is the grand total of observations, k being the number of columns, and r being the number of rows.

Cramer's *V* ranges from 0 to 1, with 0 indicating no association between variables and 1 signifying complete association. A value of 1 is achievable only when each variable is entirely determined by the other. Nevertheless, if the marginals exhibit significant inequality, the value of V tends to be less than 1.

Example 10.13

Recall the data from **Example 10.8**. We want to measure the association between smoking status and education level. Though the considered variables are categorical, i.e., not both are binary variables, we can calculate the correlation between smoking status and education level by Cramer's V using the following R-code:

R-code:

```
install.packages('rcompanion')
library('rcompanion')
table <- table(data1$currentSmoker,data1$education)
crV.correlation = cramerV(table, digits = 4)
crV.correlation
```

Output:

```
Console   Terminal ×   Background Jobs ×
R  R 4.3.2 · E:/Book - 05102023/Chapter 9 FA/
> table <- table(data1$currentSmoker,data1$education)
> crV.correlation = cramerV(table, digits = 4)
> crV.correlation
Cramer V
 0.0659
> |
```

The estimated value of Cramer's V indicates that the smoking status of the individuals is approximately not statistically correlated with their education level.

10.4.7 Odds Ratio

An odds ratio (OR) is a statistical measure that measures the strength and direction of the association between a variable (exposure) and the likelihood of a binary event (outcome). It is employed to compare the relative odds of the occurrence of a specific outcome (e.g., a disease or disorder) based on exposure to a particular variable (e.g., a health characteristic or aspect of medical history). Interpreting odds ratios commonly involves identifying risk factors by assessing the relationship between exposure to a risk factor and a medical outcome. The odds ratio is defined as the ratio of the odds of an event (binary outcome) occurring in one group to the odds of it occurring in another group (Rahman, 2016; Hossain et al., 2021; Abdulla et al., 2022). Odds ratios are particularly appropriate for case-control and cross-sectional studies.

To compute an odds ratio, we need a binary outcome or event and either a grouping variable or a continuous variable that we want to relate with the binary outcome or event of interest. When dealing with a continuous variable, the odds ratio helps to determine whether the odds of an event occurring change with changes in the continuous variable. For continuous predictors, a one-predictor logistic regression needs to be performed. The next step involves exponentiating the coefficient related to the predictor and its confidence bounds, presenting the results as the odds ratio along with its 95% confidence interval. In the case of a grouping variable, the odds ratio quantifies the likelihood of an outcome occurring in one context relative to a baseline or control condition. This assist in answering questions about whether an event is more or less likely to occur under one condition compared to another (Rahman & Rahman, 2019; Rahman et al., 2019). In the case of a grouping predictor, we can calculate the odds ratio using both contingency table and regression.

Suppose we have the following 2 × 2 contingency table:

	Event	No Event
Treatment/exposed/group-1	A	B
Control/not exposed/group-2	C	D

Then the odds ratio for treatment group compare to control group is calculated as follows:

$$\text{Odds ratio} = \frac{\text{Odds of event in treatment group}}{\text{Odds of event in control group}}$$

$$= \frac{[A/B]}{[C/D]}.$$

If we have the following 3 × 2 contingency table:

	Event	No Event
Group-1	A	B
Group-2	C	D
Group-3	E	F

Then the odds ratio is calculated as follows:

$$\text{Odds ratio for group- 1 compare to group- 3} = \frac{[A/B]}{[E/F]},$$

$$\text{Odds ratio for group- 2 compare to group- 3} = \frac{[C/D]}{[E/F]}.$$

As a rule of thumb, we can interpret the values for odds ratio:

- OR = 1: Exposure has no impact on the odds of the outcome;
- OR > 1: Exposure is associated with higher odds of the outcome; and
- OR < 1: Exposure is associated with lower odds of the outcome.

Example 10.14

Recall the data from **Example 10.8**. We want to measure the association between smoking status and heart stroke. In this case, smoking status is an exposure variable and heart stroke is an outcome variable. Both of the variables have binary categories, so we can measure the association between smoking status and heart stroke using odds ratio by considering heart stroke as outcome variable and 'no smoking' as reference category. The following R-code can be used:

R-code:

```
install.packages('epitools')
library('epitools')
table <- table(data1$currentSmoker,data1$Heart_.stroke)
table
or = oddsratio(table, method = 'wald', conf.level = 0.95)
or$measure
```

Output:

```
Console   Terminal    Background Jobs
R  R 4.3.2 · E:/Book/Exercise Solution - Supplementary/
> library("epitools")
> table <- table(data1$currentSmoker,data1$Heart_.stroke)
> table

    No  yes
0 1596  272
1 1503  285
> or = oddsratio(table, method = "wald", conf.level = 0.95)
> or$measure
 odds ratio with 95% C.I.
  estimate    lower    upper
0 1.000000       NA       NA
1 1.112628 0.9289205 1.332666
>
```

The findings reveal that compare to those individuals who are not smoker, the smoker individuals had 11.3% higher odds of suffering from heart stroke.

10.4.8 Relative Risk

Relative risk (RR) is an important indicator used to measure the strength and direction of the association between a variable (exposure) and a binary event (outcome). This measure indicates whether exposure is associated with an increase, decrease, or no change in the probability of the adverse outcome. It is defined as the ratio of the absolute risk (AR) (or probability) of an event (adverse outcome) for one group (exposed group/treatment group) to the absolute risk of the same event (adverse outcome) for a second group (unexposed group/control group). The absolute risk of an event for one group is defined as the number of occurrences of the event in that group divided by the total number of people in that group. Relative risk is suitable for cohort studies.

Suppose we have the following 2×2 contingency table:

	Event	No Event
Treatment/Exposed/Group-1	A	B
Control/Not Exposed/Group-2	C	D

Then the relative risk for treatment group compare to control group is calculated as follows:

$$RR = \frac{AR \text{ or Prob.of event in treatment group}}{AR \text{ or Prob.of event in control group}}$$

$$= \frac{\left[A/(A+B)\right]}{\left[C/(C+D)\right]}.$$

As a rule of thumb, we can interpret the values for relative risk:

- RR < 1: The event is less likely to occur in the treatment group;
- RR = 1: The event is equally likely to occur in each group; and
- RR > 1: The event is more likely to occur in the treatment group.

Example 10.15

Recall the data from **Example 10.8**. We want to measure the association between smoking status (variable: currentSmoker) and heart stroke (variable: Heart_ stroke). In this case, smoking status is an exposure variable and heart stroke is an outcome variable. Both of the variables have binary categories, so we can measure the association between smoking status and heart stroke using relative risk by considering heart stroke as outcome variable and 'no smoking' as reference category. The following R-code can be used:

R-code:

```
library('epitools')
table <- table(data1$currentSmoker,data1$Heart_.stroke)
table
rr = riskratio(table, method = 'wald', conf.level = 0.95)
rr$measure
```

Output:

```
Console   Terminal    Background Jobs

R  R 4.3.2 · E:/Book/Exercise Solution - Supplementary/
> library("epitools")
> table <- table(data1$currentSmoker,data1$Heart_.stroke)
> table

     No  yes
  0 1596  272
  1 1503  285
> rr = riskratio(table, method = "wald", conf.level = 0.95)
> rr$measure
  risk ratio with 95% C.I.
   estimate     lower    upper
 0 1.000000       NA       NA
 1 1.094675 0.9394106 1.275602
>
```

According to the estimated value of the risk ratio, the smoker individuals are 9.5% more likely to be patient of heart stroke.

10.5 Conclusion and Exercises

This section concludes the chapter with some practical problem-solving exercises. These concepts and measures are used in the subsequent chapters.

10.5.1 Concluding Remarks

The understanding of statistical association and correlation is integral to data science and biostatistics, forming a cornerstone for insightful analyses and informed decision-making. This chapter comprehensively explored various methods such as Pearson's correlation coefficient, Spearman rank correlation coefficient, Kendall's rank correlation coefficient, point-biserial correlation, phi correlation coefficient, Cramer's *V*, odds ratio, relative risk. All of these methods are employed to quantify and understand the relationships between variables. Each measure provides a unique insight into the associations within datasets from various areas, including biostatistics, medical, and public health data. These statistical data science tools enable researchers to uncover patterns, identify dependencies, and draw meaningful conclusions about the relationship between variables in the dataset. This chapter also helps the researchers select the appropriate measure based on the nature of the variables and the research question at hand. Moreover, in the next chapter, we will offer important content on regression modeling, one of the most widely used and handy tools in data analysis.

10.5.2 Practice Questions

Exercise 10.1 Consider the 'Heart Failure Clinical Records' dataset from the UCI Machine Learning Repository. The following variables are included in the dataset:

Variable Name	Description
age	Age of the patient in years [Integer]
anaemia	Decrease of red blood cells or hemoglobin [Binary]
creatinine_phosphokinase	Level of the CPK enzyme in the blood (mcg/L) [Integer]
diabetes	If the patient has diabetes [Binary]
ejection_fraction	Percentage of blood leaving the heart at each contraction (%) [Integer]
high_blood_pressure	If the patient has hypertension [Binary]
platelets	Platelets in the blood (kiloplatelets/mL) [Continuous]
serum_creatinine	Level of serum creatinine in the blood (mg/dL) [Continuous]
serum_sodium	Level of serum sodium in the blood (mEq/L) [Integer]
sex	Woman or man [Binary]
smoking	If the patient smokes or not [Binary]
time	Follow-up period in days [Integer]
death_event	If the patient died during the follow-up period [Binary]

a. Compute and interpret statistical association between age and serum creatinine using appropriate technique.

b. Estimate and interpret the measures of statistical association between sex and serum creatinine using appropriate technique.

c. Measure and interpret the statistical association between sex and diabetes using appropriate technique.

d. Measure and interpret the statistical association between sex and hypertension using appropriate technique.

e. Measure and interpret the statistical association between diabetes and hypertension using appropriate technique.

f. Compute and interpret the odds ratio between sex and diabetes (diabetes is outcome variable), sex and hypertension (hypertension is outcome variable), hypertension and diabetes (diabetes is outcome variable)

g. Compute and interpret the risk ratio between sex and diabetes (diabetes is outcome variable), sex and hypertension (hypertension is outcome variable), diabetes and hypertension (hypertension is outcome variable).

Exercise 10.2 Consider the 'Maternal Health Risk' dataset from the UCI Machine Learning Repository. The following variables are included in the dataset:

Variable Name	Description
Age	Any ages in years when a women during pregnant [Integer]
SystolicBP	Upper value of Blood Pressure in mmHg, another significant attribute during pregnancy [Integer]
DiastolicBP	Lower value of Blood Pressure in mmHg, another significant attribute during pregnancy [Integer]
BS	Blood glucose levels is in terms of a molar concentration [Integer]
BodyTemp	Body temperature measured in Fahrenheit scale [Integer]
HeartRate	A normal resting heart rate [Integer]
RiskLevel	Predicted Risk Intensity Level during pregnancy considering the previous attribute [Categorical]

a. Estimate and interpret the statistical association between systolic blood pressure and heart rate using appropriate technique.

b. Make a binary variable named *AgeCat* from the age variable considering 1 for less than age of 30 years and 2 for greater than or equal to age of 30 years. The *AgeCat* stands for age category.

c. Compute and interpret the statistical association between age category and risk level using appropriate technique.

d. Make a new risk level variable *nRiskLevel* from the risk level variable by releveling the 'mid risk' as 'low risk' and store the new variable into the existing data frame.

e. Compute and interpret the odds ratio between age category and new risk level using appropriate technique (consider high risk as event and age less than 30 years as reference category).

f. Compute and interpret the risk ratio between age category and new risk level using appropriate technique (consider high risk as event and age greater than or equal to 30 years as reference category).

Bibliography

Abdulla, F., Hossain, M. M., Karimuzzaman, M., Ali, M., & Rahman, A. (2022). Likelihood of infectious diseases due to lack of exclusive breastfeeding among infants in Bangladesh. *PLoS One*, 17(2), 1–15. e0263890. https://doi.org/10.1371/journal.pone.0263890

Ahmed, K. T., Karimuzzaman, M., Pinky, G. N., Dasgupta, D. P., Rahman, L., Hossain, M. M., & Rahman, A. (2023). Association of dietary diversity of 6–23 months aged children with prenatal and postnatal obstetric care: evidence from a nationwide cross-sectional study. *Journal of Health Population and Nutrition*, 42(1), 120. Article 120. https://doi.org/10.1186/s41043-023-00470-7

Bonett D. G., & Wright, T. A. (2000). Sample size requirements for Pearson, Kendall, and Spearman correlations. *Psychometrika*, 65, 23–28.

Bryman, A., & Cramer, D. (2002). *Quantitative data analysis with SPSS release 10 for Windows: A guide for social scientists*. Routledge.

Cohen, J., Cohen, P., West, S. G., & Aiken, L. S. (2013). *Applied multiple regression/correlation analysis for the behavioral sciences*. Routledge.

Cramér, H. (1999). *Mathematical methods of statistics* (Vol. 43). Princeton University Press.

Daniel, W. W. (1978). *Applied nonparametric statistics*. Houghton Mifflin.

Draper, N. R., & Smith, H. (1998). *Applied regression analysis* (Vol. 326). John Wiley & Sons.

Frost, J. (2019). *Regression analysis: An intuitive guide for using and interpreting linear models*. Statistics By Jim Publishing.

Gene V. Glass and Kenneth D. Hopkins (1995). *Statistical methods in education and psychology* (3rd ed.). Allyn & Bacon.

Harjule, P., Rahman, A., & Agarwal, B. (2021). A cross-sectional study of anxiety, stress, perception and mental health towards online learning of school children in India during COVID-19. *Journal of Interdisciplinary Mathematics*, 24(2), 411–424. https://doi.org/10.1080/09720502.2021.1889780

Hartshorn, S. (2017). Linear regression and correlation: A beginner's guide. Amazon Digital Services LLC.

Hill, A. B. (2015). The environment and disease: Association or causation? *Journal of the Royal Society of Medicine*, 108(1), 32–37.

Hossain, M. M., Yeasmin, S., Abdulla, F., & Rahman, A. (2021). Rural-urban determinants of vitamin a deficiency among under 5 children in Bangladesh: Evidence from National Survey 2017–18. *BMC Public Health*, 21(1), 1–10. Article 1569. https://doi.org/10.1186/s12889-021-11607-w

Kendall, M. G. (1938). A new measure of rank correlation. *Biometrika*, 30(1/2), 81–93.

Kendall, M. G. (1970). *Rank correlation methods* (4th ed.). Griffin.

Kruskal, W. H. (1958). Ordinal measures of association. *Journal of the American Statistical Association*, 53(284), 814–861.

Lachin, J. M. (2009). *Biostatistical methods: The assessment of relative risks*. John Wiley & Sons.

Liebetrau, A. M. (1983). *Measures of association* (Vol. 32). Sage.

Linacre, J. M., & Rasch, G. (2008). The expected value of a point-biserial (or similar) correlation. *Rasch Measurement Transactions*, 22(1), 1154.

Lucas, R. M., & McMichael, A. J. (2005). Association or causation: Evaluating links between 'environment and disease'. *Bulletin of the World Health Organization*, 83, 792–795.

Marjan, N., Rahman, A., Rois, R., & Rahman, A. (2021). Factors associated with coverage of vitamin a supplementation among Bangladeshi children: mixed modelling approach. *BMC Public Health*, 21(1), 1–11. Article 648. https://doi.org/10.1186/s12889-021-10735-7

Matthews, B. W. (1975). Comparison of the predicted and observed secondary structure of T4 phage lysozyme. *Biochimica et Biophysica Acta (BBA)-Protein Structure*, 405(2), 442–451.

Miles, J., & Shevlin, M. (2001). *Applying regression and correlation: A guide for students and researchers*. Sage.

Moran, P. A. P. (1948). Rank correlation and product-moment correlation. *Biometrika*, 35(1/2), 203–206.

Ng, K. Y., Hasan, M. Z., & Rahman, A. (2024). Examining the roles of meteorological variables in COVID-19 spread in Malaysia. *Aerobiologia*, 1–16. https://doi.org/10.1007/s10453-023-09804-8

Rahman, A. (2016). Significant risk factors for childhood malnutrition: Evidence from an Asian developing country. *Science Journal of Public Health*, 4(1), 16–27.

Rahman, A., & Harding, A. (2011). Social and health costs of tobacco smoking in Australia: Level, trend and determinants. *International Journal of Statistics and Systems*, 6(4), 399–411.

Rahman, A., & Hossain, M. M. (2022). Prevalence and determinants of fever, ARI and diarrhea among children aged 6–59 months in Bangladesh. *BMC Paediatrics*, 22(1), Article 117. https://doi.org/10.1186/s12887-022-03166-9

Rahman, A., & Rahman, M. S. (2019). Rural-urban differentials of childhood malnutrition in Bangladesh. *International Journal of Child Health and Nutrition*, 8, 35–42. https://doi.org/10.6000/1929-4247.2019.08.01.5

Rahman, A., Rahman, M. S., & Rahman, M. A. (2019). Determinants of infant mortality in Bangladesh: A nationally surveyed data analysis. *International Journal of Child Health and Nutrition*, 8(3), 93–102. https://doi.org/10.6000/1929-4247.2019.08.03.3

Rahman, A., Othman, N., Kuddus, M. A., & Hasan, M. Z. (2024). Impact of the COVID-19 pandemic on child malnutrition in Selangor, Malaysia: A pilot study. *Journal of Infection and Public Health*, 17(5), 833–842. https://doi.org/10.1016/j.jiph.2024.02.019

Rothman, K. J. (2012). *Epidemiology: An introduction*. Oxford University Press.

Schumacker, R. E., & Lomax, R. G. (2004). *A beginner's guide to structural equation modeling*. Psychology Press.

Sharif, O., Hasan, M. Z., & Rahman, A. (2022). Determining an effective short term COVID-19 prediction model in ASEAN countries. *Scientific Reports*, 12(1), Article 5083. https://doi.org/10.1038/s41598-022-08486-5

Shi, R., & Conrad, S. A. (2009). Correlation and regression analysis. *Annals of Allergy, Asthma & Immunology*, 103(4), S35–S41.

Sistrom, C. L., & Garvan, C. W. (2004). Proportions, odds, and risk. *Radiology*, 230(1), 12–19.

Spearman, C. (1904). The proof and measurement of association between two things. *The American Journal of Psychology*, 15(1), 72–101.

Stuart, A. (1953). The estimation and comparison of strengths of association in contingency tables. *Biometrika*, 40(1/2), 105–110.

Szumilas, M. (2010). Explaining odds ratios. *Journal of the Canadian Academy of Child and Adolescent Psychiatry*, 19(3), 227.

Wong, H. S., Hasan, M. Z., Sharif, O., & Rahman, A. (2023). Effect of total population, population density and weighted population density on the spread of Covid-19 in Malaysia. *PLoS One*, 18(4), e0284157. https://doi.org/10.1371/journal.pone.0284157

Wright, S. (1921). Correlation and causation. *Journal of Agricultural Research*, 20(7), 557–585.

Yule, G. U. (1912). On the methods of measuring association between two attributes. *Journal of the Royal Statistical Society*, 75(6), 579–652.

11

Regression Analysis

11.1 Concept and Importance of Regression Analysis

Regression analysis is a straightforward statistical technique for investigating and modeling the functional relationship among variables. This relationship is articulated through an equation or model that links the response variable to one or more predictor variables. The primary objective is to comprehend and quantify the impact of independent variables on the variability of the dependent variable, aiming to estimate and/or predict the mean or average value of the dependent variable based on the known or fixed values of the independent variables. The regression analysis involves estimating parameters that define valuable insights into the strength and direction of the relationships (Rahman & Sapkota, 2014; Rahman, 2019; Hossain et al., 2022), the significance of individual predictors (Rahman & Chowdhury, 2007; Kuddus & Rahman, 2015; Hossain et al., 2021), and the overall fit of the model to the observed data (Rahman et al., 2009). Additionally, it allows for identifying outliers and influential observations and assessing the model's assumptions. The resulting model can be used for prediction (Das et al., 2019; Karimuzzaman et al., 2020a, b; Rahman & Kuddus, 2020; Ahmed et al., 2023), hypothesis testing (Ip et al., 2021; Rahman et al., 2023), and understanding the overall pattern of the data. Regression analysis plays a crucial role in making informed decisions based on empirical and data-driven evidence.

Regression analysis serves four key purposes: description, estimation, prediction, and control. In terms of description, regression elucidates the association between dependent and independent variables. Estimation involves using observed values of independent variables to estimate the impact of explanatory variables on the dependent variable. Regression proves beneficial for predicting outcomes and variations in dependent variables based on the relationships between dependent and independent variables. Lastly, regression facilitates control over the effect of one or more independent variables while examining the relationship between one independent variable and the dependent variable.

The overall analysis involves a series of sequential steps that follow.

- **Defining the problem**: The initial step involves clearly articulating and defining the problem or research question that the regression analysis aims to address. This establishes the context and purpose of the analysis.
- **Choosing potentially pertinent variables**: Identification and selection of relevant variables that may have an impact on the dependent variable. This step involves theoretical considerations, domain knowledge, and preliminary exploration of potential predictors.

DOI: 10.1201/9781003426189-11

- **Collecting data**: The subsequent stage entails the systematic collection of data for the chosen variables. This may involve surveys, experiments, or the extraction of existing datasets, ensuring that the data collected aligns with the research objectives.

- **Specifying the model**: Defining the structure of the regression model, including the formulation of equations that express the relationships between the dependent and independent variables. This step establishes the groundwork for subsequent analyses.

- **Selecting a fitting method**: Choosing an appropriate method for fitting the model to the collected data. Common fitting methods include ordinary least squares (OLS), maximum likelihood estimation (MLE), or other specialized techniques based on the nature of the data and the assumptions of the model.

- **Fitting the model**: Implementing the chosen fitting method to estimate the regression model parameters. This involves finding the values for the coefficients that best align with the observed data and optimizing the model's fit.

- **Validating and critiquing the model**: Assessing the validity and robustness of the fitted model. This includes checking assumptions, examining residuals, and performing diagnostic tests to ensure that the model adequately represents the underlying relationships in the data.

- **Employing the selected model(s) to address the stated problem**: Utilizing the validated and critiqued model(s) to draw meaningful conclusions or make predictions about the initially defined problem. The insights gained from the analysis can inform decision-making, predictions, or further research.

Each of the abovementioned steps is crucial in conducting a thorough and effective regression analysis, ensuring the results are meaningful, reliable, and applicable to the specified research context.

Consider the following biostatistics examples where regression analysis helps to investigate the functional relationship:

Example 11.1

Examining the association between the number of hours of exercise per week and cardiovascular health indicators.

Example 11.2

Investigating the relationship between air quality indices and the prevalence of respiratory diseases.

Example 11.3

Investigating the association between certain genetic factors and the likelihood of developing a specific disease.

Example 11.4

Investigating the relationship between smoking habits, dietary patterns, and the incidence of chronic diseases such as cardiovascular diseases or diabetes.

Example 11.5

Analyzing factors influencing the frequency of hospital visits, emergency room admissions, or outpatient visits among a specific population.

Example 11.6

Examining the association between dietary habits, nutrient intake, and the prevalence of nutritional deficiencies or obesity.

Example 11.7

Modeling the spread of infectious diseases by analyzing factors like population density, vaccination rates, and environmental conditions.

Example 11.8

Investigating the impact of prenatal care, maternal nutrition, and socioeconomic status on birth outcomes such as low birth weight or preterm birth.

Example 11.9

Analyzing the relationship between geographic location, socioeconomic status, and access to healthcare services, such as the availability of primary care physicians.

Example 11.10

Examining the relationship between social support, economic factors, and mental health outcomes like depression or anxiety.

Example 11.11

Assessing the factors contributing to health disparities among different demographic groups, such as the relationship between income level and access to preventive services.

Example 11.12

Analyzing the correlation between air quality, water pollution, and the prevalence of respiratory or waterborne diseases.

Example 11.13

Investigating the association between physical activity, dietary choices, and the overall health status of a population.

Example 11.14

Assessing the impact of a specific healthcare policy or intervention on health outcomes and healthcare utilization.

Example 11.15

Analyzing factors contributing to substance abuse and addiction, such as socio-economic factors or mental health conditions.

11.2 Correlation vs Regression

Both regression analysis and correlation are two closely related statistical methods used to examine relationships between variables, but they serve distinct purposes and provide different types of information. Regression analysis is employed to model the association between a dependent variable and one or more independent variables and predict the dependent variable based on independent variables (e.g., Rahman et al., 2024a). It helps in understanding how changes in the independent variables relate to changes in the dependent variable. The coefficients in the regression equation provide information on the size and direction (ranges from $-\infty$ to $+\infty$) of the effect of each independent variable on the dependent variable.

On the other hand, correlation assesses the strength and direction of a linear relationship between two variables but does not establish predictability. It does not aim to predict one variable from another. The result of correlation is a correlation coefficient, ranging from -1 to 1, indicating the strength and direction of the linear relationship. Correlation does not distinguish between independent and dependent variables.

11.3 Regression vs Causation

Regression analysis and causation are distinct concepts, and it is important to understand the differences between them. Regression analysis is a statistical technique for modeling the relationships between variables and predictions, while causation involves establishing a direct cause-and-effect relationship between variables which implies that changes in one variable directly lead to changes in another variable.

Regression analysis is a valuable tool in observational studies to control for and assess the influence of potential confounding variables. On the other hand, causation often requires experimental designs, such as randomized controlled trials, where researchers manipulate the independent variable to observe its impact on the dependent variable while controlling for other factors. To establish causation, researchers often refer to criteria such as Bradford Hill criteria which includes many aspects like temporal precedence (the cause precedes the effect), correlation (a statistical relationship exists), and the elimination of alternative explanations (ruling out confounding variables). It is advised that readers read Section 10.1 for a clear understanding about regression versus correlation, as it discusses the differences between statistical association and causation in detail by the Bradford Hill criteria.

Example 11.16

There may be lots of factors such as sex, age, people's perception, and education that an influence the rate of vaccine coverage of COVID-19. Regression analysis helps to researcher to identify the influential factors (confounders) of the vaccine coverage of COVID-19. On the other hand, there may be a cause-and-effect relationship between hours of sleep and cognitive performance because sleep deprivation directly causes a decline in cognitive performance. The controlled design allows researchers to attribute the differences in cognitive performance directly to the manipulation of sleep, establishing a causal link. This conclusion is strengthened by the temporal precedence (sleep deprivation precedes low cognitive performance), correlation (a statistical relationship exists), and the elimination of alternative explanations through control measures.

11.4 Spurious Relationship and Regression

A spurious relationship refers to a mathematical correlation between two or more events or variables that are linked without a causal relationship. This lack of causation can be attributed to either coincidence or the presence of an unseen third factor (referred to as a 'common response variable,' 'confounding factor,' or 'lurking variable'). Suppose X and Y are two variables. A noncausal correlation can be spuriously generated by a preceding factor (W) that causes both ($W \rightarrow X$ and $W \rightarrow Y$), while variable X really doesn't affect variable Y at all.

Example 11.17

A spurious relationship becomes apparent when analyzing the sales of ice cream in a city. It is observed that sales might peak simultaneously with an increase in the rate of drownings in city swimming pools. Suggesting a causal link, implying that either ice-cream sales cause drownings or vice versa, would indicate a spurious relationship between the two. In actuality, a heat wave could be the factor influencing both phenomena. The heat wave serves as an illustration of a concealed or unobserved variable, commonly referred to as a confounding variable.

Example 11.18

The association between asthma medication use and physical exercise is spurious, as it is confounded by an unobserved variable (asthma severity). Therefore, failing to account for the confounding factor may lead to misguided public health interventions or recommendations. To accurately understand the relationship between variables, it is crucial to consider and control for relevant confounding variables in public health research and analysis.

Spurious regression refers to a situation in which a regression analysis indicates a significant relationship between two or more variables, but in reality, there is no causal or meaningful connection between them. Spurious regression can occur when the variables in the analysis are not truly related but appear to be due to random chance or the influence of common trends. Spurious regression is typically the result of a common problem in time series analysis known as autocorrelation. If two time series variables both exhibit autocorrelation, they may appear to be causally related, even if they are not.

One common cause of spurious regression is the presence of a common trend or a shared time trend in the variables. When both the dependent and independent variables have a time trend, they may show a high degree of correlation even if they are unrelated. In such cases, statistical tests may incorrectly suggest a significant relationship between the variables. However, upon closer examination, we realize that both variables are influenced by a common time trend, and there is no direct causal relationship between them. Moreover, in time series data, random fluctuations or noise can sometimes create the appearance of a relationship when there is none. In addition, failure to include relevant variables in the regression model can also lead to spurious results.

Example 11.19

Ice-cream sales and crime rate: There may be a high correlation between ice-cream sales and crime rate, which could suggest that ice-cream sales cause crime. However, this relationship is spurious, and the two variables are not causally related. Instead, the correlation may be due to a third variable, such as temperature, that affects both ice-cream sales and crime rate.

Therefore, spurious regression is a reminder of the importance of cautious interpretation in statistical analysis, especially when dealing with time series data and variables with shared trends. Researchers should be vigilant in identifying potential sources of spurious relationships and employ appropriate techniques to mitigate them. However, to avoid spurious regression, it is important to carefully select the variables to be analyzed and to account for any potential confounding variables that may be affecting the relationship between the variables of interest. This can be done using methods such as time series modeling, regression analysis, and Granger causality testing.

Variable selection methods are employed to choose a subset of predictor variables from a larger set of potential variables. There are various variable selection methods, including forward selection, backward elimination, stepwise selection, least absolute shrinkage and selection operator (LASSO), ridge regression, elastic net, recursive feature elimination (RFE), principal component regression (PCR), and best subset selection (Thevaraja et al., 2019). The choice of a specific variable selection method depends on factors such as the dataset size, the number of predictor variables, and the goals of the analysis. We briefly covered the stepwise selection method in Example 11.24.

11.5 Key Definitions and Terminology

Dependent variable: The dependent variable, also called the response or outcome variable, is a variable being predicted or explained in a regression model. It depends on the values of one or more independent variables.

Independent variable: An independent variable, also known as a predictor or explanatory variable, is a variable that is believed to influence or explain the variability in the dependent variable.

Alternative terminology of these variables: Literature has diverse terms for the dependent and independent variables. A representative list is shown in Table 11.1.

TABLE 11.1

Diverse Terms for the Dependent and Independent Variables

Dependent Variable	Independent Variable
⇕	⇕
Explained variable	Explanatory variable
⇕	⇕
Predictand variable	Predictor variable
⇕	⇕
Regressand variable	Regressor variable
⇕	⇕
Response variable	Stimulus variable
⇕	⇕
Endogenous variable	Exogenous variable
⇕	⇕
Outcome variable	Covariate variable
⇕	⇕
Controlled variable	Control variable

Regression equation: A regression equation mathematically expresses the functional relationship between the dependent variable and one or more independent variables in a regression analysis. The general form of a regression equation is:

$$Y = f(X) + \varepsilon$$

where $f(X)$ is the linear or nonlinear combination of the regression coefficients and independent variables.

Regression model: A regression model is simply a single or a set of statistical equations that explains the relationship between the dependent variable and one or more independent variables.

Suppose Y be the dependent variable and $X_1, X_2, \cdots, X_p$ be a set of independent variables then the linear regression model is defined as:

$$Y = \beta_0 + \beta_1 X_1 + \beta_2 X_2 + \cdots + \beta_p X_p + \varepsilon$$

where $\beta_0, \beta_1, \beta_2, \cdots, \beta_p$ are the regression parameters or coefficients and ε is the disturbance or random error term.

Regression coefficient: In the context of regression, coefficients represent the weights assigned to each independent variable in the regression equation. The regression coefficient provides information about the direction, strength, and statistical significance of the relationship between the independent and dependent variables in a regression model. The regression coefficient is interpreted as the change in the mean of the dependent variable for a one-unit change in the independent variable, holding other variables constant.

Intercept: The intercept is the constant term in a regression equation, representing the predicted value of the dependent variable when all independent variables are zero.

Residual: Residuals are the differences between the observed values of the dependent variable and the values predicted by the regression equation. Residuals indicate the degree of error in the predictions.

Coefficient of determination or R-squared: A measure of how well the independent variables explain the variability of the dependent variable. It represents the proportion of variability in the dependent variable explained by the independent variables in the regression model. It ranges from 0 to 1, with higher values indicating a better fit. It is mathematically defined as:

$$R^2 = 1 - \frac{\text{Sum of square of residuals}}{\text{Total sum of square}}.$$

Example 11.20

An R^2 value of 0.80 means that the independent variables account for about 80% of the variability in the dependent variable.

Sum of squares due to regression (SSR) or explained sum of squares (ESS): The sum of squares due to regression is the sum of the differences between the predicted value and the mean of the dependent variable. In other words, it describes how well our line fits the data. It is mathematically defined as:

$$\text{Sum of squares due to regression} = \sum_{i=1}^{n}\left(\hat{y}_i - \bar{y}\right)^2.$$

Sum of squares of residuals (SSR): The sum of the squared differences between the observed values of the dependent variable and the values predicted by the regression model. It is defined as:

$$\text{Sum of squares of residuals} = \sum_{i=1}^{n}\left(y_i - \hat{y}_i\right)^2.$$

Total sum of squares (SST): The sum of the squared deviations between the observed values and the mean of the dependent variable. It is defined as:

$$\text{Total sum of squares} = \sum_{i=1}^{n}\left(y_i - \bar{y}\right)^2.$$

Adjusted R-squared: An adjusted R-squared considers the number of applicable predictors in the model because the inclusion of a variable (either relevant or irrelevant) always leads to an increase in the value of the R-squared. Adjusted R-squared penalizes the inclusion of irrelevant variables in the regression model. It is expressed as:

$$R^2_{\text{Adj}} = 1 - \frac{\text{Sum of square of residuals} / \left(n - p\right)}{\text{Total sum of squares} / \left(n - 1\right)}$$

where n is number of observations in the dataset and p is number of parameters estimated including intercept. The interpretation of adjusted R-squared is same as like of R-squared.

Standard error of the estimate: The standard error of the estimate is a measure of the accuracy of the predictions made by the regression model. It represents the average amount by which the observed values differ from the predicted values.

Heteroscedasticity: Heteroscedasticity refers to the situation where the variability of the residuals is not constant across all levels of the independent variables. It violates one of the assumptions of regression analysis.

Multicollinearity: If the independent variables exhibit strong correlations between each other, they are considered multicollinear. Numerous regression techniques assume the absence of multicollinearity in the dataset, as it introduces challenges in prioritizing variables based on their significance and complicates the selection of the most crucial independent variable. Additionally, multicollinearity poses issues in accurately estimating regression coefficients and hinders the interpretation of individual variable contributions.

Autocorrelation: Autocorrelation pertains to the extent of correlation of the same variables in two consecutive time intervals. It quantifies the relationship between the lagged value of a variable and its original value within a time series. This phenomenon is also recognized as serial correlation.

Outliers: Imagine a data point in the dataset with an exceptionally high or low value compared to the other observations, indicating that it does not align with the rest of the population. This sort of data point is called an outlier, representing an extreme value. Outliers pose a challenge because they often disrupt the accuracy of the results obtained from the analysis.

Under-fitting and over-fitting model: Underfitting of the model occurs when a model struggles to fit the training set adequately. In such cases, the model fails to effectively explain or predict data relationships due to challenges of the model fitting. This situation is often described as the model having 'incorrect assumptions' and is associated with the problem of high bias.

Overfitting occurs when a model performs exceptionally well on the training set but fails to generalize effectively to testing sets. In such cases, the model becomes overly sensitive to minor fluctuations, resulting in an inability to explain or predict data relationships accurately. This phenomenon is often described as the model 'memorizing the training set' and is associated with the problem of high variance.

Fitted or predicted value: In regression analysis, a fitted or predicted value, often denoted as $\hat{Y}$ (Y-hat), represents the predicted or estimated value of the dependent variable based on the values of the independent variables in the regression model. These values are obtained by applying the estimated regression coefficients to the observed values of the independent variables. Mathematically, the fitted value is given by the regression equation:

$$\hat{Y} = \hat{f}(X).$$

For a simple regression model, the fitted value is defined as:

$$\hat{Y} = \hat{\beta}_0 + \hat{\beta}_1 X.$$

Fitted line or regression line: The fitted line is a fundamental element in regression analysis, providing a visual and quantitative representation of the modeled relationship between variables. It serves as a tool for making predictions and understanding the direction and strength of the association between the variables under consideration. In a scatterplot, the fitted line is graphed alongside the actual data points, illustrating the anticipated trend or relationship between the variables per the regression model. This line is crafted to minimize the sum of squared differences (residuals) between the observed values of the dependent variable and the corresponding predicted values from the regression equation. A fitted line is a straight line that represents the relationship between the independent variable(s) and the predicted values of the dependent variable. The term 'fitted' indicates that the line has been adjusted to best fit the observed data points in the scatterplot. The equation of the fitted line in a simple linear regression (with one independent variable) is typically expressed as:

$$\hat{Y} = \hat{\beta}_0 + \hat{\beta}_1 X.$$

In Figure 11.5, the solid line indicates the fitted line.

11.6 Types of Regression Analysis

The various classifications of regression analysis are shown in Table 11.2:

TABLE 11.2

Classifications of Regression Analysis

Types of Regression Analysis	Definition
Univariate	The regression model incorporates only one quantitative dependent variable.
Multivariate	The regression model involves two or more quantitative dependent variables.
Simple	The regression model includes only one independent variable.
Multiple	The regression model includes two or more independent variables.
Linear	Linear in variables: Variables enter the equation linearly, possibly after the transformation of the data.
	Linear in parameters: All parameters enter the equation linearly, possibly after the transformation of the data.
Nonlinear	Nonlinear in variable: All or some variables enter the equation nonlinearly, and no transformation is possible to make the variables appear linearly.
	Nonlinear in parameters: All or some parameters enter the equation nonlinearly, and no transformation is possible to make the parameters appear linearly.
Analysis of variance	All independent variables are qualitative variables.
Analysis of covariance	Some independent variables are quantitative variables, and some independent variables are qualitative variables.
Logistic	The model incorporates a dependent variable which considers the binary or multi-categories data.
Count and rate	The dependent variable considers the count and rate data.
Cox proportional hazards	The dependent variable in a Cox proportional hazards model is the survival time or time to event, representing the duration until a specific event of interest happens for each observational unit in the study.

11.7 Regression Analysis with Example

11.7.1 Simple Linear Regression

Simple linear regression is used to model the linear relationship between one independent variable and one dependent variable. The relationship is represented by a linear equation that best fits the observed data. Suppose X represent independent variable and Y represents dependent variable, then the form of simple linear regression model is:

$$Y_i = \beta_1 + \beta_2 X_i + u_i.$$

This is also called two variable regression models. The regression coefficient β_2 represents the change in Y for a one unit change in X, and the intercept β_1 is the mean value of Y when X is zero. In the model, the u_i represents the random error term.

The assumptions for simple linear regression modeling are given below:

- Assumes a normal or Gaussian distribution for the dependent data;
- Requires a linear relationship between the dependent and independent variables;
- Assumes error terms with a normal distribution, zero mean, and constant variance;
- Expects uncorrelated error terms; and
- Assumes an absence of outliers in the dataset.

Example 11.21

Suppose we have a hypothetical dataset of size 500 observations for two variables named *X* and *Y*. We can read the data file into R and see the first five observations using the following R-code:

R-code:

```
data <- read.csv('E:/Book/Chapter 11/Data SLR.csv', header = TRUE)
attach(data)
head(data, 5)
```

Output:

```
Console   Terminal ×   Background Jobs ×
R  R 4.3.2 · E:/Book/Chapter 11/
> data <- read.csv("E:/Book/Chapter 11/Data SLR.csv", header = TRUE)
> attach(data)
> head(data, 5)
  SN       X        Y
1  1 3.862647 2.314489
2  2 4.979381 3.433490
3  3 4.923957 4.599373
4  4 3.214372 2.791114
5  5 7.196409 5.596398
>
```

Now, we can check the assumption of an outlier in the dependent variable by box plot using the following R-code:

R-code:

```
install.packages('ggplot2')
library(ggplot2)
boxp <- ggplot(data = data, aes(x=factor(0), y = Y)) +
  geom_boxplot(color = 'blue') +
  stat_summary(fun = 'mean', geom = 'point', shape = 8,
               size = 6, color = 'black') +
  labs(x = ") +
  geom_jitter(position=position_jitter(0.3), size = 1, color = 'red')
+
  theme_bw()
boxp
```

The boxplot of the dependent variable *Y* is illustrated in Figure 11.1. The box plot reveals that there is no outlier observation in the dependent variable. Now, the assumption of normality can be checked by histogram using the following R-code:

R-code:

```
his<-ggplot(data, aes(x=Y)) +
  geom_histogram(color='white', fill='red') +
  labs(x='Y', y='Number of Observations') +
  theme_bw()
his
```

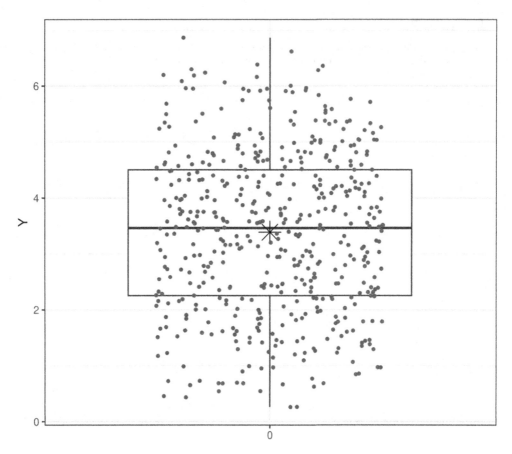

FIGURE 11.1
Box plot of the dependent variable Y.

The output of histogram is shown in Figure 11.2. The histogram indicates that the dependent variable follows approximately normal distribution. However, we can confirm by empirical test statistic such as the Anderson-Darling test using the following R-code:

R-code:

```
install.packages('nortest')
library(nortest)
ad = ad.test(Y)
ad
```

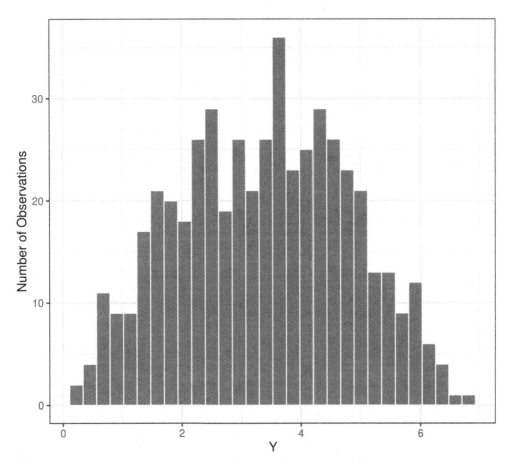

FIGURE 11.2
Histogram of the dependent variable Y.

Output:

```
Console   Terminal ×   Background Jobs ×                                    ─ □
R  R 4.3.2 · ~/
> library(nortest)
> ad = ad.test(Y)
> ad

          Anderson-Darling normality test

data:  Y
A = 1.4888, p-value = 0.0007636

> |
```

The Anderson-Darling test results concluded that the dependent variable does not follow a normal distribution with a p-value of <0.001. However, we have decided to continue further steps of simple linear regression according to the result of the histogram. Now, we can draw a scatter plot for checking the assumption of the linear relationship between the independent variable and dependent variable using the following R-code:

R-code:

```
scatter <- ggplot(data = data, aes(x=X, y=Y)) +
  geom_point(color = 'blue') +
  labs(x = 'X', y = 'Y') +
  theme_bw()
scatter
```

The scatter plot between the dependent variable Y and independent variable X is given in Figure 11.3. The scatter plot indicates that the dependent variable is linearly related to the independent variable. However, we can recheck this assumption by the Pearson correlation coefficient, which measures the linear relationship using the following R-code.

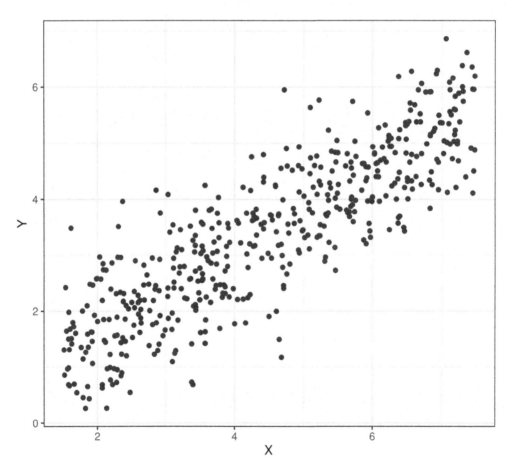

FIGURE 11.3
Scatter plot between the dependent variable *Y* and independent variable *X*.

R-code:

```
cor.coeff <- cor.test(X, Y)
cor.coeff
```

Output:

```
Console   Terminal ×   Background Jobs ×                                          ▬ ☐
R  R 4.3.2 · ~/
> cor.coeff <- cor.test(X, Y)
> cor.coeff

        Pearson's product-moment correlation

data:  X and Y
t = 38.099, df = 498, p-value < 2.2e-16
alternative hypothesis: true correlation is not equal to 0
95 percent confidence interval:
 0.8386426 0.8837007
sample estimates:
      cor
0.8628763

> |
```

The estimated correlation coefficient shows a significantly strong linear relationship between the independent and dependent variables of interest. As all of the assumptions of simple linear regression analysis meet, we can perform the modeling for regressing Y on X using the following R-code:

R-code:

```
model <- lm(Y ~ X, data = data)
summary(model)
```

Output:

```
Console   Terminal ×   Background Jobs ×                                          ▬ ☐
R  R 4.3.2 · ~/
> model <- lm(Y ~ X, data = data)
> summary(model)

call:
lm(formula = Y ~ X, data = data)

Residuals:
    Min      1Q   Median      3Q     Max
-2.36373 -0.48814  0.04383  0.45479  2.38469

Coefficients:
            Estimate Std. Error t value Pr(>|t|)
(Intercept)  0.19821    0.08975   2.209   0.0277 *
X            0.71370    0.01873  38.099   <2e-16 ***
---
Signif. codes:  0 '***' 0.001 '**' 0.01 '*' 0.05 '.' 0.1 ' ' 1

Residual standard error: 0.7257 on 498 degrees of freedom
Multiple R-squared:  0.7446,    Adjusted R-squared:  0.744
F-statistic:  1452 on 1 and 498 DF,  p-value: < 2.2e-16

> |
```

From the results, we see that the independent variable X is statistically significant (i.e., p-value <0.001) in predicting the variability of the dependent variable Y. The value of Y can be increased by 0.71 units for one unit increase the value of X. Moreover, the mean value of Y is 0.20 when the value of X is zero. We can calculate the 95% confidence interval of the coefficients using the following R-code:

R-code:

```
confint(model, level=0.95)
```

Output:

```
Console   Terminal ×   Background Jobs ×

R  R 4.3.2 · ~/
> confint(model, level=0.95)
                2.5 %      97.5 %
(Intercept) 0.02188646 0.3745399
X           0.67689705 0.7505070
>
```

The post-model assumptions on the residuals should be checked. Therefore, the assumption of heteroscedasticity of the residuals can be examined through graphical techniques using the following R-code:

R-code:

```
install.packages("ggfortyfy")
library(ggfortify)
autoplot(model, which = 1:6, ncol = 3, colour = 'black',
         ad.colour = 'black', label.n = 5, label.colour = 'black',
         smooth.colour = 'black', label.size = 3) +
   theme_bw()
```

Diagnostic plots of the residuals are illustrated in Figure 11.4. The residual plots indicate that the variance of the residuals is homoscedastic. However, empirical tests such as the Breusch–Pagan test can be used to confirm the homoscedasticity of the residuals using the following R-code:

R-code:

```
install.packages('skedastic')
library(skedastic)
skedastic::breusch_pagan(model)
```

Output:

```
Console   Terminal ×   Background Jobs ×

R  R 4.3.2 · ~/
> library(skedastic)
> skedastic::breusch_pagan(model)
# A tibble: 1 × 5
  statistic p.value parameter method              alternative
      <dbl>   <dbl>     <dbl> <chr>               <chr>
1      1.16   0.281         1 Koenker (studentised) greater
>
```

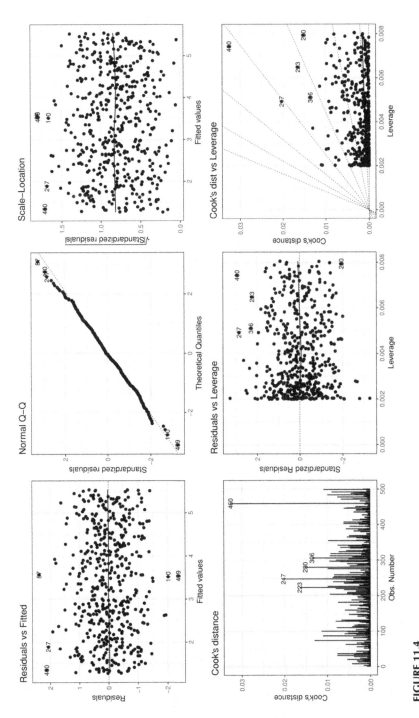

FIGURE 11.4
Diagnostic plots of the residuals.

From the results, it is seen that the Breusch–Pagan test ensures the homoscedasticity of the residuals.

Another assumption of autocorrelation of the residuals can be checked through Durbin-Watson test using the following R-code:

R-code:

```
install.packages('car')
library(car)
durbinWatsonTest(model)
```

Output:

```
Console    Terminal ×    Background Jobs ×                                    ━ ▢
R  R 4.3.2 · ~/
> library(car)
Loading required package: carData
> durbinWatsonTest(model)
 lag Autocorrelation D-W Statistic p-value
   1      0.06501821      1.865119    0.13
 Alternative hypothesis: rho != 0
>
```

According to the outputs, the Durbin-Watson test confirms that the residuals of the fitted model are not autocorrelated. The fitted model passed all the diagnostic tests, so this model can be used to predict the variability of the dependent variable of interest in terms of the known values of the independent variable. However, we need to evaluate the model performance before going to use this model for prediction purpose using the following R-code:

R-code:

```
summary(model)$r.squared
summary(model)$adj.r.squared
predictions <- predict(model, data.frame(data[,2]))
RMSE <- sqrt(mean((data$Y - predictions)^2))
print(RMSE)
```

Output:

```
Console    Terminal ×    Background Jobs ×                                    ━ ▢
R  R 4.3.2 · ~/
> summary(model)$r.squared
[1] 0.7445556
> summary(model)$adj.r.squared
[1] 0.7440426
> predictions <- predict(model, data.frame(data[,2]))
> RMSE <- sqrt(mean((data$Y - predictions)^2))
> print(RMSE)
[1] 0.7242107
>
```

From the outputs, we can conclude that the independent variable X can explain 74% variability of the dependent variable Y according to the fitted model. And the root mean square error of the fitted model is 0.72 which is significantly low. So, the performance of the fitted model is satisfactory.

Now, we can check the generalizability of the fitted model through K-fold cross-validation using the following R-code:

R-code:

```
install.packages('caret')
library(caret)
install.packages('randomForest')
library(randomForest)
train_control <- trainControl(method = 'cv', number = 10)
model <- train(Y ~ X, data = data, trControl = train_control)
print(model)
```

Output:

```
Console    Terminal    Background Jobs

R  R 4.3.2 · ~/
> library(caret)
Loading required package: lattice
> library(randomForest)
randomForest 4.7-1.1
Type rfNews() to see new features/changes/bug fixes.

Attaching package: 'randomForest'

The following object is masked from 'package:ggplot2':

    margin

> train_control <- trainControl(method = "cv", number = 10)
> model <- train(Y ~ X, data = data, trControl = train_control)
There were 11 warnings (use warnings() to see them)
> print(model)
Random Forest

500 samples
  1 predictor

No pre-processing
Resampling: Cross-Validated (10 fold)
Summary of sample sizes: 450, 450, 450, 450, 450, 451, ...
Resampling results:

  RMSE       Rsquared   MAE
  0.8293765  0.6845569  0.6694295

Tuning parameter 'mtry' was held constant at a value of 2
>
```

The 10-fold cross-validated model evaluates that the performance for generalizability of the fitted model is also satisfactory according to the performance evaluation indicators such as R-squared, root mean square error, and mean absolute error.

We can visualize the fitted model with the original data using the following R-code:

R-code:

```
install.packages('ggpubr')
library(ggpubr)
model <- lm(Y ~ X, data = data)
predictions <- predict(model, interval = 'prediction', level = 0.95)
data1 <- cbind(data, predictions)
graph<-ggplot(data1, aes(x=X, y=Y))+
  geom_point() +
  geom_smooth(method='lm', col='blue', level = 0.95) +
  geom_line(aes(y = lwr), col = 'red', linetype = 'dashed') +
  geom_line(aes(y = upr), col = 'red', linetype = 'dashed') +
  stat_regline_equation(label.x = 2, label.y = 7) +
  labs(x = 'X', y = 'Y') +
  theme_bw()
graph
```

The fitted model with original data is shown in Figure 11.5. Finally, we can predict the value of Y with 95% confidence and prediction interval for new values of X using the following R-code:

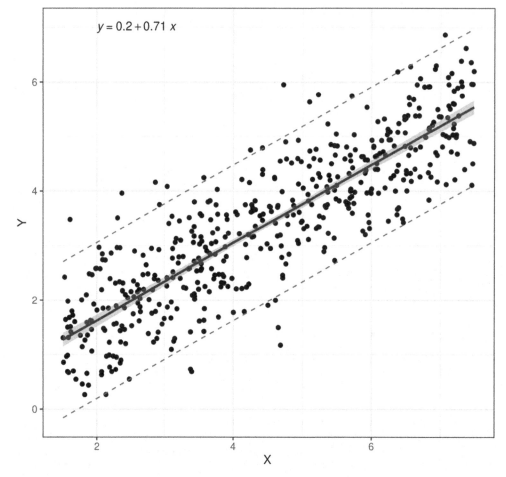

FIGURE 11.5
Fitted model with original data.

R-code:

```
newdata = data.frame(X=c(3, 5))
predict(model, newdata, interval='confidence', level = 0.95)
predict(model, newdata, interval='prediction', level = 0.95)
```

Output:

```
Console    Terminal ×    Background Jobs ×                                                          ▭☐

R  R 4.3.2 · ~/
> newdata = data.frame(X=c(3, 5))
> predict(model, newdata, interval="confidence", level = 0.95)
        fit      lwr      upr
1 2.339319 2.255777 2.422861
2 3.766723 3.700009 3.833438
> predict(model, newdata, interval="prediction", level = 0.95)
        fit       lwr      upr
1 2.339319 0.9111345 3.767504
2 3.766723 2.3394241 5.194023
>
```

The 95% confidence interval associated with a value of 5 of X is (3.7, 3.8). This means that, according to our model, the dependent variable Y with a value of 5 of X has, on average, a value ranging between 3.7 and 3.8 units. On the other hand, the 95% prediction interval associated with a value of 5 of X is (2.3, 5.2). This means that, according to our model, 95% times the dependent variable Y with a value of 5 of X have a value between 2.3 and 5.2.

11.7.2 Multiple Linear Regression

Multiple linear regression extends the concept of simple linear regression by encompassing the modeling of the association between a dependent variable and two or more independent variables. It assumes a linear connection of the dependent variable with the independent variables and aims to estimate the coefficients that best explain the variability of the dependent variable for the corresponding independent variable. Suppose Y is the dependent variable and $X_1, X_2, ..., X_p$ are independent variables, then the multiple linear regression model can be written as:

$$Y = \beta_0 + \beta_1 X_1 + \beta_2 X_2 + ... + \beta_p X_p + \varepsilon,$$

where β_0 denotes the intercept of the dependent variable and indicates the predicted value of Y when all the independent variables $X_1, X_2, ..., X_p$ are zero. Also, $\beta_1, \beta_2, ..., \beta_p$ are the regression coefficients representing the partial effect of the corresponding independent variables $(X_1, X_2, ..., X_p)$ on the dependent variable which are actually the change in Y for a one unit change in the corresponding independent $(X_1, X_2, ..., X_p)$ variable, holding other variables constant, ε is the error term, representing the variability in Y that is not explained by the linear relationship with the independent $(X_1, X_2, ..., X_p)$ variables.

The following assumptions are required to check during performing multiple linear regression modeling:

- Assumes a normal or Gaussian distribution for the dependent data;
- Requires a linear relationship between the dependent and independent variables;
- Assumes error terms with a normal distribution, zero mean, and constant variance;
- Expects uncorrelated error terms;
- Requires no multicollinearity among independent variables; and
- Assumes an absence of outliers in the dataset.

Example 11.22

Consider the birth weight data from the R-package MASS, which contains 189 observations of 10 variables. The data can be read into R and see the head of the data using the following R-code:

R-code:

```
data(birthwt, package = 'MASS')
head(birthwt, n=5)
```

Output:

```
Console   Terminal ×   Background Jobs ×
 R  R 4.3.2 · ~/
> data(birthwt, package = "MASS")
> head(birthwt, n=5)
   low age lwt race smoke ptl ht ui ftv  bwt
85   0  19 182    2     0   0  0  1   0 2523
86   0  33 155    3     0   0  0  0   3 2551
87   0  20 105    1     1   0  0  0   1 2557
88   0  21 108    1     1   0  0  1   2 2594
89   0  18 107    1     1   0  0  1   0 2600
> |
```

We can process the dataset by assigning a label for each value and variable label of the variables race and smoke using the following R-code:

R-code:

```
install.packages('tidyverse')
library(tidyverse)
install.packages('sjlabelled')
library(sjlabelled)
birthwt <- birthwt %>%
  mutate(
    smoke = factor(smoke, labels = c('Non-smoker', 'Smoker')),
    race = factor(race, labels = c('White', 'African American',
'Other'))
  ) %>%
  var_labels(
    bwt = 'Birth weight (g)',
    smoke = 'Smoking status',
    race = 'Race'
  )
head(birthwt, n = 5)
```

Output:

```
Console   Terminal ×   Background Jobs ×
R  R 4.3.2 · ~/
> library(tidyverse)
> library(sjlabelled)
> birthwt <- birthwt %>%
+   mutate(
+     smoke = factor(smoke, labels = c("Non-smoker", "Smoker")),
+     race = factor(race, labels = c("White", "African American", "Other"))
+   ) %>%
+   var_labels(
+     bwt = 'Birth weight (g)',
+     smoke = 'Smoking status',
+     race = 'Race'
+   )
> head(birthwt, n = 5)
   low age lwt             race      smoke ptl ht ui ftv  bwt
85   0  19 182 African American Non-smoker   0  0  1   0 2523
86   0  33 155            Other Non-smoker   0  0  0   3 2551
87   0  20 105            White     Smoker   0  0  0   1 2557
88   0  21 108            White     Smoker   0  0  1   2 2594
89   0  18 107            White     Smoker   0  0  1   0 2600
> |
```

First, we need to check whether the birth weight has any outlying observation. This can be done through a box plot using the following R-code:

R-code:

```
install.packages('ggplot2')
library(ggplot2)
install.packages('patchwork')
library('patchwork')
boxp1 <- ggplot(data = birthwt, aes(x = factor(0), y = bwt)) +
  geom_boxplot(color = 'red', fill = 'blue') +
  stat_summary(fun = 'mean', geom = 'point', shape = 8,
               size = 2, color = 'green') +
  labs(x=", y = get_label(birthwt$bwt)) +
  theme_bw()
boxp2 <- ggplot(data = birthwt, aes(x = race, y = bwt,
                                    fill = smoke)) +
  geom_boxplot(color = 'blue') +
  stat_summary(fun = 'mean', geom = 'point', shape = 8,
               size = 2, color = 'red') +
  guides(fill = guide_legend(title = 'Smoking Status')) +
  labs(x='Race', y = get_label(birthwt$bwt)) +
  theme_bw()
boxp1 + boxp2
```

The box plot of birth weight is shown in Figure 11.6. Although the group boxplots (right panel), according to the independent variables, indicate three outlier observations exist, the overall boxplots (left panel) reveal only one outlier point in the birth weight. Therefore, we can proceed to the next step of the regression analysis without removing the outlier observation.

The following R-code can be used to test whether the birth weight follows approximate normal distribution through histogram and Anderson-Darling test:

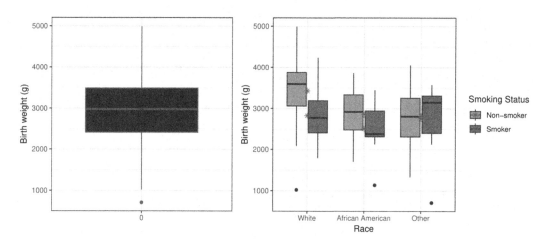

FIGURE 11.6
Box plot of birth weight.

R-code:

```
his<-ggplot(data = birthwt, aes(x=bwt)) +
  geom_histogram(color='white', fill='red') +
  labs(x = get_label(birthwt$bwt), y='Number of Observations') +
  theme_bw()
his
install.packages('nortest')
library(nortest)
ad = ad.test(bwt)
ad
```

The output of histogram of birth weight is illustrated in Figure 11.7.

In order to ensure the normality of the birth weight, the results of the Anderson-Darling test of normality is given below:

Output:

```
Console   Terminal    Background Jobs
R  R 4.3.2 · E:/Book/Chapter 11/
> library(nortest)
> ad = ad.test(bwt)
> ad

        Anderson-Darling normality test

data:  bwt
A = 0.41673, p-value = 0.3282

>
```

From the output, it is observed that both the histogram and Anderson-Darling tests indicate that the birth weight follows an approximately normal distribution. Now, we can test the assumptions of linearity and multicollinearity using the following R-code:

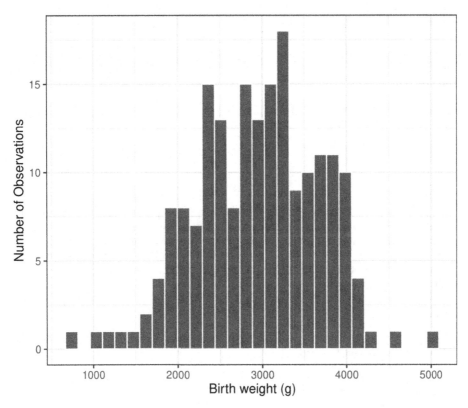

FIGURE 11.7
Histogram of birth weight.

R-code:

```
install.packages('ltm')
library('ltm')
point.biserial = biserial.cor(bwt, as.factor(smoke), use = c('all.
obs'), level = 1)
point.biserial
install.packages('rcompanion')
library('rcompanion')
table <- table(birthwt$smoke,birthwt$race)
crV.correlation = cramerV(table, digits = 4)
crV.correlation
```

Output:

```
Console   Terminal ×   Background Jobs ×                                          ▭ ▢
R  R 4.3.2 · ~/
> point.biserial = biserial.cor(bwt, as.factor(smoke), use = c("all.obs"), level = 1)
> point.biserial
[1] 0.1904481
> table <- table(birthwt$smoke,birthwt$race)
> crv.correlation = cramerv(table, digits = 4)
> crv.correlation
Cramer V
  0.3395
> |
```

The results depict that there exists a linear relationship between birth weight and smoking status. However, the correlation coefficient is low, and there is a possibility of multicollinearity between smoking status and the race of the individuals according to their correlation coefficient. However, we want to a fit multiple linear regression model without considering these results. The following R-code can be used to fit the multiple linear regression model:

R-code:

```
model <- lm(bwt ~ smoke + race, data = birthwt)
install.packages('rstatix')
library(rstatix)
model %>% Anova()
install.packages('parameters')
library(parameters)
model %>% parameters()
```

Output:

```
Console   Terminal ×   Background Jobs ×                                          ▭ ▢
R  R 4.3.2 · ~/
> # Modelling and Significance
> model <- lm(bwt ~ smoke + race, data = birthwt)
> library(rstatix)
> model %>% Anova()
Anova Table (Type II tests)

Response: bwt
          Sum Sq  Df F value     Pr(>F)
smoke     7322575   1 15.4588 0.0001191 ***
race      8712354   2  9.1964 0.0001557 ***
Residuals 87631356 185
---
Signif. codes:  0 '***' 0.001 '**' 0.01 '*' 0.05 '.' 0.1 ' ' 1
> library(parameters)
> model %>% parameters()
Parameter              | Coefficient |  SE |          95% CI | t(185) |      p
-----------------------------------------------------------------------------
(Intercept)            |     3334.95 |  91.78 | [3153.89, 3516.01] |  36.34 | < .001
smoke [Smoker]         |     -428.73 | 109.04 | [-643.86, -213.60] |  -3.93 | < .001
race [African American] |    -450.36 | 153.12 | [-752.09, -148.27] |  -2.94 | 0.004
race [Other]           |     -452.88 | 116.48 | [-682.67, -223.08] |  -3.89 | < .001

Uncertainty intervals (equal-tailed) and p-values (two-tailed) computed using a Wald
  t-distribution approximation.
> |
```

Now, we need to check the assumption of heteroscedasticity of the residuals of the fitted multiple linear regression model using the following R-code:

R-code:

```
library(ggfortify)
autoplot(model, which = 1:6, ncol = 3, colour = 'black',
         ad.colour = 'black', label.n = 5, label.colour = 'black',
         smooth.colour = 'black', label.size = 3) +
  theme_bw()
install.packages('skedastic')
library(skedastic)
skedastic::breusch_pagan(model)
skedastic::glejser(model)
```

The diagnostic plots of the residuals of the fitted model are shown in Figure 11.8. The output of the Breusch–Pagan and Glejser tests are given below:

```
Console   Terminal ×   Background Jobs ×                                                    ▭ ◻
  R 4.3.2 · E:/Book/Chapter 11/ ⟳
> library(skedastic)
> skedastic::breusch_pagan(model)
# A tibble: 1 × 5
  statistic p.value parameter method               alternative
      <dbl>   <dbl>     <dbl> <chr>                <chr>
1      1.55   0.670         3 Koenker (studentised) greater
> skedastic::glejser(model)
# A tibble: 1 × 4
  statistic p.value parameter alternative
      <dbl>   <dbl>     <dbl> <chr>
1      2.30   0.513         3 greater
> |
```

All of the graphical methods, Breusch–Pagan test and Glejser test ensured that the residuals of the fitted multiple linear regression model have homoscedastic variance. Another assumption of autocorrelation of the residuals can be checked using the following R-code:

R-code:

```
install.packages('car')
library(car)
durbinWatsonTest(model)
```

Output:

```
Console   Terminal ×   Background Jobs ×                                                    ▭ ◻
  R 4.3.2 · E:/Book/Chapter 11/ ⟳
> library(car)
> durbinWatsonTest(model)
 lag Autocorrelation D-W Statistic p-value
   1       0.8016025     0.3933732       0
Alternative hypothesis: rho != 0
> |
```

The output shows that the residuals of the fitted multiple linear regression model have autocorrelation according to the Durbin-Watson test. The performance of the fitted multiple linear regression model can be checked using the following R-code:

R-code:

```
install.packages('performance')
```

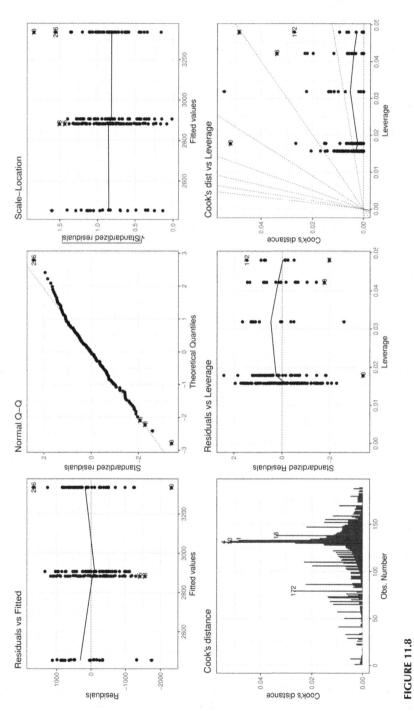

FIGURE 11.8
Diagnostic plots of the residuals of the fitted model.

```
library(performance)
model %>% performance()
```

Output:

```
Console   Terminal ×   Background Jobs ×                                    ▬ ▢
R  R 4.3.2 · ~/
> library(performance)
> model %>% performance()
# Indices of model performance

AIC       |    AICC  |      BIC |   R2 | R2 (adj.) |     RMSE |  Sigma
--------------------------------------------------------------------------
3012.223 | 3012.551 | 3028.432 | 0.123 |     0.109 | 680.924 | 688.246
> |
```

We can evaluate the generalizability of the fitted multiple linear regression model using the following R-code:

R-code:

```
install.packages('caret')
library(caret)
install.packages('randomForest')
library(randomForest)
train_control <- trainControl(method = 'cv', number = 10)
model <- train(bwt ~ smoke + race, data = birthwt, trControl =
train_control)
print(model)
```

Output:

```
Console   Terminal ×   Background Jobs ×                                    ▬ ▢
R  R 4.3.2 · ~/
> library(caret)
> library(randomForest)
> train_control <- trainControl(method = "cv", number = 10)
> model <- train(bwt ~ smoke + race, data = birthwt, trControl = train_control)
note: only 2 unique complexity parameters in default grid. Truncating the grid to 2 .

> print(model)
Random Forest

189 samples
  2 predictor

No pre-processing
Resampling: Cross-Validated (10 fold)
Summary of sample sizes: 169, 170, 170, 170, 169, 171, ...
Resampling results across tuning parameters:

  mtry  RMSE      Rsquared   MAE
  2     687.4642  0.1600866  559.9402
  3     686.6224  0.1634875  556.3686

RMSE was used to select the optimal model using the smallest value.
The final value used for the model was mtry = 3.
> |
```

Finally, we can predict the new data by our fitted multiple linear regression model using the following R-code:

R-code:

```
model <- lm(bwt ~ smoke + race, data = birthwt)
newdata = data.frame(smoke=c('Smoker', 'Non-smoker'), race=c('White',
'African American'))
predict(model, newdata, interval='confidence', level = 0.95)
predict(model, newdata, interval='prediction', level = 0.95)
```

Output:

```
Console   Terminal ×   Background Jobs ×                                                     ─□
R  R 4.3.2 · E:/Book/Chapter 11/
> # Prediction with 95% Confidence Interval
> model <- lm(bwt ~ smoke + race, data = birthwt)
> newdata = data.frame(smoke=c("Smoker", "Non-smoker"), race=c("white", "African American"))
> predict(model, newdata, interval="confidence", level = 0.95)
       fit      lwr      upr
1 2906.218 2736.139 3076.297
2 2884.588 2605.739 3163.437
> predict(model, newdata, interval="prediction", level = 0.95)
       fit      lwr      upr
1 2906.218 1537.787 4274.649
2 2884.588 1498.431 4270.746
> |
```

11.7.3 Binary Logistic Regression

Binary logistic regression is utilized to examine the association between a binary dependent variable and one or multiple independent variables to understand how changes in the independent variables relate to changes in the likelihood of an event occurring. Cases with a single independent variable are referred to as simple binary logistic regression, while cases with two or more independent variables are called multiple binary logistic regression (Rahman et al., 2009). This modeling technique is employed to predict the probability of a binary outcome, typically coded as 0 and 1. Logistic regression is suitable when predicting the presence or absence of a characteristic based on predictor variables. The coefficients derived from logistic regression help the estimation of odds ratios for each independent variable included in the model (e.g., Rahman et al., 2024a).

Suppose X represent independent variable and Y represents binary dependent variable, then the binary logistic regression model (LRM) can be written as,

$$\Pr(Y = 1 \mid X) = \frac{\exp(X_i\beta)}{1 + \exp(X_i\beta)}.$$

A convenient form of this model can be expressed as:

$$\ln\left[\frac{P_i}{1 - P_i}\right] = \beta_0 + \beta_1 X_1 + \beta_2 X_2 + \ldots + \beta_k X_k$$

where p_i = indicates the probability of $Y = 1$.
For a sample of size n, the likelihood for a binary logistic regression is given by:

$$L(\beta; y, X) = \prod_{i=1}^{n} p_i^{y_i} (1-p_i)^{1-y_i} = \prod_{i=1}^{n} \left(\frac{\exp(X_i\beta)}{1+\exp(X_i\beta)} \right)^{y_i} \left(\frac{1}{1+\exp(X_i\beta)} \right)^{1-y_i}.$$

This generates the log likelihood:

$$l(\beta) = \sum_{i=1}^{n} \left[y_i X_i \beta - \log\left(1 + \exp(X_i\beta)\right) \right],$$

where, $\exp(X_i\beta) = \beta_0 ++ \beta_1 X_1 + \beta_2 X_2 + \ldots + \beta_k X_k$.

In order to perform binary logistic regression model, the following assumptions are required to meet:

- Assumes a binomial distribution for the dependent variable;
- Requires a linear relationship between the independent variables and the log-odds of the dependent variable;
- Expects no outliers in the data;
- Requires independent variables to be uncorrelated;
- Assumes independence of observations in the dataset;
- Expects independence of errors or residuals in the logistic regression model; and
- Performs better with larger sample sizes.

Example 11.23

Consider the 'PimaIndiansDiabetes2' dataset from the R-package 'mlbench' that contains 768 observations of 9 variables. We can read this data file from mlbench package and see the first five observations using the following R-code:

R-code:

```
install.packages('mlbench')
library(mlbench)
data(PimaIndiansDiabetes2, package = 'mlbench')
attach(PimaIndiansDiabetes2)
head(PimaIndiansDiabetes2, n=5)
```

Output:

```
Console   Terminal ×   Background Jobs ×                                          ▭ □
 R 4.3.2 · ~/
> library(mlbench)
> data(PimaIndiansDiabetes2, package = "mlbench")
> attach(PimaIndiansDiabetes2)
The following object is masked from package:datasets:

    pressure

> head(PimaIndiansDiabetes2, n=5)
  pregnant glucose pressure triceps insulin mass pedigree age diabetes
1        6     148       72      35      NA 33.6    0.627  50      pos
2        1      85       66      29      NA 26.6    0.351  31      neg
3        8     183       64      NA      NA 23.3    0.672  32      pos
4        1      89       66      23      94 28.1    0.167  21      neg
5        0     137       40      35     168 43.1    2.288  33      pos
> |
```

We see that *diabetes* is the binary variable, which is our target dependent variable with categories – pos/neg (Positive/Negative). In addition, we have the following eight independent variables in the dataset:

- Pregnant – Number of times pregnant;
- Glucose – Plasma glucose concentration (glucose tolerance test);
- Pressure – Diastolic blood pressure (mm Hg);
- Triceps – Skinfold thickness (mm);
- Insulin – 2-Hr serum insulin (mu U/ml);
- Mass – Body mass index (weight in Kg/ (height in m)^2);
- Pedigree – Diabetes pedigree function; and
- Age – Age (years).

First, we need to analyze the descriptive statistics for this dataset in order to check whether any missing observations in the dataset using the following R-code:

R-code:

```
summary(PimaIndiansDiabetes2)
```

Output:

```
Console   Terminal ×   Background Jobs ×                                          ▭ □
 R 4.3.2 · ~/
> summary(PimaIndiansDiabetes2)
    pregnant         glucose         pressure         triceps          insulin
 Min.   : 0.000   Min.   : 44.0   Min.   : 24.00   Min.   : 7.00   Min.   : 14.00
 1st Qu.: 1.000   1st Qu.: 99.0   1st Qu.: 64.00   1st Qu.:22.00   1st Qu.: 76.25
 Median : 3.000   Median :117.0   Median : 72.00   Median :29.00   Median :125.00
 Mean   : 3.845   Mean   :121.7   Mean   : 72.41   Mean   :29.15   Mean   :155.55
 3rd Qu.: 6.000   3rd Qu.:141.0   3rd Qu.: 80.00   3rd Qu.:36.00   3rd Qu.:190.00
 Max.   :17.000   Max.   :199.0   Max.   :122.00   Max.   :99.00   Max.   :846.00
                  NA's   :5       NA's   :35       NA's   :227     NA's   :374
      mass           pedigree           age          diabetes
 Min.   :18.20   Min.   :0.0780   Min.   :21.00   neg:500
 1st Qu.:27.50   1st Qu.:0.2437   1st Qu.:24.00   pos:268
 Median :32.30   Median :0.3725   Median :29.00
 Mean   :32.46   Mean   :0.4719   Mean   :33.24
 3rd Qu.:36.60   3rd Qu.:0.6262   3rd Qu.:41.00
 Max.   :67.10   Max.   :2.4200   Max.   :81.00
 NA's   :11
> |
```

We see that some missing observations in the dataset, and they need to be estimated or discarded from the dataset for performing regression analysis. There are several methods for estimating the missing observations such as mean/median/mode imputation, linear regression imputation, k-nearest neighbors (KNN) imputation, multiple imputation, expectation-maximization (EM) algorithm, hot deck imputation, regression-based imputation, interpolation and extrapolation, stochastic imputation, and deep learning imputation. The choice of imputation method should be made carefully, considering the characteristics of the data and the assumptions underlying each method. It is essential to assess the impact of imputation on the validity and reliability of subsequent analyses. However, we discarded the missing observations in this example using the following R-code:

R-code:

```
data <- na.omit(PimaIndiansDiabetes2)
summary(data)
```

Output:

```
Console  Terminal ×  Background Jobs ×

R  R 4.3.2 · ~/
> data <- na.omit(PimaIndiansDiabetes2)
> summary(data)
    pregnant         glucose         pressure         triceps          insulin
 Min.   : 0.000   Min.   : 56.0   Min.   : 24.00   Min.   : 7.00   Min.   : 14.00
 1st Qu.: 1.000   1st Qu.: 99.0   1st Qu.: 62.00   1st Qu.:21.00   1st Qu.: 76.75
 Median : 2.000   Median :119.0   Median : 70.00   Median :29.00   Median :125.50
 Mean   : 3.301   Mean   :122.6   Mean   : 70.66   Mean   :29.15   Mean   :156.06
 3rd Qu.: 5.000   3rd Qu.:143.0   3rd Qu.: 78.00   3rd Qu.:37.00   3rd Qu.:190.00
 Max.   :17.000   Max.   :198.0   Max.   :110.00   Max.   :63.00   Max.   :846.00
      mass           pedigree           age          diabetes
 Min.   :18.20   Min.   :0.0850   Min.   :21.00   neg:262
 1st Qu.:28.40   1st Qu.:0.2697   1st Qu.:23.00   pos:130
 Median :33.20   Median :0.4495   Median :27.00
 Mean   :33.09   Mean   :0.5230   Mean   :30.86
 3rd Qu.:37.10   3rd Qu.:0.6870   3rd Qu.:36.00
 Max.   :67.10   Max.   :2.4200   Max.   :81.00
> |
```

We can check the multicollinearity among the independent variables through a matrix of scatter plots using the following R-code:

R-code:

```
install.packages('GGally')
library(GGally)
plots <- ggpairs(data[,1:8]) +
  theme_bw()
plots
```

The correlation matrix as well as the matrix of scatter plots among the variables is shown in Figure 11.9. Now we can fit the binary logistic regression model using the following R-code:

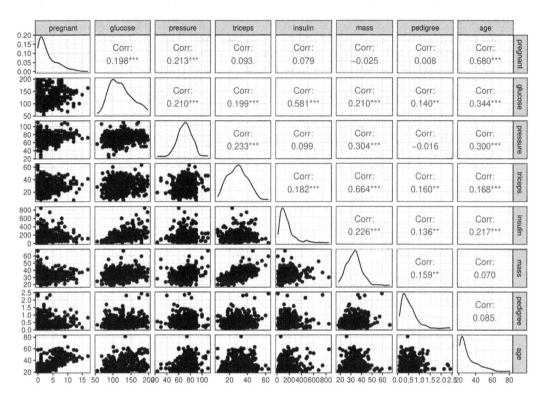

FIGURE 11.9
Correlation matrix as well as matrix of scatter plots among the variables.

R-code:

```
logit <- glm(diabetes~., family = binomial, data = data)
install.packages('parameters')
library(parameters)
library(tidyverse)
logit %>% parameters()
logit %>% parameters(exponentiate = TRUE)
```

Output:

```
Console   Terminal ×   Background Jobs ×                                          ▬ ☐
ℝ R 4.3.2 · ~/
> # Fitting Binary Logistic Regression
> logit <- glm(diabetes~., family = binomial, data = data)
> library(parameters)
> logit %>% parameters()
Parameter   | Log-Odds |       SE |          95% CI |     z |      p
--------------------------------------------------------------------------
(Intercept) |   -10.04 |     1.22 | [-12.55, -7.76] | -8.25 | < .001
pregnant    |     0.08 |     0.06 | [ -0.03,  0.19] |  1.48 | 0.138
glucose     |     0.04 | 5.77e-03 | [  0.03,  0.05] |  6.64 | < .001
pressure    | -1.42e-03 |    0.01 | [ -0.02,  0.02] | -0.12 | 0.904
triceps     |     0.01 |     0.02 | [ -0.02,  0.04] |  0.66 | 0.511
insulin     | -8.25e-04 | 1.31e-03 | [ -0.00,  0.00] | -0.63 | 0.528
mass        |     0.07 |     0.03 | [  0.02,  0.13] |  2.58 | 0.010
pedigree    |     1.14 |     0.43 | [  0.32,  2.00] |  2.67 | 0.008
age         |     0.03 |     0.02 | [ -0.00,  0.07] |  1.85 | 0.065

Uncertainty intervals (profile-likelihood) and p-values (two-tailed) computed using a
  Wald z-distribution approximation.
> logit %>% parameters(exponentiate = TRUE)
Parameter   | Odds Ratio |       SE |       95% CI |     z |      p
--------------------------------------------------------------------------
(Intercept) |  4.36e-05 | 5.31e-05 | [0.00, 0.00] | -8.25 | < .001
pregnant    |     1.09 |     0.06 | [0.97, 1.21] |  1.48 | 0.138
glucose     |     1.04 | 5.99e-03 | [1.03, 1.05] |  6.64 | < .001
pressure    |     1.00 |     0.01 | [0.98, 1.02] | -0.12 | 0.904
triceps     |     1.01 |     0.02 | [0.98, 1.05] |  0.66 | 0.511
insulin     |     1.00 | 1.31e-03 | [1.00, 1.00] | -0.63 | 0.528
mass        |     1.07 |     0.03 | [1.02, 1.13] |  2.58 | 0.010
pedigree    |     3.13 |     1.34 | [1.38, 7.37] |  2.67 | 0.008
age         |     1.03 |     0.02 | [1.00, 1.07] |  1.85 | 0.065

Uncertainty intervals (profile-likelihood) and p-values (two-tailed) computed using a
  Wald z-distribution approximation.
> |
```

The performance of the fitted model can be checked using the following R-code:

R-code:

```
install.packages('performance')
library(performance)
logit %>% performance()
```

Output:

```
Console   Terminal ×   Background Jobs ×                                          ▬ ☐
ℝ R 4.3.2 · ~/
> library(performance)
> logit %>% performance()
# Indices of model performance

AIC     |    AICc |     BIC | Tjur's R2 | RMSE | Sigma | Log_loss | Score_log | Score_spherical |   PCP
------------------------------------------------------------------------------------------------------
362.021 | 362.492 | 397.763 |     0.364 | 0.376 | 1.000 |    0.439 |  -74.015 |           0.009 | 0.718
> |
```

We can fit the model again, excluding the insignificant variables, and check the model performance using the following R-code:

R-code:

```
logit1 <- glm(diabetes~ glucose+mass+pedigree, family = binomial, data
= data)
logit1 %>% parameters()
logit1 %>% parameters(exponentiate = TRUE)
logit1 %>% performance()
```

Output:

```
Console   Terminal ×   Background Jobs ×                                                    ─ □
R  R 4.3.2 · ~/
> logit1 <- glm(diabetes~ glucose+mass+pedigree, family = binomial, data = data)
> logit1 %>% parameters()
Parameter   | Log-Odds |      SE |         95% CI |     z |      p
-------------------------------------------------------------------
(Intercept) |    -8.79 |    0.97 | [-10.80, -6.98] | -9.04 | < .001
glucose     |     0.04 | 4.89e-03 | [  0.03,  0.05] |  8.32 | < .001
mass        |     0.07 |    0.02 | [  0.03,  0.11] |  3.43 | < .001
pedigree    |     1.17 |    0.41 | [  0.37,  2.00] |  2.83 | 0.005

Uncertainty intervals (profile-likelihood) and p-values (two-tailed) computed using a Wald
  z-distribution approximation.
> logit1 %>% parameters(exponentiate = TRUE)
Parameter   | Odds Ratio |      SE |      95% CI |     z |      p
-------------------------------------------------------------------
(Intercept) |  1.52e-04 | 1.48e-04 | [0.00, 0.00] | -9.04 | < .001
glucose     |      1.04 | 5.09e-03 | [1.03, 1.05] |  8.32 | < .001
mass        |      1.07 |    0.02 | [1.03, 1.11] |  3.43 | < .001
pedigree    |      3.23 |    1.34 | [1.45, 7.38] |  2.83 | 0.005

Uncertainty intervals (profile-likelihood) and p-values (two-tailed) computed using a Wald
  z-distribution approximation.
> logit1 %>% performance()
# Indices of model performance

AIC     |    AICc |     BIC | Tjur's R2 | RMSE | Sigma | Log_loss | Score_log | Score_spherical |  PCP
--------------------------------------------------------------------------------------------------------
371.702 | 371.805 | 387.587 |     0.324 | 0.387 | 1.000 |    0.464 |   -70.965 |           0.006 | 0.700
> |
```

Now, we can compute the confusion matrix and different performance indicators such as accuracy, precision, sensitivity, specificity, and F1 score of the fitted binary logistic regression model using the following R-code:

R-code:

```
data$Predict = ifelse(logit1$fitted.values >0.5,'pos','neg')
mytable <- table(data$diabetes,data$Predict)
rownames(mytable) <- c('Obs. neg','Obs. pos')
colnames(mytable) <- c('Pred. neg','Pred. pos')
mytable
accuracy <- (sum(diag(mytable))/sum(mytable))*100
accuracy
precision <- (mytable[2,2]/(mytable[2,2] + mytable[1,2]))*100
precision
sensitivity_logit1 <- (mytable[2,2]/(mytable[2,2] + mytable[2,1]))*100
sensitivity_logit1
Specificity_logit1 <- (mytable[1,1]/(mytable[1,1] + mytable[1,2]))*100
Specificity_logit1
F1_Score <- 2 * ((precision * sensitivity_logit1) / (precision +
sensitivity_logit1))
F1_Score
```

Output:

```
Console   Terminal ×   Background Jobs ×                                    ___ □
 R  R 4.3.2 · E:/Book/Chapter 11/
> data$Predict = ifelse(logit1$fitted.values >0.5,"pos","neg")
> mytable <- table(data$diabetes,data$Predict)
> rownames(mytable) <- c("Obs. neg","Obs. pos")
> colnames(mytable) <- c("Pred. neg","Pred. pos")
> mytable

          Pred. neg Pred. pos
  Obs. neg       234        28
  Obs. pos        57        73
> accuracy <- (sum(diag(mytable)))/sum(mytable))*100
> accuracy
[1] 78.31633
> precision <- (mytable[2,2]/(mytable[2,2] + mytable[1,2]))*100
> precision
[1] 72.27723
> sensitivity_logit1 <- (mytable[2,2]/(mytable[2,2] + mytable[2,1]))*100
> sensitivity_logit1
[1] 56.15385
> Specificity_logit1 <- (mytable[1,1]/(mytable[1,1] + mytable[1,2]))*100
> Specificity_logit1
[1] 89.31298
> F1_Score <- 2 * ((precision * sensitivity_logit1) / (precision + sensitivity_logit1))
> F1_Score
[1] 63.20346
>
```

The Receiver Operating Characteristic (ROC) curve assesses a model's performance by comparing sensitivity versus specificity, and the area under the ROC Curve serves as an accuracy index. A greater area under the curve indicates better predictive model performance, with a perfect predictive model having an AUC of 1. We can make the ROC curve with an AUC value for our fitted model using the following R-code:

R-code:

```
install.packages('pROC')
library(pROC)
roc(diabetes~logit1$fitted.values, data = data, plot = TRUE, print.auc
= TRUE, col= 'blue')
```

The ROC curve of the fitted model is illustrated in Figure 11.10.

Finally, the generalizability of the fitted binary logistic regression model can be checked through cross-validation using the following R-code:

R-code:

```
install.packages('caret')
library(caret)
install.packages('randomForest')
library(randomForest)
train_control <- trainControl(method = 'cv', number = 10)
logit2 <- train(diabetes~ glucose+mass+pedigree, data = data,
trControl = train_control, method = 'glm', family = binomial())
print(logit2)
```

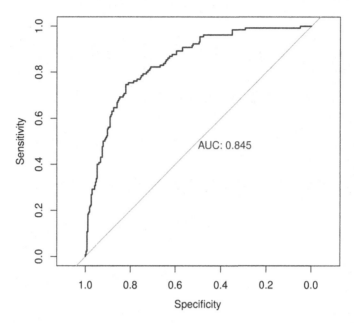

FIGURE 11.10
ROC curve of the fitted model.

Output:

```
Console   Terminal ×   Background Jobs ×                                          ⊟ ☐
ℝ  R 4.3.2 · ~/
> library(caret)
> library(randomForest)
> train_control <- trainControl(method = "cv", number = 10)
> logit2 <- train(diabetes~ glucose+mass+pedigree, data = data, trControl = train_control, method = "glm",
family = binomial())
> print(logit2)
Generalized Linear Model

392 samples
  3 predictor
  2 classes: 'neg', 'pos'

No pre-processing
Resampling: Cross-Validated (10 fold)
Summary of sample sizes: 353, 353, 353, 353, 353, 353, ...
Resampling results:

  Accuracy   Kappa
  0.7778205  0.4673541

>
```

11.7.4 Poisson Regression

Poisson regression is a form of generalized linear regression model where the random component follows the Poisson distribution. It is a useful statistical technique for studying the response variable of count and rate data such as the occurrences of an event or phenomenon within a specific time or region. Examples of count variables encompass various scenarios, such as the number of heart attacks, alcoholic drink usage days, days from outbreak to infection, people in an emergency department queue, yearly lung cancer incidents, hospital admissions, cancerous lesions, asthmatic attacks, and more. Suppose n independent observations y_1, y_2, ..., y_n follows a Poisson distribution. If it is assumed that the natural log of the conditional mean of y_1, y_2, ..., y_n, i.e., the natural log of the expected count (λ) is a linear combination of a set of p categorical or numerical covariates x_j; $j = 1, 2,$ ..., p. Then the Poisson regression model is given by the following equation

$$\ln\left(E[y_i \mid x_i]\right) = \ln\left(\lambda\right) = \beta_0 + \beta_1 x_1 + \ldots + \beta_p x_p ; i = 1, 2, \ldots, n.$$

The natural link function is the log link. It ensures that $\lambda \geq 0$ and can be written as

$$E[y_i \mid x_i] = \lambda = \exp\left(\beta_0 + \beta_1 x_1 + \ldots + \beta_p x_p\right)$$

where n is the sample size; β_0 is the intercept; and β_1, β_2, ..., β_p are the regression coefficients associated with the predictor variables x_1, x_2, ..., x_p.

While linear regression might be tempting to use for handling count outcomes by assuming discrete numerical data as continuous, it can lead to nonsensical predictions. For instance, employing linear regression to predict the number of heart attacks or strokes in the past year may result in a negative count, which does not make any clinical sense! Therefore, it is advisable to opt for Poisson regression when dealing with count variables. Another circumstance that warrants Poisson regression usage is when the number of cases (e.g., deaths, accidents) is relatively small compared to nonevents (e.g., alive, no accident). In such cases, focusing solely on information from the cases within a population of interest makes more sense than obtaining data from both cases and noncases, as is common in cohort and case-control studies. For example, in publicly available COVID-19 data, only the number of reported deaths, along with basic sociodemographic and clinical details for cases, may be available. The logistic regression is recommended when information for noncases is available. Logistic regression determines the occurrence of an event, while Poisson regression assesses the frequency of occurrences when the event transpires.

Poisson regression requires the following assumptions to make inferences:

- The response variable is a count (nonnegative integers) per unit of time or space.
- The observations must be independent of one another.
- The distribution of counts (conditional on the model) follows a Poisson distribution.
- The factors affecting the mean are also affecting the variance. The variance is assumed to be equal to the mean. When this assumption does not hold, we should consider a Quasi-Poisson generalized linear model (GLM) for over-dispersed (or under-dispersed) data or a Negative Binomial GLM instead.
- Independent variables can be measured on a continuous, ordinal, or nominal/ dichotomous scale.
- The log of the mean rate, $\log(\lambda)$, must be a linear function of x.

Example 11.24

The dataset on the number of lung cancer cases (Doll, 1971) contains four variables which are described in Table 11.3.

TABLE 11.3

Description of the Variables Included in the Dataset on the Number of
Lung Cancer Cases (Doll, 1971)

Variable Name	Description
smoke_yrs	Years of smoking (categorical) {15–19, 20–24, 25–29, 30–34, 35–39, 40–44, 45–49, 50–54, 55–59}
cigar_day	Cigarettes per day (numerical)
person_yrs	Person-years at risk of lung cancer (numerical)
case	Number of lung cancer cases (count) among doctors

The following R-code can be used for importing the dataset into R and see the head of
the dataset:

R-code:

```
smoke <- read.csv('E:/Book/Chapter 11/Smoke.csv', header = TRUE)
attach(smoke)
head(smoke)
```

Output:

```
Console   Terminal   Background Jobs
R  R 4.3.2 · ~/
> smoke <- read.csv('E:/Book/Chapter 11/Smoke.csv', header = TRUE)
> attach(smoke)
> head(smoke)
  smoke_yrs cigar_day person_yrs case
1    15-19         0      10366    1
2    20-24         0       8162    0
3    25-29         0       5969    0
4    30-34         0       4496    0
5    35-39         0       3512    0
6    40-44         0       2201    0
>
```

Suppose we want to predict the rate of lung cancer cases based on the predictors such
as daily cigarette consumption and duration of smoking in years considering their interaction terms. In order to declare the rate of lung cancer, we need to provide the log of
the number of person-years at risk of lung cancer as offset or denominator in the glm
function in R. We should bear in mind that there is no need to declare offset in the glm
function in case of dealing with count data. The following R-code can be used:

R-code:

```
set.seed(1234)
pois_model = glm(case ~ cigar_day * smoke_yrs, data = smoke,
                 family = 'poisson', offset = log(person_yrs))
summary(pois_model)
```

Output:

```
Console   Terminal ×   Background Jobs ×                                          ─ □
R 4.3.2 · E:/icddrb/ELISA Results/
> set.seed(1234)
> pois_model = glm(case ~ cigar_day * smoke_yrs, data = smoke,
+                   family = "poisson", offset = log(person_yrs))
warning message:
glm.fit: fitted rates numerically 0 occurred
> summary(pois_model)

Call:
glm(formula = case ~ cigar_day * smoke_yrs, family = "poisson",
    data = smoke, offset = log(person_yrs))

Coefficients:
                          Estimate Std. Error z value Pr(>|z|)
(Intercept)                -9.2463     1.0000  -9.246  < 2e-16 ***
cigar_day                  -2.8137   319.6351  -0.009 0.992976
smoke_yrs20-24             -0.7514     1.5114  -0.497 0.619084
smoke_yrs25-29             -0.2539     1.3518  -0.188 0.851030
smoke_yrs30-34              1.2493     1.1032   1.132 0.257489
smoke_yrs35-39              0.5856     1.1660   0.502 0.615498
smoke_yrs40-44              1.9706     1.0790   1.826 0.067800 .
smoke_yrs45-49              2.1070     1.0982   1.919 0.055044 .
smoke_yrs50-54              3.0710     1.0761   2.854 0.004318 **
smoke_yrs55-59              3.5704     1.0706   3.335 0.000853 ***
cigar_day:smoke_yrs20-24    2.8609   319.6351   0.009 0.992859
cigar_day:smoke_yrs25-29    2.8736   319.6351   0.009 0.992827
cigar_day:smoke_yrs30-34    2.8736   319.6351   0.009 0.992827
cigar_day:smoke_yrs35-39    2.8997   319.6351   0.009 0.992762
cigar_day:smoke_yrs40-44    2.8845   319.6351   0.009 0.992800
cigar_day:smoke_yrs45-49    2.8886   319.6351   0.009 0.992790
cigar_day:smoke_yrs50-54    2.8669   319.6351   0.009 0.992844
cigar_day:smoke_yrs55-59    2.8641   319.6351   0.009 0.992851
---
Signif. codes:  0 '***' 0.001 '**' 0.01 '*' 0.05 '.' 0.1 ' ' 1

(Dispersion parameter for poisson family taken to be 1)

    Null deviance: 445.099  on 62  degrees of freedom
Residual deviance:  60.597  on 45  degrees of freedom
AIC: 216.44

Number of Fisher Scoring iterations: 18

> |
```

In the outputs, the p-value indicates that the interaction terms are not statistically significant for predicting the rate of lung cancer cases. Thus, we can use the stepwise model selection procedure for choosing the best set of predictors as it can provide a best set of relevant covariates based on Akaike Information Criterion (AIC) or Bayesian Information Criterion (BIC). The AIC quantifies the amount of information loss due to adding or removing any variable from the model. Hence, AIC provides a means for model selection and in case of multiple models, the one which has lower AIC value suggest a better model. In R, stepAIC is a feature selection function using the stepwise algorithm for a specified model in the case of both linear and logistic regression. It begins with an empty set of predictors and iteratively adds or removes predictors based on their AIC score until no additional enhancement in the AIC score is observed. The chosen features are returned when there is no further improvement in the AIC score. We can use either backward or forward, or both method in stepAIC function in R. The following R-code can be used to produce the results:

R-code:

```
library(MASS)
stepAIC(pois_model, direction = 'both')
```

Output:

```
Console  Terminal ×  Background Jobs ×                                    ▭ ☐
R R 4.3.2 · E:/icddrb/ELISA Results/
> library(MASS)
> stepAIC(pois_model, direction = 'both')
Start:  AIC=216.44
case ~ cigar_day * smoke_yrs

                        Df Deviance    AIC
- cigar_day:smoke_yrs  8    68.015 207.86
<none>                       60.597 216.44

Step:  AIC=207.86
case ~ cigar_day + smoke_yrs

                        Df Deviance    AIC
<none>                       68.02 207.86
+ cigar_day:smoke_yrs  8    60.60 216.44
- cigar_day            1   170.17 308.01
- smoke_yrs            8   324.71 448.55

Call:  glm(formula = case ~ cigar_day + smoke_yrs, family = "poisson",
    data = smoke, offset = log(person_yrs))

Coefficients:
    (Intercept)        cigar_day   smoke_yrs20-24   smoke_yrs25-29   smoke_yrs30-34   smoke_yrs35-39
      -11.31716          0.06436          0.96242          1.71089          3.20768          3.24278
  smoke_yrs40-44   smoke_yrs45-49   smoke_yrs50-54   smoke_yrs55-59
        4.20836          4.44897          4.89367          5.37260

Degrees of Freedom: 62 Total (i.e. Null); 53 Residual
Null Deviance:      445.1
Residual Deviance: 68.02       AIC: 207.9
warning message:
glm.fit: fitted rates numerically 0 occurred
> |
```

From the outputs, we can conclude that both of the variables daily cigarette consumption and duration of smoking in years are potential for predicting the rate of lung cancer cases. Now, we can fit the model again and then check the assumption of equality of mean and variance. The variance of y_i is approximated by $(y_i - \lambda_i)^2$. The following R-code can be used for investigating the equality of the variance and mean by graphical methods:

R-code:

```
pois_model_rate = glm(case ~ cigar_day + smoke_yrs, data = smoke,
                family = 'poisson', offset = log(person_yrs))
# Assumption of equality of mean and variance by graph
lambdahat <-fitted(pois_model_rate)
app.var <- (smoke$case-lambdahat)^2
Pearson_res <- resid(pois_model_rate, type='pearson')
data <- data.frame(lambdahat, app.var, Pearson_res)
library('ggplot2')
library('patchwork')
plot1 <- ggplot(data = data, aes(x=lambdahat, y=app.var)) +
  geom_point(color = 'blue') +
  labs(x = expression(hat(lambda)), y = expression((y-hat(lambda))^2
)) +
  theme_bw()
plot2 <- ggplot(data = data, aes(x=lambdahat, y=Pearson_res)) +
  geom_point(color = 'red') +
  labs(x = expression(hat(lambda)), y = 'Pearson Residuals') +
  theme_bw()
plot1 + plot2
```

The graphical investigation of the assumption of equality of mean and variance is shown in Figure 11.11. The left panel of Figure 11.11 reveals no significant difference between the range of variance and range of mean. Additionally, the right panel illustrates no noticeable pattern in the residuals, indicating that the expected rate of lung cancer cases is approximately equal to its variance. To further explore this, we can estimate the dispersion parameter ϕ, where $\phi > 1$ indicates that the data are over-dispersed and $\phi < 1$ indicates that the data are under-dispersed. We can use the following R-code for estimating the dispersion parameter:

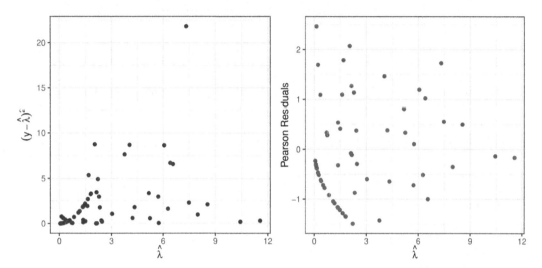

FIGURE 11.11
Graphical investigation of the assumption of equality of mean and variance: $\hat{\lambda}$ versus $\left(y - \hat{\lambda}\right)^2$ (left panel) and $\hat{\lambda}$ versus Pearson residuals (right panel)

R-code:

```
# Assumption of equality of mean and variance by dispersion parameter
Pearson_rate <- sum((smoke$case - pois_model_rate$fitted.values)^2 /
pois_model_rate$fitted.values)
dispersion_rate <- Pearson_rate / pois_model_rate$df.residual
dispersion_rate
```

Output:

```
Console   Terminal   Background Jobs
R 4.3.2 · E:/icddrb/ELISA Results/
> # Assumption of equality of mean and variance by dispersion parameter
> Pearson_rate <- sum((smoke$case - pois_model_rate$fitted.values)^2 / pois_model_rate$fitted.values)
> dispersion_rate <- Pearson_rate / pois_model_rate$df.residual
> dispersion_rate
[1] 1.072472
>
```

The estimated value of dispersion was found to be approximately 1, which indicates that the mean and variance are approximately equal; hence, Poisson regression model is appropriate to predict the rate of lung cancer cases based on daily cigarette consumption and duration of smoking in years. Now, we can see the diagnostic plots of the fitted Poisson regression model using the following R-code:

R-code:

```
library(ggfortify)
autoplot(pois_model_rate, which = 1:6, ncol = 3, colour = 'black',
         ad.colour = 'black', label.n = 5, label.colour = 'black',
         smooth.colour = 'black', label.size = 3) +
   theme_bw()
```

The diagnostic plots of the fitted Poisson regression model are shown in Figure 11.12. It is observed that there are three influential data points based on the Cook's distance. However, we can consider this model as final without removing these influential data points using the following R-code:

R-code:

```
final.pois_model_rate <- pois_model_rate
summary(final.pois_model_rate)
```

Output:

```
Console  Terminal ×  Background Jobs ×

R  R 4.3.2 · ~/
> final.pois_model_rate <- pois_model_rate
> summary(final.pois_model_rate)

Call:
glm(formula = case ~ cigar_day + smoke_yrs, family = "poisson",
    data = smoke, offset = log(person_yrs))

Coefficients:
                Estimate Std. Error z value Pr(>|z|)
(Intercept)    -11.317162   1.007492 -11.233  < 2e-16 ***
cigar_day        0.064361   0.006401  10.054  < 2e-16 ***
smoke_yrs20-24   0.962417   1.154694   0.833  0.40457
smoke_yrs25-29   1.710894   1.080420   1.584  0.11330
smoke_yrs30-34   3.207676   1.020378   3.144  0.00167 **
smoke_yrs35-39   3.242776   1.024187   3.166  0.00154 **
smoke_yrs40-44   4.208361   1.013726   4.151 3.30e-05 ***
smoke_yrs45-49   4.448972   1.017054   4.374 1.22e-05 ***
smoke_yrs50-54   4.893674   1.019945   4.798 1.60e-06 ***
smoke_yrs55-59   5.372600   1.023404   5.250 1.52e-07 ***
---
Signif. codes:  0 '***' 0.001 '**' 0.01 '*' 0.05 '.' 0.1 ' ' 1

(Dispersion parameter for poisson family taken to be 1)

    Null deviance: 445.099  on 62  degrees of freedom
Residual deviance:  68.015  on 53  degrees of freedom
AIC: 207.86

Number of Fisher Scoring iterations: 5

> |
```

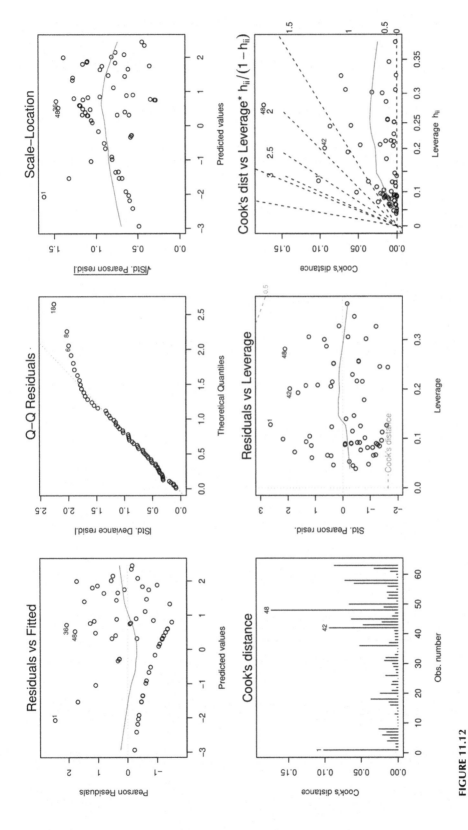

FIGURE 11.12
Diagnostic plots of the fitted Poisson regression model.

Now, we can check the goodness-of-fit and performance of the model using the following R-code:

R-code:

```
deviance_rate <- deviance(final.pois_model_rate)
p_value_dev_rate <- 1 - pchisq(deviance_rate, df = final.pois_model_
rate$df.residual)
p_value_dev_rate
Pearson_rate <- sum((smoke$case - final.pois_model_rate$fitted.
values)^2 / final.pois_model_rate$fitted.values)
p_value_Pear_rate <- 1 - pchisq(Pearson_rate, df = final.pois_model_
rate$df.residual)
p_value_Pear_rate
install.packages('tidyverse')
library(tidyverse)
install.packages('performance')
library(performance)
final.pois_model_rate %>% performance()
install.packages('caret')
library(caret)
install.packages('randomForest')
library(randomForest)
train_control <- trainControl(method = 'cv', number = 10)
final.pois_model_rate_cv <- train(case ~ cigar_day + smoke_yrs, data =
smoke, trControl = train_control, method = 'glm', family = poisson())
print(final.pois_model_rate_cv)
```

Output:

```
Console   Terminal ×   Background Jobs ×                                    ─□
 R R 4.3.2 · E:/icddrb/ELISA Results/
> deviance_rate <- deviance(final.pois_model_rate)
> p_value_dev_rate <- 1 - pchisq(deviance_rate, df = final.pois_model_rate$df.residual)
> p_value_dev_rate
[1] 0.0802866
> Pearson_rate <- sum((smoke$case - final.pois_model_rate$fitted.values)^2 / final.pois_model_rate$fitt
ed.values)
> p_value_Pear_rate <- 1 - pchisq(Pearson_rate, df = final.pois_model_rate$df.residual)
> p_value_Pear_rate
[1] 0.334015
> library(tidyverse)
> library(performance)
> final.pois_model_rate %>% performance()
# Indices of model performance

AIC     |   AICc |    BIC | Nagelkerke's R2 | RMSE | Sigma | Score_log | Score_spherical
--------------------------------------------------------------------------------------------
207.857 | 212.088 | 229.289 |          0.998 | 1.448 | 1.000 |    -1.491 |           0.097
> library(caret)
> library(randomForest)
> train_control <- trainControl(method = "cv", number = 10)
> final.pois_model_rate_cv <- train(case ~ cigar_day + smoke_yrs, data = smoke,
+                     trControl = train_control, method = "glm", family = poisson())
> print(final.pois_model_rate_cv)
Generalized Linear Model

63 samples
 2 predictor

No pre-processing
Resampling: Cross-Validated (10 fold)
Summary of sample sizes: 55, 58, 56, 56, 57, 57, ...
Resampling results:

  RMSE      Rsquared   MAE
  2.927536  0.4282541  2.220782

> |
```

We can see from the output that the considered Poisson regression model is good fit at 5% level of significance based on the deviance and Pearson goodness-of-fit test as the *p*-value of both tests is greater than 0.05. We also found lowest RMSE value of 2.93 and the model can explain 42.8% variability of the outcome variable. Now, we can see and save the estimates of the regression coefficients along with their significance and 95% confidence interval using the following R-code:

R-code:

```
install.packages('parameters')
library(parameters)
results <- final.pois_model_rate %>% parameters(exponentiate = TRUE,
conf.int = TRUE)
write.csv(results, 'E:/Book/Chapter 11/Pois results.csv', row.names =
FALSE)
install.packages('gtsummary')
library(gtsummary)
tbl_regression(final.pois_model_rate, exponentiate = TRUE)
```

Output:

Characteristic	IRR[1]	95% CI[1]	p-value
cigar_day	1.07	1.05, 1.08	<0.001
smoke_yrs			
15-19	—	—	
20-24	2.62	0.34, 52.9	0.4
25-29	5.53	0.94, 105	0.11
30-34	24.7	5.23, 442	0.002
35-39	25.6	5.35, 459	0.002
40-44	67.2	14.6, 1,195	<0.001
45-49	85.5	18.3, 1,525	<0.001
50-54	133	28.3, 2,385	<0.001
55-59	215	45.1, 3,862	<0.001

[1] IRR = Incidence Rate Ratio, CI = Confidence Interval

From the output, we can conclude that the increasing daily cigarette consumption by one raises the likelihood of developing lung cancer by 1.07 when adjusting for other factors. We found that the years of smoking from 20 to 29 are not significantly associated with the risk of lung cancer. Moreover, the individuals who smoked for 30 or more years have a significantly heightened risk of lung cancer, where the individuals of aged 30–34 years have an IRR of 24.7 when controlling for other variables. The risk of lung cancer is observed as increasing with the years of smoking.

Moreover, there are other types of modeling techniques utilize for biological and health sciences data analysis, including multinomial logistic regression (Rahman et al., 2008), multilevel modeling (Das et al., 2019), biophysical modeling (Rahman & Hakim, 2016, 2017), microsimulation modeling (Hakim & Rahman, 2018), and mathematical modeling (Karimuzzaman et al., 2020a, b; Kuddus et al., 2021; Rahman & Kuddus, 2014a, b, 2021; Kuddus & Rahman, 2021, 2022; Rahman et al., 2021, 2024b). Although they could be handy tools for data analysis, further discussion of those methods are beyond the scope of this book.

11.7.5 Analysis of Variance (ANOVA)

Analysis of variance is a statistical method used to analyze the differences among group means in a sample. It is often employed when comparing three or more groups to determine whether there are statistically significant differences between them. The basic idea behind ANOVA is to partition the total variance observed in a dataset into different components, namely:

- **Between-group variance**: This represents the variability among the means of different groups. If this variance is significantly larger than what would be expected by chance, it suggests that there are significant differences between the groups.
- **Within-group variance**: This represents the variability within each group. It is the variance that is expected to occur due to random chance.

The ANOVA test then compares the ratio of between-group variance to within-group variance using F-test by setting the null hypothesis that there is no significance difference among the group means. If the ratio is sufficiently large and the observed differences between group means are unlikely to have occurred by chance, the result is deemed statistically significant.

ANOVA relies on several assumptions to be valid. Violations of these assumptions may affect the accuracy and reliability of the ANOVA results. Here are the key assumptions of ANOVA:

- **Normality**: The data within each group should be approximately normally distributed.
- **Homogeneity of variance (homoscedasticity)**: The variances of the groups being compared should be approximately equal. In other words, the spread of scores in one group should be roughly the same as the spread in another group.
- **Independence**: Observations within each group should be independent of each other.
- **Random sampling**: Data should be collected through a random sampling process to ensure that the results generalize to the larger population.
- **Variable types**: The dependent variable (the one being measured) should be measured on an interval or ratio scale. ANOVA is not appropriate for nominal or ordinal data. The independent variables should be measured in qualitative scale with at least two levels.
- **Outlier**: There should be no significant outliers in the different groups
- **Homogeneity of regression slopes**: This assumption is specific to two-way ANOVA. It assumes that the relationship between the independent variables and the dependent variable is consistent across all levels of the other independent variable.

There are nonparametric alternatives to ANOVA, such as the Kruskal-Wallis test, which can be used when assumptions of normality and homogeneity of variance are not met. If other assumptions are violated, alternative statistical methods or transformations of the data might be necessary.

There are different types of ANOVA tests, such as:

a. **One-way ANOVA**: For comparing means across one factor. Mathematically, the basic model for a one-way ANOVA can be written as:

$$Y_{ij} = \mu + \alpha_i + \varepsilon_{ij}$$

where Y_{ij} is the observation in the jth level (group) of the ith factor (independent variable), μ is the overall mean of all observations, α_i is the effect of the ith level of the factor, and ε_{ij} is the random error term.

The null hypothesis for a one-way ANOVA is that there is no significant difference between the group means, i.e., $\alpha_1 = \alpha_2 = \ldots = \alpha_k = 0$, where k is the number of groups. The alternative hypothesis is that at least one group mean is significantly different from the others.

Example 11.25

A pharmaceutical company is testing the effectiveness of three different drugs for reducing blood pressure. They measure the change in blood pressure after a specified treatment period for three groups of patients: Group A receives Drug X, Group B receives Drug Y, and Group C receives Drug Z. In this case, one-way ANOVA can be applied to determine if there are any statistically significant differences in the mean reduction of blood pressure among the three drug groups.

b. **Two-way ANOVA**: For comparing means across two factors. The general model for a two-way ANOVA can be stated mathematically below

$$Y_{ijk} = \mu + \alpha_i + \beta_j + (\alpha\beta)_{ij} + \varepsilon_{ijk}$$

where Y_{ijk} is the observation in the kth level of the jth factor of the ith factor combination, μ is the overall mean, α_i is the effect of the ith level of the first factor, β_j is the effect of the jth level of the second factor, $(\alpha\beta)_{ij}$ is the interaction effect between the ith level of the first factor and the jth level of the second factor, and ε_{ijk} is the random error term.

The factors can be referred to as the 'main effects', and the interaction term represents the combined effect of both factors. The null hypothesis for each main effect and the interaction effect is that there is no significant difference. The hypotheses can be expressed as follows:

- Null hypothesis for the first factor (α): $H_0 : \alpha_1 = \alpha_2 = \ldots = \alpha_a = 0$, where a is the number of levels for the first factor.
- Null hypothesis for the second factor (β): $H_0 : \beta_1 = \beta_2 = \ldots = \beta_b = 0$, where b is the number of levels for the second factor.
- Null hypothesis for the interaction ($\alpha\beta$): $H_0 : (\alpha\beta)_{ij} = 0$ for all i and j.

Example 11.26

An education researcher is investigating the effects of teaching method (Traditional vs. Online) and class size (Small vs. Large) on students' performance in a mathematics course. Students are randomly assigned to one of the four groups formed by the combination of teaching method and class size. Two-way ANOVA can be used to examine the main effects of teaching method and class size, as well as their interaction effect on students' math scores.

c. **Repeated measures ANOVA**: When measurements are taken on the same subjects over multiple points in time or conditions.

Example 11.27

A psychologist is studying the impact of three different therapies on anxiety levels within the same group of participants. Anxiety levels are measured before therapy, after four weeks of therapy, and after eight weeks of therapy. Repeated measures ANOVA can be applied to assess whether there are statistically significant differences in anxiety levels across the three time points and to determine if these changes are associated with the type of therapy.

Example 11.28

Let us consider the poison dataset from https://raw.githubusercontent.com/guru99-edu/R-Programming/master/poisons.csv. The dataset contains 48 rows of 3 variables:

- Time: Survival time of the animal,
- poison: Type of poison used: factor level: 1, 2, and 3, and
- treat: Type of treatment used: factor level: A, B, C, and D.

Suppose we want to test the significance of mean difference of survival time among poison levels. This can be done by performing one-way ANOVA test because the dependent variable is a continuous variable and the independent variable is a categorical variable with three levels. Now, we can import this dataset into R using the following R-code:

R-code:

```
library(dplyr)
data <- read.csv('https://raw.githubusercontent.com/guru99-edu/R--
Programming/master/poisons.csv') %>%
  select(-X) %>%
  mutate(poison = factor(poison, ordered = TRUE))
  mutate(treat = factor(treat, ordered = FALSE))
glimpse(data)
```

Output:

```
Console   Terminal   Background Jobs                                          
R  R 4.3.2 · C:/Users/faruq.abdulla/Desktop/
> data <- read.csv("https://raw.githubusercontent.com/guru99-edu/R-Programming/master/poisons.c
sv") %>%
+   select(-X) %>%
+   mutate(poison = factor(poison, ordered = TRUE)) %>%
+   mutate(treat = factor(treat, ordered = FALSE))
> glimpse(data)
Rows: 48
Columns: 3
$ time   <dbl> 0.31, 0.45, 0.46, 0.43, 0.36, 0.29, 0.40, 0.23, 0.22, 0.21, 0.18, 0.23, 0...
$ poison <ord> 1, 1, 1, 1, 2, 2, 2, 2, 3, 3, 3, 3, 1, 1, 1, 1, 2, 2, 2, 2, 3, 3, 3, 3, 1...
$ treat  <fct> A, A, A, A, A, A, A, A, A, A, A, A, B, B, B, B, B, B, B, B, B, B, B, B, C...
> |
```

We can see the factor levels using the following R-code:

R-code:

```
levels(data$poison)
levels(data$treat)
```

Output:

```
Console   Terminal ×   Background Jobs ×
R  R 4.3.2 · C:/Users/faruq.abdulla/Desktop/
> levels(data$poison)
[1] "1" "2" "3"
> levels(data$treat)
[1] "A" "B" "C" "D"
>
```

There are three levels in 'poison' variable and four levels in the 'treat' variable. We can calculate the mean and standard deviation of the survival time for each level of the poison using the following R-code:

R-code:

```
data %>%
  group_by(poison) %>%
  summarise(
    count_poison = n(),
    mean_time = mean(time, na.rm = TRUE),
    sd_time = sd(time, na.rm = TRUE)
  )
```

Output:

```
Console   Terminal ×   Background Jobs ×
R  R 4.3.2 · C:/Users/faruq.abdulla/Desktop/
> data %>%
+   group_by(poison) %>%
+   summarise(
+     count_poison = n(),
+     mean_time = mean(time, na.rm = TRUE),
+     sd_time = sd(time, na.rm = TRUE)
+   )
# A tibble: 3 × 4
  poison count_poison mean_time sd_time
  <ord>         <int>     <dbl>   <dbl>
1 1                16     0.618   0.209
2 2                16     0.544   0.289
3 3                16     0.276   0.0623
>
```

Now, we can check the assumption of outliers and equal variance of survival time among different levels of poison using the following R-commands:

R-code:

```
install.packages('ggplot2')
library(ggplot2)
plot1 = ggplot(data) +
   aes(x = poison, y = time, color = poison) +
   geom_jitter() +
   theme(legend.position = 'none')
plot2 = ggplot(data, aes(x = poison, y = time, fill = poison)) +
   geom_boxplot() +
   geom_jitter(shape = 15,
               color = 'steelblue',
               position = position_jitter(0.21)) +
   theme_classic()
install.packages('patchwork')
library(patchwork)
plot1 + plot2
```

The output of grouped scatter plot and box plot are shown in Figure 11.13. From the box plot in Figure 11.13, we can conclude that the variance of the survival time among three different levels of poison is not equal. However, we can also check the equality of variance among groups using dot plot using the following R-code:

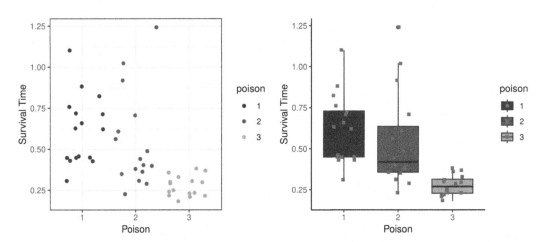

FIGURE 11.13
Scatter plot and box plot of the survival time grouped by the levels of poison.

R-code:

```
install.packages('lattice')
library('lattice')
dotplot(time ~ poison,
        data = data, col = 'red', xlab = 'Poison', ylab = 'Survival
Time')
```

The output of dot plot is illustrated in Figure 11.14.

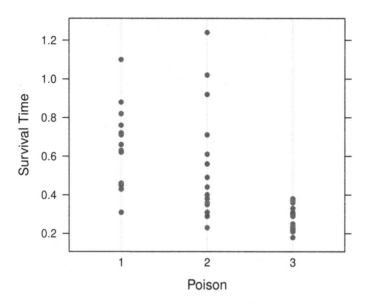

FIGURE 11.14
Dot plot of the survival time (grouped by the levels of poison).

The dot plot also depicted that the variance of the survival time among the three levels of poison are not equal. We can also use formally test for equality of the variances with a Levene's test by setting the null hypothesis that the variances are equal. The following R-code can be used:

R-code:

```
install.packages('car')
library(car)
leveneTest(time ~ poison, data = data)
```

Output:

```
Console   Terminal    Background Jobs

R  R 4.3.2 · C:/Users/faruq.abdulla/Desktop/
> library(car)
> leveneTest(time ~ poison, data = data)
Levene's Test for Homogeneity of Variance (center = median)
      Df F value  Pr(>F)
group  2  4.1964 0.02133 *
      45
---
Signif. codes:  0 '***' 0.001 '**' 0.01 '*' 0.05 '.' 0.1 ' ' 1
>
```

The result of Levene's test confirms that the variance of the survival time among the three levels of poison is not equal.

Now, we can fit one-way ANOVA modeling using the following R-code:

R-code:

```
anova1 <- aov(time~poison, data = data)
```

Now, we can check the assumption of independence of the observations using the Durbin-Watson test of autocorrelation of the residuals of the fitted model. The following R-code can be used:

R-code:

```
install.packages('car')
library(car)
durbinWatsonTest(anova1)
```

Output:

```
Console   Terminal ×   Background Jobs ×                                      ___ ☐

  R  R 4.3.2 · C:/Users/faruq.abdulla/Desktop/ ⇗
> library(car)
> durbinWatsonTest(anova1)
 lag Autocorrelation D-W Statistic p-value
   1      0.3958906     1.158806   0.004
 Alternative hypothesis: rho != 0
>
```

From the results, we observed that there is some degree of autocorrelation among the residuals of the fitted model. Therefore, we can conclude that the observations are not independent.

After fitting the model, normality assumption can be checked using the following R-code:

R-code:

```
par(mfrow = c(1, 2))
hist(anova1$residuals, xlab = 'Residuals', ylab = 'Frequency', main =
'')
library(car)
qqPlot(anova1$residuals, xlab = 'Normal Quantiles', ylab =
'Residuals',
      id = FALSE)
```

The output of histogram and Q-Q plot of the residuals are illustrated in Figure 11.15. Histogram and Q-Q plot depicted that the residuals of the fitted model do not follow normal distribution. We can also test normality of the residuals with the Shapiro-Wilk test using the following R-code:

R-code:

```
shapiro.test(anova1$residuals)
```

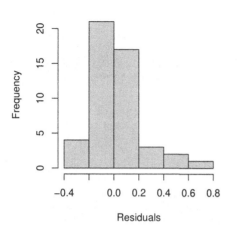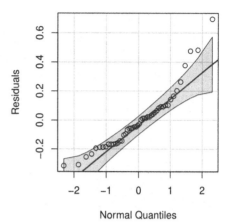

FIGURE 11.15
Histogram and Q-Q plot of the residuals.

Output:

```
Console   Terminal ×   Background Jobs ×
R  R 4.3.2 · C:/Users/faruq.abdulla/Desktop/
> shapiro.test(anoval$residuals)

        Shapiro-Wilk normality test

data:  anoval$residuals
W = 0.90648, p-value = 0.001025

>
```

The Shapiro-Wilk test also indicated that the distribution of the residuals of the fitted model is not normal.

To check whether the model fits the assumption of homoscedasticity, look at the model diagnostic plots in R using the following R-code:

R-code:

```
library(ggfortify)
autoplot(anoval, which = 1:6, ncol = 3, colour = 'black',
        ad.colour = 'black', label.n = 5, label.colour = 'black',
        smooth.colour = 'black', label.size = 3) +
    theme_bw()
```

The residual diagnostic plots are shown in Figure 11.16.
Now, we can see the results of the fitted ANOVA model using the following R-code:

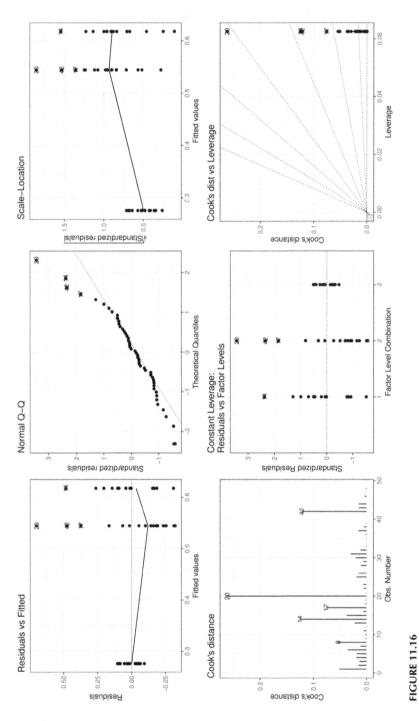

FIGURE 11.16
Residual diagnostic plots.

R-code:

```
summary(anova1)
```

Output:

```
Console  Terminal ×  Background Jobs ×                                    ─ □
R  R 4.3.2 · C:/Users/faruq.abdulla/Desktop/
> summary(anova1)
            Df Sum Sq Mean Sq F value  Pr(>F)
poison       2  1.033  0.5165   11.79 7.66e-05 ***
Residuals   45  1.972  0.0438
---
Signif. codes:  0 '***' 0.001 '**' 0.01 '*' 0.05 '.' 0.1 ' ' 1
>
```

The p-value is lower than the usual threshold of 0.05. Therefore, we are confident to say there is a statistical difference between the groups. Now, we are interested to seek which pairs of levels of poison have difference of means of survival time. To find out which groups are statistically different from one another, we can perform a post-hoc test. Post-hoc tests are a family of statistical tests; the most common ones are:

- Tukey's Honestly Significant Difference (Tukey's HSD), used to compare all groups to each other (so all possible comparisons of two groups).
- Dunnett, used to make comparisons with a reference group. For example, consider two treatment groups and one control group. If you only want to compare the two treatment groups with respect to the control group, and you do not want to compare the two treatment groups to each other, the Dunnett's test is preferred.
- Bonferroni correction if one has a set of planned comparisons to do.

We can perform Tukey's HSD for pairwise comparisons using the following R-code:

R-code:

```
TukeyHSD(anova1)
```

Output:

```
Console  Terminal ×  Background Jobs ×                                    ─ □
R  R 4.3.2 · C:/Users/faruq.abdulla/Desktop/
> TukeyHSD(anova1)
  Tukey multiple comparisons of means
    95% family-wise confidence level

Fit: aov(formula = time ~ poison, data = data)

$poison
         diff        lwr         upr      p adj
2-1 -0.073125 -0.2525046  0.10625464 0.5881654
3-1 -0.341250 -0.5206296 -0.16187036 0.0000971
3-2 -0.268125 -0.4475046 -0.08874536 0.0020924

>
```

From the results, we observed that the mean survival time of level 3 is significantly differs from the mean survival time of levels 1 and 2.

We can also check the pairwise difference using Bonferroni method. The following R-code can be used:

R-code:

```
pairwise.t.test(data$time, data$poison, p.adjust.method =
    'bonferroni')
```

Output:

```
Console   Terminal ×   Background Jobs ×                                           ─ □
R  R 4.3.2 · C:/Users/faruq.abdulla/Desktop/
> pairwise.t.test(data$time, data$poison, p.adjust.method = "bonferroni")

        Pairwise comparisons using t tests with pooled SD

data:  data$time and data$poison

  1       2
2 0.9853  -
3 1e-04   0.0022

P value adjustment method: bonferroni
>
```

The findings of Bonferroni method also indicate the similar conclusion as like of Tukey's HSD method.

Now, if we want to add another factor variable such as 'treat' in the ANOVA model then two-way ANOVA is appropriate to perform ANOVA. The procedure of checking assumptions for two-way ANOVA is same as like of one-way ANOVA. The following R-code can be used to perform two-way ANOVA model:

R-code:

```
# Two-way ANOVA without mixed effect
anova2 <- aov(time~poison + treat, data = data)
summary(anova2)
# Two-way ANOVA with mixed effect
anova3 <- aov(time~poison * treat, data = data)
summary(anova3)
```

Output:

```
Console   Terminal ×   Background Jobs ×                                           ─ □
R  R 4.3.2 · C:/Users/faruq.abdulla/Desktop/
> # Two-way ANOVA without mixed effect
> anova2 <- aov(time~poison + treat, data = data)
> summary(anova2)
            Df Sum Sq Mean Sq F value  Pr(>F)
poison       2 1.0330  0.5165   20.64 5.7e-07 ***
treat        3 0.9212  0.3071   12.27 6.7e-06 ***
Residuals   42 1.0509  0.0250
---
Signif. codes:  0 '***' 0.001 '**' 0.01 '*' 0.05 '.' 0.1 ' ' 1
> # Two-way ANOVA with mixed effect
> anova3 <- aov(time~poison * treat, data = data)
> summary(anova3)
              Df Sum Sq Mean Sq F value   Pr(>F)
poison         2 1.0330  0.5165  23.222 3.33e-07 ***
treat          3 0.9212  0.3071  13.806 3.78e-06 ***
poison:treat   6 0.2501  0.0417   1.874    0.112
Residuals     36 0.8007  0.0222
---
Signif. codes:  0 '***' 0.001 '**' 0.01 '*' 0.05 '.' 0.1 ' ' 1
>
```

The results depicted that there is a significant mean difference of survival time among the groups of poison and treatment. However, no significant difference of mean survival time of the groups of poison from the groups of treatment. Now, we can compare both of these models to select the best model using the following R-code:

R-code:

```
install.packages('AICcmodavg')
library(AICcmodavg)
model.set <- list(anova2, anova3)
model.names <- c('Two-way ANOVA without Mixed Effect', 'Two-way ANOVA
with Mixed Effect')
aictab(model.set, modnames = model.names)
```

Output:

```
Console   Terminal ×   Background Jobs ×

R  R 4.3.2 · C:/Users/faruq.abdulla/Desktop/
> library(AICcmodavg)
> model.set <- list(anova2, anova3)
> model.names <- c("Two-way ANOVA without Mixed Effect", "Two-way ANOVA with Mixed Effect")
> aictab(model.set, modnames = model.names)

Model selection based on AICc:

                                      K    AICc  Delta_AICc  AICcWt  Cum.Wt    LL
Two-way ANOVA without Mixed Effect    7  -30.42        0.00    0.97    0.97  23.61
Two-way ANOVA with Mixed Effect      13  -23.56        6.86    0.03    1.00  30.13

>
```

In the result table, AICc is a measure of how well a statistical model fits the data. Lower AICc values indicate better-fitting models; Delta_AICc represents the difference in AICc between each model and the best-fitting model. Smaller values are better, and a difference of 2 or more is considered significant; AICcWt indicates the probability that a given model is the best among the ones considered; Cum.Wt shows the cumulative probability that any model up to a particular row is the best-fitting model; and LL (Log-likelihood) measures how well a model explains the observed data. Higher values mean a better fit. From the results, we can conclude that the two-way ANOVA model without mixed effect is the best model. Now, we can see the pairwise comparison by Tukey's HSD method using the following R-code:

R-code:

```
TukeyHSD(anova2)
```

Output:

```
Console   Terminal ×   Background Jobs ×                                        ▬ ☐
R   R 4.3.2 · C:/Users/faruq.abdulla/Desktop/ ∞
> TukeyHSD(anova2)
  Tukey multiple comparisons of means
    95% family-wise confidence level

Fit: aov(formula = time ~ poison + treat, data = data)

$poison
         diff        lwr        upr     p adj
2-1 -0.073125 -0.2089936  0.0627436 0.3989657
3-1 -0.341250 -0.4771186 -0.2053814 0.0000008
3-2 -0.268125 -0.4039936 -0.1322564 0.0000606

$treat
           diff        lwr        upr     p adj
B-A  0.36250000  0.18976135  0.53523865 0.0000083
C-A  0.07833333 -0.09440532  0.25107198 0.6221729
D-A  0.22000000  0.04726135  0.39273865 0.0076661
C-B -0.28416667 -0.45690532 -0.11142802 0.0004090
D-B -0.14250000 -0.31523865  0.03023865 0.1380432
D-C  0.14166667 -0.03107198  0.31440532 0.1416151

>
```

The results indicate significant differences in the mean survival time among the poison groups, with group 3 differing significantly from groups 1 and 2. Furthermore, there are significant differences in the mean survival time between treatment A and treatments B and D. Additionally, the mean survival time of treatment B significantly differs from that of treatment C.

11.8 Conclusion and Exercises

This section concludes the chapter with some practical problem-solving exercises. These concepts and measures are used in the subsequent chapters.

11.8.1 Concluding Remarks

In conclusion, the regression analysis conducted in this chapter has provided valuable insights into the relationships between the variables under investigation. The findings of the regression analysis have not only confirmed or refuted our initial hypotheses but have also unearthed nuanced patterns and trends within the data. This analysis has enabled us to quantify the strength and direction of the relationships, offering a quantitative basis for predicting the outcome variable based on the independent variables. In this chapter, our exploration began with a comprehensive overview of the theoretical foundations of regression analysis, highlighting its significance in understanding the complexities of real-world phenomena. We delved into the different types of regression models through real-life examples with the application of R-programming, from simple linear regression to more advanced methods such as multiple linear regression, logistic regression, Poisson regression, and analysis of variance, each offering unique perspectives on the relationships under investigation. Furthermore, our discussion extended to the interpretation of coefficients, addressing the practical implications of our findings and their relevance in the context of our research objectives. As we move forward, the lessons learned from this chapter will undoubtedly inform subsequent chapters, shaping a comprehensive and robust narrative that adds meaningful contributions to the academic discourse in our chosen field of study.

11.8.2 Practice Questions

Exercise 11.1 Consider the 'Cardiovascular Diseases Risk Prediction Dataset' dataset from **Example 3.4**. Check whether simple linear regression is appropriate to estimate the effect of fruit consumption on the body mass index.

Exercise 11.2 Consider the 'nsduh2019_adult_sub_rmph.RData' dataset from the book entitled *Introduction to Regression Methods for Public Health Using R* written by Ramzi W. Nahhas (2024). This dataset was processed from National Survey on Drug Use and Health 2019 (NSDUH-2019), which is freely available at SAMHSA (U.S. Department of Health and Human Services, Substance Abuse and Mental Health Services Administration, Center for Behavioral Health Statistics and Quality 2019) for research and statistical purposes. This dataset includes a random subset of 1,000 observations of adults, and variables that have been renamed from NSDUH-2019.

    a.

- Download NSDUH-2019-DS0001-bndl-data-r.zip file from https://www.datafiles.samhsa.gov/sites/default/files/field-uploads-protected/studies/NSDUH-2019/NSDUH-2019-datasets/NSDUH-2019-DS0001/NSDUH-2019-DS0001-bundles-with-study-info/NSDUH-2019-DS0001-bndl-data-r.zip that contains NSDUH_2019 as .RData file. Extract the NSDUH_2019.RData file from the downloaded .zip file and paste to your working directory.

- Download the R script files NSDUH_2019 Process.R from https://github.com/rwnahhas/RMPH_Resources/blob/main/NSDUH_2019%20Process.R.

- Save the 'nsduh2019_adult_sub_rmph.RData' dataset by running the downloaded R script with changing the working directory.

    b. Perform a binary logistic regression for checking the association of lifetime marijuana use (*mj_lifetime*) with sex (*demog_sex*), age category with 6 categories (*demog_age_cat6*), age at first use of alcohol (*alc_age-first*), and marital status (*demog_marital*). Consider female, 65+, and married as reference category for sex, age category, and marital status.

    c. Check the performance of the fitted model

    d. Check the validity of the fitted model by 10-fold cross-validation

    e. Visualize the odds ratio using forest plot

    f. Predict the lifetime marijuana use using the fitted model for sex: Female, age category: 18–25, marital status: Married, and age at first use of alcohol: 20

Exercise 11.3 The dataset named 'Affairs' in the package 'AER' contains 601 observations of the following 9 variables:

Variable Name	Description
affairs	How often engaged in extramarital sexual intercourse during the past year? 0 = none, 1 = once, 2 = twice, 3 = 3 times, 7 = 4–10 times, 12 = monthly, 12 = weekly, 12 = daily
gender	Factor indicating gender categorized as female and male
age	Numeric variable coding age in years: 17.5 = under 20, 22 = 20–24, 27 = 25–29, 32 = 30–34, 37 = 35–39, 42 = 40–44, 47 = 45–49, 52 = 50–54, 57 = 55 or over
yearsmarried	Numeric variable coding number of years married: 0.125 = 3 months or less, 0.417 = 4–6 months, 0.75 = 6 months–1 year, 1.5 = 1–2 years, 4 = 3–5 years, 7 = 6–8 years, 10 = 9–11 years, 15 = 12 or more years
children	Are there children in the marriage? categorized as no and yes
religiousness	Numeric variable coding religiousness: 1 = anti, 2 = not at all, 3 = slightly, 4 = somewhat, 5 = very
education	Numeric variable coding level of education: 9 = grade school, 12 = high school graduate, 14 = some college, 16 = college graduate, 17 = some graduate work, 18 = master's degree, 20 = Ph.D., M.D., or other advanced degree
occupation	Numeric variable coding occupation according to Hollingshead classification (reverse numbering)
rating	Numeric variable coding self-rating of marriage: 1 = very unhappy, 2 = somewhat unhappy, 3 = average, 4 = happier than average, 5 = very happy

Fit an appropriate regression model in order to predict the expected times of engaging in extramarital sexual intercourse based on the covariates available in the dataset.

Exercise 11.4 Consider the built-in R dataset named 'ToothGrowth' that contains data from a study evaluating the effect of vitamin C on tooth growth in Guinea pigs. The experiment has been performed on 60 pigs, where each animal received one of three dose levels of vitamin C (0.5, 1, and 2 mg/day) by one of two delivery methods, (orange juice, coded as OJ or ascorbic acid (a form of vitamin C and coded as VC). In the dataset, dose levels of vitamin C are stored under the variable *dose* and the delivery methods are stored under the variable *supp*. Tooth length was measured and stored under the variable *len*. Call the dataset under the data name 'mydata' and solve the following problems:

a. Make the *dose* variable as a factor variable.

```
[Hints: mydata$dose <- factor(mydata$dose,
          levels = c(0.5, 1, 2),
          labels = c('D0.5', 'D1', 'D2'))]
```

b. Calculate the means and standard deviations of tooth growth for each level of dose and for each delivery method.

c. Check whether any difference of tooth growth among different dose levels.

    d. If significant difference found in question 'c', find which pair of doses has significantly different tooth growth.

    e. Check whether any difference of tooth growth over both of dose levels and delivery method without considering their combined effect.

    f. Check whether any difference of tooth growth over both of dose levels and delivery method with considering their combined effect.

Bibliography

Abuse, S., & Administration, M. H. S. (2014). *National survey on drug use and health*. Substance Abuse and Mental Health Services Administration. Available at https://www.samhsa.gov/data/data-we-collect/nsduh-national-survey-drug-use-and-health

Ahmed, K. T., Karimuzzaman, M., Afroz, S., Hossain, M. M., Huq, S. S., Abdulla, F., & Rahman, A. (2023). Trends and long-term variation explaining nutritional determinants of child linear growth: Analysis of Bangladesh Demographic and Health Surveys 1996–2018. *Public Health Nutrition*, 2758–2770. https://doi.org/10.1017/S1368980023002288

Allen, M. P. (2004). *Understanding regression analysis*. Springer Science & Business Media.

Arkes, J. (2023). *Regression analysis: A practical introduction*. Taylor & Francis.

Atkinson, A. C., Riani, M., & Riani, M. (2000). *Robust diagnostic regression analysis* (Vol. 2). Springer.

Berry, W. D. (1993). *Understanding regression assumptions* (Vol. 92). Sage.

Brook, R. J., & Arnold, G. C. (2018). *Applied regression analysis and experimental design*. CRC Press.

Cameron, A. C., & Trivedi, P. K. (1986). Econometric models based on count data. Comparisons and applications of some estimators and tests. *Journal of Applied Econometrics*, 1(1), 29–53.

Cameron, A. C., & Trivedi, P. K. (2009). *Advances in Count Data Regression Talk for the Applied Statistics Workshop*, March 28, 2009. Retrieved from (12/21/2015) http://cameron.econ.ucdavis.edu/racd/count.html

Cameron, A. C., & Trivedi, P. K. (2013). *Regression analysis of count data* (Vol. 53). Cambridge University Press.

Chatterjee, S., & Hadi, A. S. (2015). *Regression analysis by example*. John Wiley & Sons.

Chatterjee, S., & Simonoff, J. S. (2013). *Handbook of regression analysis* (Vol. 5). John Wiley & Sons.

Chen, D. G. D., & Chen, J. K. (2021). *Statistical regression modeling with R*. Springer International Publishing.

Christensen, R. (1996). *Analysis of variance, design, and regression: Applied statistical methods*. CRC Press.

Ciaburro, G. (2018). *Regression Analysis with R: Design and develop statistical nodes to identify unique relationships within data at scale*. Packt Publishing Ltd.

Cohen, J., Cohen, P., West, S. G., & Aiken, L. S. (2013). *Applied multiple regression/correlation analysis for the behavioral sciences*. Routledge.

Cook, R. D., & Weisberg, S. (2009). *An introduction to regression graphics* (Vol. 405). John Wiley & Sons.

Daniel, W. W., & Cross, C. L. (2018). *Biostatistics: A foundation for analysis in the health sciences*. Wiley.

Darlington, R. B., & Hayes, A. F. (2016). *Regression analysis and linear models: Concepts, applications, and implementation*. Guilford Publications.

Das, S., Rahman, A., Ahamed, A., & Rahman, S. T. (2019). Multi-level models can benefit from minimizing higher-order variations: An illustration using child malnutrition data. *Journal of Statistical Computation and Simulation*, 89(6), 1090–1110. https://doi.org/10.1080/00949655.2018.1553242

Dobson, A. J., & Barnett, A. G. (2018). *An introduction to generalized linear models*. Chapman and Hall/CRC.

Doll, R. (1971). The age distribution of cancer: Implications for models of carcinogenesis. *Journal of the Royal Statistical Society. Series A (General)*, 134 (2): 133–55. https://doi.org/10.2307/2343871

Draper, N. R., & Smith, H. (1998). *Applied regression analysis* (Vol. 326). John Wiley & Sons.

Dupont, W. D. (2009). *Statistical modeling for biomedical researchers: A simple introduction to the analysis of complex data.* Cambridge University Press.

Faraway, J. J. (2016). *Extending the linear model with R: generalized linear, mixed effects and nonparametric regression models.* Chapman and Hall/CRC.

Ferreira, E. B., Cavalcanti, P. P., & Nogueira, D. A. (2014). ExpDes: An R package for ANOVA and experimental designs. *Applied Mathematics*, 5(19), 2952.

Fleiss, J. L., Levin, B., & Paik, M. C. (2013). *Statistical methods for rates and proportions.* John Wiley & Sons.

Fox, J., & Weisberg, S. (2018). *An R companion to applied regression.* Sage Publications.

Franzese, R., & Kam, C. (2009). *Modeling and interpreting interactive hypotheses in regression analysis.* University of Michigan Press.

Freund, R. J., Wilson, W. J., & Sa, P. (2006). *Regression analysis.* Elsevier.

Frome, E. L. (1983). The analysis of rates using Poisson regression models. *Biometrics*, 39(3), 665–674.

Frost, J. (2019). *Regression analysis: An intuitive guide for using and interpreting linear models.* Statistics By Jim Publishing.

Golberg, M. A., & Cho, H. A. (2004). *Introduction to regression analysis.* WIT Press.

Gunst, R. F., & Mason, R. L. (2018). *Regression analysis and its application: a data-oriented approach.* CRC Press.

Hakim, M. A., & Rahman, A. (2018). Simulating the nutritional traits of populations at the small area levels using spatial microsimulation modelling approach. *Computational Biology and Bioinformatics*, 6(1), 25–30.

Hocking, R. R. (2013). *Methods and applications of linear models: regression and the analysis of variance.* John Wiley & Sons.

Hoffmann, J. P. (2016). *Regression models for categorical, count, and related variables: An applied approach.* University of California Press.

Hoffmann, J. P. (2021). *Linear regression models: applications in R.* CRC Press.

Hossain, M. M. (2022). Statistical regression modeling with R. *Journal of the Royal Statistical Society Series A: Statistics in Society*, 185(2), 743–744.

Hossain, M. M., Abdulla, F., Rahman, A., & Khan, H. T. A. (2022). Prevalence and determinants of wife-beating in Bangladesh: Evidence from a nationwide survey. *BMC Psychiatry*, 22(1), 9. https://doi.org/10.1186/s12888-021-03652-x

Hossain, M. M., Sobhan, M. A., Rahman, A., Flora, S. S., & Irin, Z. S. (2021). Trends and determinants of vaccination among children aged 06–59 months in Bangladesh: country representative survey from 1993 to 2014. *BMC Public Health*, 21(1), 1–11. 1578. https://doi.org/10.1186/s12889-021-11576-0

Hsu, J. (1996). *Multiple comparisons: Theory and methods.* CRC Press.

Ip, H. L., Demskoi, D., Rahman, A., & Zheng, L. (2021). Evaluation of COVID-19 mitigation policies in Australia using generalised space-time autoregressive intervention models. *International Journal of Environmental Research and Public Health*, 18(14), 1–17. 7474. https://doi.org/10.3390/ijerph18147474

Kahane, L. H. (2007). *Regression basics.* Sage Publications.

Karimuzzaman, M., Afroz, S., Hossain, M. M., & Rahman, A. (2020a). Forecasting the COVID-19 pandemic with climate variables for top five burdening and three south Asian countries. https://doi.org/10.1101/2020.05.12.20099044

Karimuzzaman, M., Hossain, M. M., & Rahman, A. (2020b). Finite mixture modelling approach to identify factors affecting children ever born for 15–49 year old women in Asian country. In A. Rahman (Ed.), *Statistics for data science and policy analysis* (pp. 221–236). Springer. https://doi.org/10.1007/978-981-15-1735-8_17

Kuddus, A., & Rahman, A. (2015). Affect of urbanization on health and nutrition. *International Journal of Statistics and Systems*, 10(2), 165–175.

Kuddus, M. A., Mohiuddin, M., & Rahman, A. (2021). Mathematical analysis of a measles transmission dynamics model in Bangladesh with double dose vaccination. *Scientific Reports*, 11(1), 1–16. 16571. https://doi.org/10.1038/s41598-021-95913-8

Kuddus, M. A., & Rahman, A. (2021). Analysis of COVID-19 using a modified SLIR model with nonlinear incidence. *Results in Physics*, 27, 104478. https://doi.org/10.1016/j.rinp.2021.104478

Kuddus, M. A., & Rahman, A. (2022). Modelling and analysis of human–mosquito malaria transmission dynamics in Bangladesh. *Mathematics and Computers in Simulation*, 193, 123–138. Article MATCOM5498. https://doi.org/10.1016/j.matcom.2021.09.021

Kuddus, M. A., Rahman, A., Alam, F., & Mohiuddin, M. (2023). Analysis of the different interventions scenario for programmatic measles control in Bangladesh: A modelling study. *PLoS One*, 18(6), e0283082. https://doi.org/10.1371/journal.pone.0283082

MacFarland, T. W. (2011). *Two-way analysis of variance: Statistical tests and graphics using R*. Springer Science & Business Media.

McCullagh, P. (2019). *Generalized linear models*. Routledge.

Menard, S. (2002). *Applied logistic regression analysis* (No. 106). Sage.

Montgomery, D. C. (2017). *Design and analysis of experiments*. John Wiley & Sons.

Montgomery, D. C., Peck, E. A., & Vining, G. G. (2021). *Introduction to linear regression analysis*. John Wiley & Sons.

Musa, K. I., Mansor, W. N. A. W., & Hanis, T. M. (2023). *Data analysis in medicine and health using R*. CRC Press.

Nahhas, R. W. (2024). *Introduction to regression methods for public health using R*. Creative Common.

Rahman, A. (2019). Geospatial, socioeconomic, demographic and health determinants of childhood mortality in Bangladesh. *Demography India*, 48(2), 74–85.

Rahman, A., & Chowdhury Biswas, S. (2009). Nutritional status of under-5 children in Bangladesh. *South Asian Journal of Population and Health*, 2(1), 1–11.

Rahman, A., & Chowdhury, S. (2007). Determinants of chronic malnutrition among preschool children in Bangladesh. *Journal of Biosocial Science*, 39(2), 161–173. https://doi.org/10.1017/S0021932006001295

Rahman, A., Chowdhury, S., & Hossain, D. (2009). Acute malnutrition in Bangladeshi children levels and determinants. *Asia-Pacific Journal of Public Health*, 21(3), 294–302. https://doi.org/10.1177/1010539509335399

Rahman, A., Chowdury, S., Karim, A., & Ahmed, S. (2008). Factors associated with nutritional status of children in Bangladesh: A multivariate analysis. *Demography India*, 37(1), 95–109.

Rahman, A., & Hakim, M. A. (2016). Measuring modified mass energy equivalence in nutritional epidemiology: A proposal to adapt the biophysical modelling approach. *International Journal of Statistics in Medical Research*, 5(3), 219–223.

Rahman, A., & Hakim, M. A. (2017). Modeling health status using the logarithmic biophysical modulator. *Journal of Public Health and Epidemiology*, 9(5), 145–150. DE4D0AF63797. https://doi.org/10.5897/JPHE2016.0867

Rahman, A., & Kuddus, A. (2014a). A new model to study on physical behaviour among susceptible infective removal population. *Far East Journal of Theoretical Statistics*, 46(2), 115–135.

Rahman, A., & Kuddus, A. (2014b). Effects of some sociological factors on the outbreak of chickenpox disease. *JP Journal of Biostatistics*, 11(1), 37–53.

Rahman, A., & Kuddus, M. A. (2020). Cost-effective modeling of the transmission dynamics of malaria: A case study in Bangladesh. *Communications in Statistics: Case Studies, Data Analysis and Applications*, 6(2), 270–286. https://doi.org/10.1080/23737484.2020.1731724

Rahman, A., & Kuddus, M. A. (2021). Modelling the transmission dynamics of COVID-19 in six high-burden countries. *BioMed Research International*, 2021, 1–17, 5089184. https://doi.org/10.1155/2021/5089184

Rahman, A., Kuddus, M. A., Ip, H. L., & Bewong, M. (2021). A review of COVID-19 modelling strategies in three countries to develop a research framework for regional areas. *Viruses*, 13(11), 2185. https://doi.org/10.3390/v13112185

Rahman, A., Kuddus, M. A., Ip, R. H. L., & Bewong, M. (2023). Modelling COVID-19 pandemic control strategies in metropolitan and rural health districts in New South Wales, Australia. *Scientific Reports*, 13, 1–18, 10352. https://doi.org/10.1038/s41598-023-37240-8

Rahman, A., Othman, N., Kuddus, M. A., & Hasan, M. Z. (2024a). Impact of the COVID-19 pandemic on child malnutrition in Selangor, Malaysia: A pilot study. *Journal of Infection and Public Health*, 17(5), 833–842. https://doi.org/10.1016/j.jiph.2024.02.019

Rahman, A., Kuddus, M. A., Paul, A. K., & Hasan, M. Z. (2024b). The impact of triple doses vaccination and other interventions for controlling the outbreak of COVID-19 cases and mortality in Australia: A modelling study. *Heliyon*, *10*(4), 1–12. e25945. https://doi.org/10.1016/j.heliyon.2024.e25945

Rahman, A., & Sapkota, M. (2014). Knowledge on vitamin A rich foods among mothers of preschool children in Nepal: Impacts on public health and policy concerns. *Science Journal of Public Health*, 2(4), 316–322. https://doi.org/10.11648/j.sjph.20140204.22

Rawlings, J. O., Pantula, S. G., & Dickey, D. A. (Eds.). (1998). *Applied regression analysis: A research tool*. Springer New York.

Roback, P., & Legler, J. (2021). *Beyond multiple linear regression: Applied generalized linear models and multilevel models in R*. Chapman and Hall/CRC.

Roberts, M., & Russo, R. (2014). *A student's guide to analysis of variance*. Routledge.

Ryan, T. P. (2008). *Modern regression methods* (Vol. 655). John Wiley & Sons.

Schroeder, L. D., Sjoquist, D. L., & Stephan, P. E. (2016). *Understanding regression analysis: An introductory guide* (Vol. 57). Sage Publications.

Sheather, S. (2009). *A modern approach to regression with R*. Springer Science & Business Media.

Shoukri, M. M., & Cihon, C. (1998). *Statistical methods for health sciences*. CRC Press.

Thevaraja, M., Rahman, A., & Gabirial, M. (2019). Recent developments in data science: Comparing linear, ridge and lasso regressions techniques using wine data. In F. Hidoussi (Ed.), *Proceedings of the International Conference on Digital Image & Signal Processing* (pp. 1–6). 217, University of Oxford.

Thrane, C. (2019). *Applied regression analysis: Doing, interpreting, and reporting*. Routledge.

Tukey, J. W. (1949). Comparing individual means in the analysis of variance. *Biometrics*, 5(2), 99–114.

Turner, J. R., & Thayer, J. (2001). *Introduction to analysis of variance: Design, analysis & interpretation*. Sage.

Wang, G. C., & Jain, C. L. (2003). *Regression analysis: Modeling & forecasting*. Institute of Business Forecasting.

Westfall, P. H., & Arias, A. L. (2020). *Understanding regression analysis: A conditional distribution approach*. CRC Press.

Wright, D. B., & London, K. (2009). *Modern regression techniques using R: A practical guide*. Sage.

Yan, X., & Su, X. (2009). *Linear regression analysis: Theory and computing*. World Scientific.

Zelterman, D. (2022). *Regression for health and social science: Applied linear models with R*. Cambridge University Press.

12

Survival Analysis and Factor Analysis

Survival analysis is a statistical approach applied in many domains, including biostatistics, social sciences, data science, engineering, health, agriculture, and economics, to determine how long an event of interest will take to occur. The term 'event' in survival analysis need not always be negative; it can be any relevant event, such as a disease developing, a mechanical component failing, or the amount of time before a client makes a purchase. Survival analysis is a powerful tool for modeling time-to-event data (e.g. Pandey et al., 2018), providing valuable insights into the factors influencing the timing of events.

Factor analysis is a statistical method used to explore the underlying structure of a set of variables. The primary goal of factor analysis is to identify latent factors that can explain patterns of correlations among observed variables. Factor analysis helps reduce the dimensionality of the data by identifying a smaller number of factors that can explain the observed patterns of correlations. It is widely used in psychometrics to develop and validate tests, in market research to identify underlying consumer preferences, and in various other disciplines where understanding latent structures is essential.

In this chapter, different survival models with their applications, including biostatistics and health sciences, are discussed. This chapter also includes the concept of factor analysis with practical applications using real-world scientific data. Moreover, the relevant R-code and output are presented throughout the chapter so that learners can practice as they study.

12.1 Key Concept Associated with Survival Analysis

This section presents some key concepts that need to be understood for survival analysis.

Survival function: The survival function is denoted by $S(t)$ is the probability that a subject will survive beyond a certain time t, or in other words, the probability that the event of interest, e.g., death, occurrence of a disease, system failure, etc. has not occurred by time t. Mathematically, the survival function is defined as follows:

$$S(t) = P(T > t)$$

where T is the random variable representing the time until the event of interest, and t is a specific time point.

The survival function provides a cumulative perspective on the survival experience of a group of subjects. For example, $S(15) = 0.75$ means that the probability of surviving beyond 15 units of time is 0.75 or 75%.

The survival function is associated with the cumulative distribution function (CDF). The CDF, denoted as $F(t)$, represents the probability that the event occurs by time t, which is complementary to the survival function. Thus, we can write as,

$$F(t) = 1 - S(t).$$

Hazard function: The hazard function is denoted by $h(t)$ or $\lambda(t)$ and is the instantaneous rate of occurrence of an event at a specific time t, given that the individual has survived up to that time. In other words, the hazard function represents the probability per unit of time that an event will occur at time t, given survival up to t. Mathematically, the hazard function is defined as the ratio of the probability density function (pdf) to the survival function:

$$h(t) = \frac{f(t)}{S(t)}.$$

Alternatively, the hazard function can be defined in terms of the negative logarithm of the survival function:

$$h(t) = -\frac{d}{dt}\ln(S(t)),$$

where $f(t)$ is the probability density function, representing the probability of the event occurring at time t, and $S(t)$ is the survival function.

The hazard function provides insights into the changing risk of an event over time. A high hazard function indicates an increased risk of the event occurring, while a low hazard function suggests a lower risk.

Censoring: Censoring is a concept frequently encountered in survival analysis, where it refers to the incomplete observation of the time until an event of interest occurs. Alternatively, censoring occurs when the exact time of the event is not known for some subjects in the study. It occurs when the event of interest has not occurred for some individuals by the end of the study. These individuals are still part of the analysis, and their data is considered in estimating survival probabilities. There are three types of censoring which follow.

i. *Left censoring*: This occurs when the event of interest has already occurred before the study begins, but the exact time of occurrence is not known. Left-censored data may arise in cases where individuals enter a study already having experienced the event.

ii. *Right censoring*: This is the most common type of censoring in survival analysis. Right censoring occurs when a subject is still under observation at the end of the study period, and the event of interest has not occurred by that time. The observed time is then treated as a lower bound for the time until the event, but the exact time is unknown. Right censoring is denoted by a vertical bar ' | ' on a survival time, indicating that the event did not occur up to that time.

iii. *Interval censoring*: In some cases, the event of interest is only known to have occurred within a certain time interval. This is referred to as interval censoring.

12.2 Models Used in Survival Analysis

Several models are used in survival analysis to analyze time-to-event data and understand the factors influencing the occurrence of events. Choosing the appropriate model depends on the characteristics of the data, the assumptions that can be reasonably made, and the research question. It is common to use a combination of exploratory data analysis and statistical tests to guide the selection of the most suitable model for a particular study. Moreover, researchers should be aware of the assumptions associated with each model and assess their validity in the context of the data being analyzed. The commonly used models are given below.

    i. Kaplan-Meier Estimator.
   ii. Cox Proportional Hazards Model.
  iii. Parametric Survival Models.
  iv. Frailty Models.

Each of these techniques is discussed below.

12.2.1 Kaplan-Meier Estimator

The Kaplan-Meier estimator is a nonparametric method used in survival analysis to estimate the probability of survival beyond a certain time point. It is particularly useful when dealing with time-to-event data in the presence of censoring. The Kaplan-Meier estimator provides a step-function estimate of the survival function, representing the probability that an individual survives beyond a specified time-period.

The Kaplan-Meier estimate of the survival function at time t can be written as:

$$S(t) = \prod_{i:t_i \leq t} \left(1 - \frac{d_i}{n_i}\right)$$

where d_i is the number of events at time t_i, and n_i is the number of individuals at risk just before time t_i.

Example 12.1

We consider the 'lung' dataset in the `survival` package of R that contains information about survival times and censoring status for patients with advanced lung cancer. In this dataset, the following variables are available:

- inst: Institution code.
- time: This variable represents the survival time or the time until death (measured in days).
- status: This variable indicates the censoring status. A value of 1 represents an observed event (death), and a value of 0 represents censoring (individuals who were still alive at the end of the study).
- sex: The gender of the patient, coded as 1 for male and 2 for female.
- age: The age of the patient at the time of diagnosis.
- ph.ecog: The performance status of the patient, measured on the ECOG scale (Eastern Cooperative Oncology Group). It is a categorical variable representing the overall health

and activity level of the patient. Common values include 0 (fully active), 1 (restricted activity but ambulatory), 2 (ambulatory but unable to work), and so on.

- ph.karno: The Karnofsky performance score, another measure of the patient's ability to perform normal daily activities.
- pat.karno: The Karnofsky performance score for the patient's spouse or partner.
- meal.cal: The number of calories consumed during a meal.
- wt.loss: Weight loss in the last six months.

To perform a Kaplan-Meier survival analysis in R, we use the survival package. In order to install and load the survival package as well as to see the structure of the 'lung' dataset from the survival package, the following R-code can be used:

R-code:

```
# Install and load the survival package
install.packages('survival')
library(survival)
# Explore the structure of the lung dataset
head(lung)
```

Output:

```
Console  Terminal   Background Jobs                                          _ □
R  R 4.3.2 · ~/
> library(survival)
> # Explore the structure of the lung dataset
> head(lung)
  inst time status age sex ph.ecog ph.karno pat.karno meal.cal wt.loss
1    3  306      2  74   1       1       90       100     1175      NA
2    3  455      2  68   1       0       90        90     1225      15
3    3 1010      1  56   1       0       90        90       NA      15
4    5  210      2  57   1       1       90        60     1150      11
5    1  883      2  60   1       0      100        90       NA       0
6   12 1022      1  74   1       1       50        80      513       0
>
```

We use the following code in R to get the Kaplan-Meier estimate and plot:

R-code:

```
# Create a Surv object with survival times and event/censoring
indicator
surv_object <- with(lung, Surv(time, status))
# Fit the Kaplan-Meier estimator
fit_km <- survfit(surv_object ~ 1) #The formula ~ 1 indicates a single
group
summary(fit_km)
# Plot the Kaplan-Meier curve
plot(fit_km, main = 'Kaplan-Meier Survival Curve', xlab = 'Time', ylab
= 'Survival Probability', col='blue')
```

Output:

```
Console   Terminal ×   Background Jobs ×                                              ─ □
 R  R 4.3.2 · ~/
> library(survival)
> # Explore the structure of the lung dataset
> head(lung)
  inst time status age sex ph.ecog ph.karno pat.karno meal.cal wt.loss
1    3  306      2  74   1       1       90       100     1175      NA
2    3  455      2  68   1       0       90        90     1225      15
3    3 1010      1  56   1       0       90        90       NA      15
4    5  210      2  57   1       1       90        60     1150      11
5    1  883      2  60   1       0      100        90       NA       0
6   12 1022      1  74   1       1       50        80      513       0
> # Create a Surv object with survival times and event/censoring indicator
> surv_object <- with(lung, Surv(time, status))
> # Fit the Kaplan-Meier estimator
> fit_km <- survfit(surv_object ~ 1) #The formula ~ 1 indicates a single group
> summary(fit_km)
Call: survfit(formula = surv_object ~ 1)

 time n.risk n.event survival std.err lower 95% CI upper 95% CI
    5    228       1   0.9956 0.00438       0.9871        1.000
   11    227       3   0.9825 0.00869       0.9656        1.000
   12    224       1   0.9781 0.00970       0.9592        0.997
   13    223       2   0.9693 0.01142       0.9472        0.992
   15    221       1   0.9649 0.01219       0.9413        0.989
  735     12       1   0.0979 0.02660       0.0575        0.167
  765     10       1   0.0881 0.02568       0.0498        0.156
  791      9       1   0.0783 0.02462       0.0423        0.145
  814      7       1   0.0671 0.02351       0.0338        0.133
  883      4       1   0.0503 0.02285       0.0207        0.123
> # Plot the Kaplan-Meier curve
> plot(fit_km, main = "Kaplan-Meier Survival Curve", xlab = "Time", ylab = "Survival Probabilit
y", col="blue")
> |
```

The Kaplan-Meier survival curve is presented in Figure 12.1.

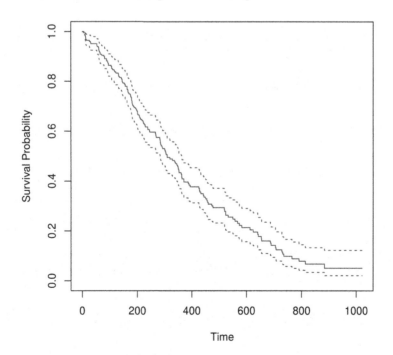

FIGURE 12.1
Kaplan-Meier survival curve.

We may also consider multiple groups, and we can compare survival curves. Suppose we consider the *ph.ecog* variable to compare the survival curves for different treatment groups. For this purpose, we use the following code in R:

R-code:

```
# Fit Kaplan-Meier estimator for different treatment groups
# Convert 'ph.ecog' to a factor
lung$ph.ecog <- as.factor(lung$ph.ecog)
# Check the levels of 'ph.ecog'
print(levels(lung$ph.ecog))
fit_km_group <- survfit(surv_object ~ ph.ecog, data = lung)
summary(fit_km_group)
# Plot the Kaplan-Meier curves for each group
plot(fit_km_group, main = 'Kaplan-Meier Survival Curves by Treatment
Group', col = 1:length(levels(lung$ph.ecog)), xlab = 'Time', ylab =
'Survival Probability')
legend('topright', legend = levels(lung$ph.ecog), col =
1:length(levels(lung$ph.ecog)), bty = 'n', lty = 1)
```

Output:

```
Console   Terminal ×   Background Jobs ×
R 4.3.2 · E:/Book/Chapter 12/
> # Fit Kaplan-Meier estimator for different treatment groups
> # Convert 'ph.ecog' to a factor
> lung$ph.ecog <- as.factor(lung$ph.ecog)
> # Check the levels of 'ph.ecog'
> print(levels(lung$ph.ecog))
[1] "0" "1" "2" "3"
> fit_km_group <- survfit(surv_object ~ ph.ecog, data = lung)
> summary(fit_km_group)
Call: survfit(formula = surv_object ~ ph.ecog, data = lung)

1 observation deleted due to missingness
                ph.ecog=0
 time n.risk n.event survival std.err lower 95% CI upper 95% CI
    5     63       1   0.9841  0.0157       0.9537        1.000
   11     62       1   0.9683  0.0221       0.9259        1.000
   15     61       1   0.9524  0.0268       0.9012        1.000
   31     60       1   0.9365  0.0307       0.8782        0.999
   53     59       1   0.9206  0.0341       0.8562        0.990
  524      6       1   0.1392  0.0533      0.06566        0.295
  533      5       1   0.1113  0.0494      0.04666        0.266
  654      3       1   0.0742  0.0448      0.02277        0.242
  707      2       1   0.0371  0.0345      0.00601        0.229
  814      1       1   0.0000     NaN           NA           NA

            ph.ecog=3
       time     n.risk      n.event     survival     std.err lower 95% CI
        118          1            1            0         NaN           NA
upper 95% CI
         NA

> # Plot the Kaplan-Meier curves for each group
> plot(fit_km_group, main = "Kaplan-Meier Survival Curves by Treatment Group", col = 1:lengt
h(levels(lung$ph.ecog)) , xlab = "Time", ylab = "Survival Probability")
> legend("topright", legend = levels(lung$ph.ecog), col = 1:length(levels(lung$ph.ecog)), bt
y = "n", lty = 1)
> |
```

The Kaplan-Meier survival curves by different treatment groups are illustrated in Figure 12.2.

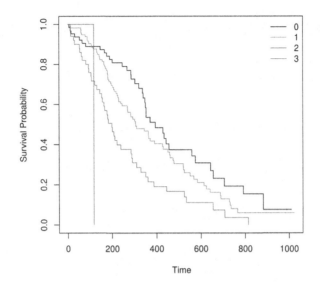

FIGURE 12.2
Kaplan-Meier survival curves by different treatment groups.

12.2.2 Cox Proportional Hazards Model

The Cox Proportional Hazards model, introduced by David R. Cox in 1972, is a widely used statistical model in survival analysis. It is a semi-parametric model that examines the association between the survival time of individuals and one or more predictor variables. The key assumption of the Cox model is that the hazard function for any individual is a constant multiple of a baseline hazard function that changes with time.

The hazard function for the Cox Proportional Hazards model is written as:

$$h(t) = h_0(t) \times \exp(\beta_1 X_1 + \beta_2 X_2 + \ldots + \beta_k X_k),$$

where $h_0(t)$ is the baseline hazard function that changes with time, $\beta_1, \beta_2, \ldots, \beta_k$ are the coefficients associated with predictor variables $X_1, X_2, \ldots, X_k$ respectively.

Example 12.2

Recall the dataset used in **Example 12.1**. Suppose we want to include the variables *age*, *sex*, the performance status of the patient (*ph.ecog*), the Karnofsky performance score of the patient, number of calories consumed during a meal, and weight loss in the last six months to fit the Cox Proportional Hazards model.

We use coxph function to fit the Cox Proportional Hazards model and the commands used in R, are given below:

R-code:

```
# Fit Cox Proportional Hazards model
fit_cox_model <- coxph(Surv(time, status) ~ age + sex + ph.ecog +
ph.karno + meal.cal + wt.loss, data = lung)
# Display summary of the model
```

```
summary(fit_cox_model)
# Create survival curves
surv_curves <- survfit(fit_cox_model)
# Plot survival curves
plot(surv_curves, col = c('blue', 'red', 'green'), lty = 1, lwd = 2,
     main = 'Survival Curves by Category', xlab = 'Time', ylab =
'Survival Probability')
```

Output:

```
Console  Terminal ×  Background Jobs ×                                    □

R R4.3.2 · ~/
> # Fit Cox Proportional Hazards model
> fit_cox_model <- coxph(Surv(time, status) ~ age + sex + ph.ecog + ph.karno + meal.cal + wt.lo
ss, data = lung)
> # Display summary of the model
> summary(fit_cox_model)
Call:
coxph(formula = Surv(time, status) ~ age + sex + ph.ecog + ph.karno +
    meal.cal + wt.loss, data = lung)

  n= 170, number of events= 123
   (58 observations deleted due to missingness)

              coef   exp(coef)  se(coef)     z Pr(>|z|)
age       1.197e-02  1.012e+00  1.169e-02  1.023 0.306103
sex      -5.510e-01  5.763e-01  1.996e-01 -2.761 0.005756 **
ph.ecog1  6.172e-01  1.854e+00  2.781e-01  2.219 0.026465 *
ph.ecog2  1.570e+00  4.807e+00  4.318e-01  3.637 0.000276 ***
ph.ecog3  2.771e+00  1.598e+01  1.118e+00  2.478 0.013214 *
ph.karno  1.983e-02  1.020e+00  1.116e-02  1.777 0.075524 .
meal.cal -3.492e-05  1.000e+00  2.590e-04 -0.135 0.892730
wt.loss  -1.209e-02  9.880e-01  7.727e-03 -1.565 0.117527
---
Signif. codes:  0 '***' 0.001 '**' 0.01 '*' 0.05 '.' 0.1 ' ' 1

         exp(coef) exp(-coef) lower .95 upper .95
age        1.0120    0.98811    0.9891    1.0355
sex        0.5763    1.73507    0.3898    0.8522
ph.ecog1   1.8538    0.53943    1.0748    3.1974
ph.ecog2   4.8072    0.20802    2.0624   11.2046
ph.ecog3  15.9783    0.06258    1.7848  143.0464
ph.karno   1.0200    0.98036    0.9980    1.0426
meal.cal   1.0000    1.00003    0.9995    1.0005
wt.loss    0.9880    1.01217    0.9731    1.0031

Concordance= 0.636  (se = 0.029 )
Likelihood ratio test= 26.77  on 8 df,   p=8e-04
Wald test            = 27.23  on 8 df,   p=6e-04
Score (logrank) test = 29.14  on 8 df,   p=3e-04

> # Create survival curves
> surv_curves <- survfit(fit_cox_model)
> # Plot survival curves
> plot(surv_curves, col = c("blue", "red", "green"), lty = 1, lwd = 2,
+      main = "Survival Curves by Category", xlab = "Time", ylab = "Survival Probability")
> |
```

Findings depict that age is insignificant at the 5% level of significance; however, sex and the performance status of the patient are statistically significant at the 1% level of significance. It is observed that sex has a negative influence, and the performance status of the patient has a positive effect on the survival of the event.

The survival curves for the Cox Proportional Hazards model is shown in Figure 12.3. From the survival curve, it is observed that the median survival time is about 550.

Suppose we want to estimate the survival curve for the specific covariates, i.e., $sex = 1$, $age = 55$, $ph.ecog = '1'$, $ph.karno = 50$, $meal.cal = 1150$, and $wt.loss = 11$. The following R-code produced the survival curve:

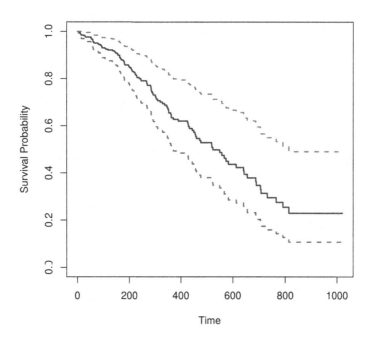

FIGURE 12.3
Survival curves by category using cox proportional hazards model.

R-code:

```
#create a survival curve given your specified covariates
newdata <- data.frame(age = 57,
                      sex= 1,
                      ph.ecog = '1',
                      ph.karno = 50,
                      meal.cal = 1175,
                      wt.loss = 11)
estimate <- survfit(fit_cox_model, newdata = newdata)

plot(survfit(fit_cox_model), ylab = 'Probability of Survival',
     xlab = 'Time', col = c('red', 'black', 'orange'))
lines(estimate$time, estimate$surv, col = 'blue', type = 's')
legend('topright', legend = c('mean', '95% Lower', '95% Upper',
'estimate'), col = c('red', 'black', 'orange', 'blue'), lty=1)
```

We get the curve illustrated in Figure 12.4.

12.2.3 Parametric Survival Models

Parametric survival models are a class of models in survival analysis that make explicit assumptions about the distribution of the survival times. Unlike nonparametric methods (e.g., Kaplan-Meier estimator) or semi-parametric methods (e.g., Cox Proportional Hazards model), parametric models specify a functional form for the survival distribution. This allows for more specific modeling of the underlying survival time distribution (e.g. Rahman & Asaduzzaman, 2019). Commonly used parametric survival models include Exponential, Weibull, Log-Normal, Gompertz, and Gamma models.

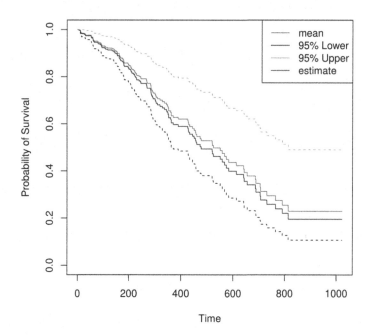

FIGURE 12.4
Survival curve for specific value of the covariates, i.e., *sex*= 1, *age*= 55, *ph.ecog* = '1', *ph.karno* = 50, *meal.cal* = 1150, and *wt.loss* = 11 using cox proportional hazards model.

Exponential model: In this model, we assume a constant hazard rate over time, and the survival function $S(t)$ can be written as $S(t) = e^{-\lambda t}$, where λ is the hazard rate.

Weibull model: It allows the hazard rate to change over time and the survival function is $S(t) = e^{-(\lambda t)^\gamma}$, where λ is the scale parameter and γ is the shape parameter.

Log-normal model: It is assumed that the logarithm of survival time follows a normal distribution. The survival function involves the cumulative distribution function of the normal distribution.

Gompertz model: In this model, the hazard rate increases exponentially with age. The survival function is expressed in terms of the Gompertz function.

Gamma model: It is assumed that survival times follow a gamma distribution. The survival function involves the incomplete gamma function.

Example 12.3

Recall the dataset used in **Example 12.1**. Now, we fit the Exponential, Weibull, Log-Normal, Gompertz, and Gamma survival models successively considering the variables used in **Example 12.1**. We use the following code in R:

Exponential survival model:

R-code:

```
# Fit an exponential survival model
fit_exponential_model <- survreg(Surv(time, status) ~ age + sex
+ ph.ecog + ph.karno + meal.cal + wt.loss, data = lung, dist =
'exponential')
# Display summary
summary(fit_exponential_model)
```

Output:

```
Console   Terminal    Background Jobs                                          — □
R 4.3.2 · ~/
> # Fit an exponential survival model
> fit_exponential_model <- survreg(Surv(time, status) ~ age + sex + ph.ecog + ph.karno + meal.
cal + wt.loss, data = lung, dist = "exponential")
> # Display summary
> summary(fit_exponential_model)

call:
survreg(formula = Surv(time, status) ~ age + sex + ph.ecog +
    ph.karno + meal.cal + wt.loss, data = lung, dist = "exponential")
              Value Std. Error     z       p
(Intercept)  7.62e+00  1.47e+00   5.17 2.3e-07
age         -9.91e-03  1.14e-02  -0.87 0.3836
sex          5.15e-01  2.00e-01   2.57 0.0101
ph.ecog1    -4.79e-01  2.76e-01  -1.74 0.0823
ph.ecog2    -1.28e+00  4.23e-01  -3.03 0.0024
ph.ecog3    -1.99e+00  1.10e+00  -1.81 0.0702
ph.karno    -1.52e-02  1.13e-02  -1.35 0.1774
meal.cal     5.18e-05  2.48e-04   0.21 0.8344
wt.loss      8.73e-03  7.39e-03   1.18 0.2375

Scale fixed at 1

Exponential distribution
Loglik(model)= -857   Loglik(intercept only)= -867.5
        chisq= 21.03 on 8 degrees of freedom, p= 0.0071
Number of Newton-Raphson Iterations: 4
n=170 (58 observations deleted due to missingness)

>
```

Weibull survival model:

R-code:

```
# Fit a Weibull survival model
fit_weibull_model <- survreg(Surv(time, status) ~ age + sex + ph.ecog
+ ph.karno + meal.cal + wt.loss, data = lung, dist = 'weibull')
# Display summary
summary(fit_weibull_model)
```

Output:

```
Console   Terminal    Background Jobs                                          — □
R 4.3.2 · ~/
> # Fit a weibull survival model
> fit_weibull_model <- survreg(Surv(time, status) ~ age + sex + ph.ecog + ph.karno + meal.cal
+ wt.loss, data = lung, dist = "weibull")
> # Display summary
> summary(fit_weibull_model)

call:
survreg(formula = Surv(time, status) ~ age + sex + ph.ecog +
    ph.karno + meal.cal + wt.loss, data = lung, dist = "weibull")
              Value Std. Error     z       p
(Intercept)  7.53e+00  1.02e+00   7.37 1.7e-13
age         -7.29e-03  8.19e-03  -0.89 0.37362
sex          3.98e-01  1.43e-01   2.78 0.00548
ph.ecog1    -4.26e-01  1.96e-01  -2.18 0.02954
ph.ecog2    -1.14e+00  2.99e-01  -3.80 0.00014
ph.ecog3    -1.96e+00  7.79e-01  -2.51 0.01204
ph.karno    -1.48e-02  7.70e-03  -1.93 0.05383
meal.cal     3.36e-05  1.82e-04   0.19 0.85304
wt.loss      8.08e-03  5.43e-03   1.49 0.13645
Log(scale)  -3.40e-01  7.21e-02  -4.72 2.3e-06

Scale= 0.712

weibull distribution
Loglik(model)= -847.5   Loglik(intercept only)= -861
        chisq= 26.97 on 8 degrees of freedom, p= 0.00072
Number of Newton-Raphson Iterations: 6
n=170 (58 observations deleted due to missingness)

>
```

Log-normal survival model:

R-code:

```
# Fit a log-normal survival model
fit_lognormal_model <- survreg(Surv(time, status) ~ age + sex
+ ph.ecog + ph.karno + meal.cal + wt.loss, data = lung, dist =
'lognormal')
# Display summary
summary(fit_lognormal_model)
```

Output:

```
Console   Terminal ×   Background Jobs ×
R R 4.3.2 · ~/
> # Fit a log-normal survival model
> fit_lognormal_model <- survreg(Surv(time, status) ~ age + sex + ph.ecog + ph.karno + meal.ca
l + wt.loss, data = lung, dist = "lognormal")
> # Display summary
> summary(fit_lognormal_model)

call:
survreg(formula = Surv(time, status) ~ age + sex + ph.ecog +
    ph.karno + meal.cal + wt.loss, data = lung, dist = "lognormal")
                 Value Std. Error     z        p
(Intercept)   6.861994   1.503375   4.56  5e-06
age          -0.017280   0.010281  -1.68  0.0928
sex           0.512223   0.184798   2.77  0.0056
ph.ecog1     -0.212515   0.256766  -0.83  0.4079
ph.ecog2     -0.944929   0.418571  -2.26  0.0240
ph.ecog3     -1.231016   1.123374  -1.10  0.2732
ph.karno     -0.009521   0.011845  -0.80  0.4215
meal.cal      0.000268   0.000224   1.20  0.2302
wt.loss       0.005991   0.006447   0.93  0.3528
Log(scale)    0.027046   0.064720   0.42  0.6760

Scale= 1.03

Log Normal distribution
Loglik(model)= -860.2   Loglik(intercept only)= -874.4
        Chisq= 28.39 on 8 degrees of freedom, p= 0.00041
Number of Newton-Raphson Iterations: 4
n=170 (58 observations deleted due to missingness)

>
```

Gompertz survival model:

R-code:

```
# Fit a Gompertz survival model
# Install and load the flexsurv package
install.packages('flexsurv')
library(flexsurv)
# Fit a Gompertz survival model
fit_gompertz_model <- flexsurvreg(Surv(time, status) ~ age + sex
+ ph.ecog + ph.karno + meal.cal + wt.loss, data = lung, dist =
'gompertz')
# Display summary
summary(fit_gompertz_model)
fit_gompertz_model$coefficients
fit_gompertz_model$AIC
```

Output:

```
Console   Terminal ×   Background Jobs ×                                              ⎯ ☐
 R  R 4.3.2 · ~/
> library(flexsurv)
> # Fit a Gompertz survival model
> fit_gompertz_model <- flexsurvreg(Surv(time, status) ~ age + sex + ph.ecog + ph.karno + me
al.cal + wt.loss, data = lung, dist = "gompertz")
> # Display summary
> summary(fit_gompertz_model)
age=62.7411764705882,sex=1.38235294117647,ph.ecog1=0.476470588235294,ph.ecog2=0.235294117647
059,ph.ecog3=0.00588235294117647,ph.karno=81.9411764705882,meal.cal=924.388235294118,wt.loss
=9.76470588235294
      time        est         lcl          ucl
1        5 0.992842959 0.989358150 0.99500539
2       11 0.984230836 0.976631079 0.98896798
3       12 0.982793023 0.974514564 0.98795710
4       13 0.981354518 0.972399397 0.98694489
5       15 0.978475443 0.968173135 0.98491650
147    814 0.058139625 0.023037392 0.11094251
148    821 0.055355400 0.021398160 0.10692272
149    840 0.048286786 0.017726912 0.09641983
150    965 0.017129008 0.004271564 0.04311755
151   1022 0.009754968 0.001996665 0.02829327
> fit_gompertz_model$coefficients
        shape          rate           age           sex       ph.ecog1      ph.ecog2
 1.944971e-03 -9.187697e+00  1.161597e-02 -5.413610e-01  6.419393e-01  1.690979e+00
     ph.ecog3      ph.karno       meal.cal      wt.loss
 2.752043e+00  2.512697e-02 -2.254047e-05 -1.035909e-02
> fit_gompertz_model$AIC
[1] 1715.584
> |
```

Gamma survival model:

R-code:

```
# Fit a gamma survival model
fit_gamma_model <- flexsurvreg(Surv(time, status) ~ age + sex +
ph.ecog + ph.karno  + wt.loss, data = lung, dist = 'gamma')
# Display summary
summary(fit_gamma_model)
fit_gamma_model$coefficients
fit_gamma_model$AIC
```

Output:

```
Console    Terminal ×    Background Jobs ×
R  R 4.3.2 · ~/
> # Fit a gamma survival model
> fit_gamma_model <- flexsurvreg(Surv(time, status) ~ age + sex + ph.ecog + ph.karno  + wt.l
oss, data = lung, dist = "gamma")
> # Display summary
> summary(fit_gamma_model)
age=62.5023474178404,sex=1.4037558685446,ph.ecog1=0.497652582159624,ph.ecog2=0.2112676056338
03,ph.ecog3=0.00469483568075117,ph.karno=82.1596244131455,wt.loss=9.72769953051643
    time        est          lcl          ucl
1       5 0.99897976  0.99705031  0.99971859
2      11 0.99627645  0.99139772  0.99863102
3      12 0.99570823  0.99032545  0.99837171
4      13 0.99511024  0.98922306  0.99809052
5      15 0.99382925  0.98691159  0.99746263
174   840 0.09664191  0.07208869  0.12031129
175   883 0.08327248  0.06048074  0.10637355
176   965 0.06251939  0.04292331  0.08389162
177  1010 0.05334659  0.03546415  0.07368652
178  1022 0.05112882  0.03369212  0.07117627
> fit_gamma_model$coefficients
        shape         rate          age          sex      ph.ecog1      ph.ecog2
  0.507571209  -6.537342250  0.009890963  -0.465749975  0.428972854  1.000110739
      ph.ecog3      ph.karno      wt.loss
  1.664281226  0.008481681  -0.006275392
> fit_gamma_model$AIC
[1] 2112.201
>
```

12.2.4 Frailty Models

Frailty models are statistical models commonly used in survival analysis to account for unobserved heterogeneity among individuals that may influence their survival times. The concept of frailty refers to an individual-specific random effect that captures this unobserved variability. There are two main types of frailty models:

Shared frailty models: In shared frailty models, a random frailty term is assumed to be shared among a group of correlated individuals, such as members of the same family or subjects within the same cluster. The shared frailty term allows for the modeling of correlation among individuals within the same group, acknowledging that they may share common characteristics that affect their survival.

Unshared frailty models (population frailty models): In unshared frailty models, each individual has their own independent frailty term. These models are suitable when there is no clear clustering of individuals, and each person is assumed to have a unique frailty that influences their survival time.

For a shared frailty model, the survival function $S(t)$ can be written as:

$$S(t) = \prod_{i=1}^{n} S_0(t)^{\exp(\beta X_i + \theta)}.$$

For an unshared frailty model, the survival function $S(t)$ is given by:

$$S(t) = \prod_{i=1}^{n} S_0(t)^{\exp(\beta X_i + \theta_i)}$$

where $S_0(t)$ is the baseline survival function, β represents the fixed effects associated with covariates, X_i represents the covariates for the i-th individual, θ represents the frailty term, and θ_i is the frailty term specific to the i-th individual.

The functions coxph is used for shared frailty models and coxme is used for unshared frailty models in R.

Example 12.4

Consider the 'lung' dataset available in the survival R package. Now we fit both the shared and unshared frailty models. In R, we use the following commands:

Shared frailty models:

Here, we are assuming that patients within the same institution (inst) share a common frailty effect. So, we use the frailty term as frailty(inst).

R-code:

```
# load package for data
install.packages('survival')
library(survival)
#It is assuming that patients within the same institution share a
common frailty effect.
fit_shared_frailty <- coxph(Surv(time, status) ~ age + sex + ph.ecog +
ph.karno  + wt.loss + frailty(inst), data = lung)
# Display the summary of the model
summary(fit_shared_frailty)
```

Output:

```
Console   Terminal    Background Jobs
R R 4.3.2 · ~/
> library(survival)
> #It is assuming that patients within the same institution share a common frailty effect.
> fit_shared_frailty <- coxph(Surv(time, status) ~ age + sex + ph.ecog + ph.karno  + wt.loss +
frailty(inst), data = lung)
> # Display the summary of the model
> summary(fit_shared_frailty)
Call:
coxph(formula = Surv(time, status) ~ age + sex + ph.ecog + ph.karno +
    wt.loss + frailty(inst), data = lung)

  n= 212, number of events= 150
   (16 observations deleted due to missingness)

                coef     se(coef)  se2       Chisq DF   p
age           0.014951 0.009818 0.009809   2.32 1.00 0.13000
sex          -0.628483 0.178366 0.178297  12.42 1.00 0.00043
ph.ecog1      0.676829 0.248950 0.248692   7.39 1.00 0.00660
ph.ecog2      1.479570 0.390056 0.389278  14.39 1.00 0.00015
ph.ecog3      2.693241 1.095788 1.095097   6.04 1.00 0.01400
ph.karno      0.015359 0.009918 0.009904   2.40 1.00 0.12000
wt.loss      -0.009361 0.006730 0.006726   1.93 1.00 0.16000
frailty(inst)                             0.26 0.21 0.35000

          exp(coef) exp(-coef) lower .95 upper .95
age        1.0151    0.98516    0.9957    1.0348
sex        0.5334    1.87476    0.3760    0.7566
ph.ecog1   1.9676    0.50823    1.2079    3.2052
ph.ecog2   4.3911    0.22774    2.0443    9.4316
ph.ecog3  14.7795    0.06766    1.7255  126.5889
ph.karno   1.0155    0.98476    0.9959    1.0354
wt.loss    0.9907    1.00940    0.9777    1.0038

Iterations: 5 outer, 23 Newton-Raphson
    Variance of random effect= 0.001678629   I-likelihood = -653.1
Degrees of freedom for terms= 1.0 1.0 3.0 1.0 1.0 0.2
Concordance= 0.643  (se = 0.026 )
Likelihood ratio test= 34.35  on 7.2 df,    p=2e-05

>
```

Unshared frailty models:

In this example, we are allowing each institution (inst) to have its own unique frailty effect. In R, the unshared frailty term considers (1|inst). Now, we use the following commands in R:

R-code:

```
# load package
install.packages('coxme')
library(coxme)
# Fit an unshared frailty Cox model
fit_unshared_frailty <- coxme(Surv(time, status) ~ age + sex + ph.ecog
+ ph.karno  + wt.loss + (1|inst), data = lung)
# Display the summary of the model
summary(fit_unshared_frailty)
```

Output:

```
Console  Terminal ×  Background Jobs ×                                           ━ ☐
R R 4.3.2 · ~/
> library(coxme)
> # Fit an unshared frailty Cox model
> fit_unshared_frailty <- coxme(Surv(time, status) ~ age + sex + ph.ecog + ph.karno  + wt.loss
+ (1|inst), data = lung)
> # Display the summary of the model
> summary(fit_unshared_frailty)
Cox mixed-effects model fit by maximum likelihood
  Data: lung
  events, n = 150, 212 (16 observations deleted due to missingness)
  Iterations= 9 75
                   NULL Integrated    Fitted
Log-likelihood -669.984   -652.9935 -651.2101

                  Chisq  df         p   AIC   BIC
Integrated loglik 33.98 8.00 4.0949e-05 17.98 -6.10
 Penalized loglik 37.55 8.62 1.5383e-05 20.31 -5.63

Model:  Surv(time, status) ~ age + sex + ph.ecog + ph.karno + wt.loss +      (1 | inst)
Fixed coefficients
                 coef  exp(coef)     se(coef)    z       p
age       0.014912756  1.0150245 0.009883420  1.51 0.13000
sex      -0.631480280  0.5318040 0.178792208 -3.53 0.00041
ph.ecog1  0.685200469  1.9841696 0.250635323  2.73 0.00630
ph.ecog2  1.520897652  4.5763313 0.394514617  3.86 0.00012
ph.ecog3  2.742669296 15.5283797 1.100363642  2.49 0.01300
ph.karno  0.015590561  1.0157127 0.009980785  1.56 0.12000
wt.loss  -0.009474504  0.9905702 0.006741213 -1.41 0.16000

Random effects
 Group Variable Std Dev     Variance
 inst  Intercept 0.12144853 0.01474975
> |
```

12.3 Factor Analysis

Factor analysis (FA) is used to explore the underlying structure of a set of variables. The primary goal of factor analysis is to identify latent factors that can explain patterns of correlations among observed variables. Factor analysis helps reduce the dimensionality of the data by identifying a smaller number of factors that can explain the observed patterns of correlations. It is commonly employed in many fields, such as medical sciences, psychology, sociology, economics, and other social sciences.

An FA model assumes that the observed variables are linear combinations of these latent factors plus unique variance or error. The basic equation for the observed variables X in a factor analysis model is:

$$X = AF + \varepsilon$$

where X is a matrix of observed variables, A is the matrix of factor loadings that represent the strength and direction of the relationship between the latent factors and the observed variables, F is a matrix of latent factors, and ε is a matrix of unique variance or error terms.

The goal of factor analysis is to estimate the factor loadings (A), and latent factors (F).

Remember that when conducting a factor analysis, we usually do it on the correlation matrix or covariance matrix of the data. This is because factor loadings are influenced by the scale of the variables, and using correlation or covariance matrices helps standardize the analysis. It is often a good idea to use the correlation matrix if the dataset includes variables with various scales or units.

Example 12.5

We consider a dataset named 'Student Stress Factors: A Comprehensive Analysis', which is available on the 'Kaggle' database. This dataset includes 21 variables that create the most impact on the Stress of a Student. Perform a factor analysis considering this dataset.

First, download the dataset from the 'Kaggle' database using the login credentials and save it to your computer. Load the dataset from your computer to R and employ the following command to perform the factor analysis:

R-code:

```
# Load data into R
data_fa<-read.csv('E:/Book/Chapter 12/StressLevelDataset.csv',
header=TRUE)
attach(data_fa)
# To see the structure of the data
str(data_fa)
```

Output:

```
Console   Terminal ×   Background Jobs ×

R R 4.3.2 · ~/
> # Load data into R
> data_fa<-read.csv("E:/Book/Chapter 12/StressLevelDataset.csv", header=TRUE)
> attach(data_fa)
> # To see the structure of the data
> str(data_fa)
'data.frame':   1100 obs. of  21 variables:
 $ anxiety_level               : int  14 15 12 16 16 20 4 17 13 6 ...
 $ self_esteem                 : int  20 8 18 12 28 13 26 3 22 8 ...
 $ mental_health_history       : int  0 1 1 1 0 1 0 1 1 0 ...
 $ depression                  : int  11 15 14 15 7 21 6 22 12 27 ...
 $ headache                    : int  2 5 2 4 2 3 1 4 3 4 ...
 $ blood_pressure              : int  1 3 1 3 3 3 2 3 1 3 ...
 $ sleep_quality               : int  2 1 2 1 5 1 4 1 2 1 ...
 $ breathing_problem           : int  4 4 2 3 1 4 1 5 4 2 ...
 $ noise_level                 : int  2 3 2 4 3 3 1 3 3 0 ...
 $ living_conditions           : int  3 1 2 2 2 2 4 1 3 5 ...
 $ safety                      : int  3 2 3 2 4 2 4 1 3 2 ...
 $ basic_needs                 : int  2 2 2 2 3 1 4 1 3 2 ...
 $ academic_performance        : int  3 1 2 2 4 2 5 1 3 2 ...
 $ study_load                  : int  2 4 3 4 3 5 1 3 3 2 ...
 $ teacher_student_relationship: int  3 1 3 1 1 2 4 2 2 1 ...
 $ future_career_concerns      : int  3 5 2 4 2 5 1 4 3 5 ...
 $ social_support              : int  2 1 2 1 1 1 3 1 3 1 ...
 $ peer_pressure               : int  3 4 3 4 5 4 2 4 3 5 ...
 $ extracurricular_activities  : int  3 5 2 4 0 4 2 4 2 3 ...
 $ bullying                    : int  2 5 2 5 5 5 1 5 2 4 ...
 $ stress_level                : int  1 2 1 2 1 2 0 2 1 1 ...
>
```

To see the structure of the data frame, we use the `str()` function in R. Using this command, we can see the number of observations, number of variables, and details about each variable, such as its name, type, and other.

Eigenvalues play a crucial role in factor analysis and they are associated with the factors extracted from the dataset. Higher eigenvalues indicate more important factors. In practice, factors with eigenvalues greater than 1 are often considered significant. To get the eigenvalues, we use the following commands in R:

R-code:

```
# Calculate the correlation matrix
cor_matrix <- cor(data_fa)
# Use the eigen() function to obtain eigenvalues
eigenvalues <- eigen(cor_matrix)$values
# Print the eigenvalues
print(eigenvalues, digits = 3)
```

Output:

```
Console   Terminal ×   Background Jobs ×
R  R 4.3.2 · ~/
> # Calculate the correlation matrix
> cor_matrix <- cor(data_fa)
> # Use the eigen() function to obtain eigenvalues
> eigenvalues <- eigen(cor_matrix)$values
> # Print the eigenvalues
> print(eigenvalues, digits = 3)
 [1] 12.703  1.199  0.694  0.595  0.559  0.526  0.474  0.458  0.406  0.386  0.364  0.348
[13]  0.329  0.313  0.312  0.282  0.273  0.266  0.233  0.175  0.102
>
```

In this example, we may decide to retain the first two factors with eigenvalues 12.703 and 1.119, as they are above a chosen threshold e.g., 1.

Now, we use the following code in R, to get the output of factor analysis:

R-code:

```
# Perform factor analysis
fa_result <- factanal(
   factors = 2,           # Number of factors to extract
   covmat = cor(data_fa),  # Input the correlation matrix
   rotation = 'varimax'   # Use varimax rotation for better
interpretability
)
# Print the results
print(fa_result, digits = 2)
```

Output:

```
Console   Terminal ×   Background Jobs ×                                              ─ □
 R  R 4.3.2 · ~/
> # Perform factor analysis
> fa_result <- factanal(
+    factors = 2,          # Number of factors to extract
+    covmat = cor(data_fa),  # Input the correlation matrix
+    rotation = "varimax"   # Use varimax rotation for better interpretability
+ )
> # Print the results
> print(fa_result, digits = 2)

Call:
factanal(factors = 2, covmat = cor(data_fa), rotation = "varimax")

Uniquenesses:
             anxiety_level            self_esteem   mental_health_history
                      0.30                   0.30                    0.45
                depression               headache          blood_pressure
                      0.31                   0.39                    0.00
             sleep_quality      breathing_problem             noise_level
                      0.30                   0.56                    0.50
         living_conditions                 safety             basic_needs
                      0.56                   0.38                    0.39
      academic_performance             study_load teacher_student_relationship
                      0.36                   0.53                    0.36
    future_career_concerns         social_support           peer_pressure
                      0.29                   0.25                    0.41
   extracurricular_activities             bullying             stress_level
                      0.41                   0.31                    0.19

Loadings:
                             Factor1 Factor2
anxiety_level                 0.80    0.24
self_esteem                  -0.71   -0.44
mental_health_history         0.71    0.22
depression                    0.75    0.36
headache                      0.73    0.28
blood_pressure                0.11    0.99
sleep_quality                -0.81   -0.21
breathing_problem             0.65
noise_level                   0.65    0.28
living_conditions            -0.63   -0.21
safety                       -0.76   -0.21
basic_needs                  -0.76   -0.20
academic_performance         -0.78   -0.18
study_load                    0.63    0.28
teacher_student_relationship -0.76   -0.27
future_career_concerns        0.77    0.35
social_support               -0.51   -0.70
peer_pressure                 0.69    0.33
extracurricular_activities    0.68    0.35
bullying                      0.78    0.29
stress_level                  0.85    0.30

               Factor1 Factor2
SS loadings     10.49    2.95
Proportion Var   0.50    0.14
Cumulative Var   0.50    0.64

The degrees of freedom for the model is 169 and the fit was 0.8246
> |
```

The eigenvalues suggest only two factors for this dataset. However, the two factors collectively explain 64% of the variance in the original variables. The factor loadings represent the correlation between each variable and each factor. Strong loading indicates which factor the item is included in. Moreover, it is essential to consider both statistical results and theoretical reasoning for making the factors and ensure that the identified factors make sense in the context of the studied phenomenon.

12.4 Conclusion and Exercises

In this section, the summary of the chapter is discussed with some problem-solving exercises in biostatistics.

12.4.1 Concluding Remarks

This chapter served as a comprehensive exploration of the intricacies of handling time-to-event data. From the foundational notions of survival and hazard functions to the practical applications in health research, this chapter provides a solid grounding for researchers and practitioners. The Kaplan-Meier estimator emerges as a reliable ally in estimating survival functions, especially when dealing with incomplete information. The Cox Proportional Hazards model stands tall as a versatile and widely adopted method for assessing the impact of covariates on survival outcomes. As an example of a practical application, each model is estimated in this chapter using a dataset. It is crucial, however, for practitioners to tread carefully, acknowledging the assumptions and limitations that underpin survival analysis.

Factor analysis allows researchers to identify and understand the latent factors that drive patterns in the data. The practical applications of factor analysis span a wide range of disciplines, from psychology and education to finance and marketing. Uncovering the latent structure in data provides valuable insights for decision-making.

By reading this chapter, the reader gains a solid understanding of survival analysis and their practical applications, and they will be capable of using these concepts in their statistical analysis and biostatistics research. Readers can also perform factor analysis using their dataset. This concluding chapter offered valuable resources for researchers, analysts, biostatisticians, data scientists, and statisticians, providing a comprehensive knowledge and understanding of survival and factor analyses along with their pivotal role in modern data science, including medical statistics and bioinformatics.

12.4.2 Practice Questions

Exercise 12.1 Use the dataset named 'cancer' available in `survival` R-package. This dataset contains 686 observations and following 11 variables.

- pid: patient identifier.
- age: age, years.
- meno: menopausal status (0 = premenopausal, 1 = postmenopausal).
- size: tumor size, mm.
- grade: tumor grade.
- nodes: number of positive lymph nodes.
- pgr: progesterone receptors (fmol/l).
- er: estrogen receptors (fmol/l).
- hormon: hormonal therapy, 0 = no, 1 = yes.
- rfstime: recurrence free survival time; days to first of reccurence, death or last follow-up.
- status: 0= alive without recurrence, 1= recurrence or death.

Now, solve the following problems:

a. Print the structure of the data frame.

b. Find the Kaplan-Meier estimates using the suitable variable. Also, draw the survival curve.

c. Fit the Cox Proportional Hazards model considering the relevant variables and draw the survival curve.

d. Fit different parametric models.

e. Fit both shared and unshared frailty models.

Exercise 12.2 Consider a dataset named 'kidney' available in `survival` R-package. Now, fit the following models.

a. Cox Proportional Hazards model considering the relevance covariates.

b. Shared and unshared frailty models.

Exercise 12.3 Download a dataset named 'Predicting Divorce' from the 'Kaggle' database and save it into your computer.

a. Load the dataset in R.

b. Find the eigenvalues and determine the number of factors needed in factor analysis.

c. Perform a factor analysis.

Bibliography

Bandalos, D. L., & Finney, S. J. (2018). Factor analysis: Exploratory and confirmatory. In Hancock, G. R., Stapleton, L. M., & Mueller, R. O. (Eds.), *The reviewer's guide to quantitative methods in the social sciences* (pp. 98–122). Routledge.

Bartholomew, D. J., Knott, M., & Moustaki, I. (2011). *Latent variable models and factor analysis: A unified approach* (Vol. 904). John Wiley & Sons.

Bogaerts, K., Komarek, A., & Lesaffre, E. (2017). *Survival analysis with interval-censored data: A practical approach with examples in R, SAS, and BUGS*. CRC Press.

Brown, T. A. (2015). *Confirmatory factor analysis for applied research*. Guilford Publications.

Denis, D. J. (2021). *Applied univariate, bivariate, and multivariate statistics: Understanding statistics for social and natural scientists, with applications in SPSS and R*. John Wiley & Sons.

Diez, D. (2021). *Survival analysis in R*. OpenIntro.

Fabrigar, L. R., & Wegener, D. T. (2011). *Exploratory factor analysis*. Oxford University Press.

Flores, A. Q. (2022). *Survival analysis: A new guide for social scientists*. Cambridge University Press.

Garson, G. D. (2022). *Factor analysis and dimension reduction in R: A social scientist's toolkit*. Taylor & Francis.

Gorsuch, R. L. (2014). *Factor analysis* (Classic ed.). Routledge.

Hosmer Jr, D. W., Lemeshow, S., & May, S. (2011). *Applied survival analysis: Regression modeling of time-to-event data*. John Wiley & Sons.

Hossain, M. M. (2022). Applied univariate, bivariate, and multivariate statistics: Understanding statistics for social and natural scientists, with applications in SPSS and R. *Journal of the Royal Statistical Society Series A: Statistics in Society*, 185(2), 727–728.

James, G., Witten, D., Hastie, T., & Tibshirani, R. (2013). *An introduction to statistical learning*. Springer.

Klein, J. P., & Moeschberger, M. L. (2003). *Survival analysis: Techniques for censored and truncated data* (Vol. 1230). Springer.

Kleinbaum, D. G., & Klein, M. (1996). *Survival analysis a self-learning text*. Springer.

Kleinbaum, D. G., Klein, M., Kleinbaum, D. G., & Klein, M. (2012). *Introduction to survival analysis*. *Survival analysis: A self-learning text*. Springer.

Lee, E. T., & Wang, J. (2003). *Statistical methods for survival data analysis* (Vol. 476). John Wiley & Sons.

Li, J., & Ma, S. (2013). *Survival analysis in medicine and genetics*. CRC Press.

Loprinzi, C. L., Laurie, J. A., Wieand, H. S., Krook, J. E., Novotny, P. J., Kugler, J. W., Bartel, J., Law, M., Bateman, M., Klatt, N. E., et al. (1994). Prospective evaluation of prognostic variables from patient-completed questionnaires. North Central Cancer Treatment Group. *Journal of Clinical Oncology*. 12(3), 601–607.

Machin, D., Cheung, Y. B., & Parmar, M. (2006). *Survival analysis: A practical approach*. John Wiley & Sons.

McGilchrist, C. A., Aisbett, C. W. (1991). Regression with frailty in survival analysis. *Biometrics*, 47, 461–466.

Moore, D. F. (2016). *Applied survival analysis using R*. Springer.

O'Quigley, J. (2008). *Proportional hazards regression*. Springer.

Pagès, J. (2014). *Multiple factor analysis by example using R*. CRC Press.

Pandey, S., Rahman, A., & Gurr, G. M. (2018). Australian native flowering plants enhance the longevity of three parasitoids of brassica pests. *Entomologia Experimentalis et Applicata*, 166(4), 265–276. https://doi.org/10.1111/eea.12668

Pett, M. A., Lackey, N. R., & Sullivan, J. J. (2003). *Making sense of factor analysis: The use of factor analysis for instrument development in health care research*. Sage.

Rahman, A., & Asaduzzaman, M. (2019). Statistical modelling of seed germination and seedlings root response of Annual Ryegrass (Lolium rigidum) to different stress. *Agricultural Research*, 8(2), 262–269. https://doi.org/10.1007/s40003-018-0379-6

Royston, P. and Douglas Altman, D. (2013). External validation of a Cox prognostic model: principles and methods. *BMC Medical Research Methodology*, 13, 33.

Therneau, T. M., Lumley, T., Elizabeth, A., & Cynthia, C. (2024). Package Survival: A Package for Survival Analysis in R. R Package. Version 3.7-0. https://CRAN.R-project.org/package=survival

Watkins, M. (2020). *A step-by-step guide to exploratory factor analysis with R and RStudio*. Routledge.

Yöntem, M., Adem, K, İlhan, T, Kılıçarslan, S. (2019). Divorce prediction using correlation based feature selection and artificial neural networks. *Nevşehir Hacı Bektaş Veli University SBE Dergisi*, 9(1), 259–273.

Index

Pages in *italics* refer to figures and pages in **bold** refer to tables.

Printed in the United States
by Baker & Taylor Publisher Services